T0237785

Lecture Notes in Computer Scier

Edited by G. Goos, J. Hartmanis, and J. van Leeuwen

Springer
Berlin
Heidelberg
New York
Barcelona
Hong Kong
London
Milan
Paris
Singapore
Tokyo

Leo Bachmair (Ed.)

Rewriting Techniques and Applications

11th International Conference, RTA 2000
Norwich, UK, July 10-12, 2000
Proceedings

Springer

Series Editors

Gerhard Goos, Karlsruhe University, Germany
Juris Hartmanis, Cornell University, NY, USA
Jan van Leeuwen, Utrecht University, The Netherlands

Volume Editor

Leo Bachmair
SUNY at Stony Brook, Department of Computer Science
Stony Brook, New York 11794, USA
E-mail: leo@cs.sunysb.edu

Cataloging-in-Publication Data applied for

Die Deutsche Bibliothek - CIP-Einheitsaufnahme

Rewriting techniques and applications : 11th international conference ;
proceedings / RTA 2000, Norwich, UK, July 10 - 12, 2000. Leo
Bachmair (ed.). - Berlin ; Heidelberg ; New York ; Barcelona ; Hong
Kong ; London ; Milan ; Paris ; Singapore ; Tokyo : Springer, 2000
   (Lecture notes in computer science ; Vol. 1833)
   ISBN 3-540-67778-X

CR Subject Classification (1998): F.4, F.3.2, D.3, I.2.2-3, I.1

ISSN 0302-9743
ISBN 3-540-67778-X Springer-Verlag Berlin Heidelberg New York

Springer-Verlag is a company in the BertelsmannSpringer publishing group.
© Springer-Verlag Berlin Heidelberg 2000
Printed in Germany

Typesetting: Camera-ready by author, data conversion by Christian Grosche, Hamburg
Printed on acid-free paper   SPIN: 10721975   06/3142   5 4 3 2 1 0

Preface

This volume contains the proceedings of the *11th International Conference on Rewriting Techniques and Applications*. The conference was held July 10-12, 2000, at the University of East Anglia, Norwich, U.K. It is the major forum for the presentation of research on all theoretical and practical aspects of rewriting. Information about previous RTA conferences can be found at

http://rewriting.loria.fr/rta/

and information about the general research area of rewriting at

http://www.loria.fr/ vigneron/RewritingHP/

The program committee selected 18 papers, including three system descriptions, from a total of 44 submissions. In addition the program included invited talks by Jose Meseguer, Dale Miller, and Andrei Voronkov; and an invited tutorial by Sophie Tison.

Many people contributed to RTA-2000 and I would like to express my sincere thanks to all of them. I am grateful to the program committee members and the external referees for reviewing the submissions and maintaining the high standard of the RTA conferences; to Richard Kennaway, who was responsible for the local arrangements for the conference; and to José Meseguer, the RTA publicity chair. It is a particular pleasure to thank Ashish Tiwari for his extensive assistance in many of my tasks as the program chair. Finally, I wish to thank the School of Information Systems at the University of East Anglia both for financial support and for providing the facilities.

May 2000   Leo Bachmair

Conference Organization

Program Chair

Leo Bachmair (Stony Brook)

Program Committee

Franz Baader (Aachen)
Gilles Dowek (Rocquencourt)
Neil Ghani (Leicester)
Juergen Giesl (Albuquerque)
Jean-Pierre Jouannaud (Orsay)
Chris Lynch (Potsdam, New York)
Aart Middeldorp (Tsukuba)
Mitsuhiro Okada (Tokyo)
Femke van Raamsdonk (Amsterdam)
Albert Rubio (Barcelona)
Yoshihito Toyama (Tatsunokuchi)
Rakesh Verma (Houston)

Local Arrangements Chair

Richard Kennaway (Norwich)

Publicity Chair

José Meseguer (Menlo Park)

RTA Organizing Committee

Hubert Comon (Cachan)
Tobias Nipkow (Munich)
Nachum Dershowitz, chair (Urbana)
Michael Rusinowitch (Nancy)
José Meseguer (Menlo Park)
Yoshihito Toyama (Tatsunokuchi)

External Referees

Y. Akama
T. Arts
R. Bloo
J. Chrzaszcz
D. Dougherty
M. Fernandez
G. Godoy
T. Ida
S. Kahrs
D. Kesner
F. Lang
M. Leucker
R. Matthes
B. Monate
M. Nagayama
J. Niehren
K. Ogata
S. Okui
V. van Oostrom
J. van de Pol
M. R. K. Krishna Rao
M. Rittri
M. Rusinowitch
M. Sakai
M. Schmidt-Schauss
T. Suzuki
A. Tiwari
F.-J. de Vries
C. Walther

T. Aoto
F. Blanqui
A. Boudet
E. Contejean
K. Erk
W. Fokkink
B. Gramlich
F. Jacquemard
Y. Kaji
K. Kusakari
C. Lüth
J. Levy
R. Mayr
J. J. Moreno Navarro
P. Narendran
R. Nieuwenhuis
E. Ohlebusch
C.-H. L. Ong
F. Otto
C. Prehofer
E. Ritter
K. Rose
K. Sakai
G. Schmidt
K. U. Schulz
R. Thomas
R. Treinen
J. Waldmann
T. Yamada

Table of Contents

Invited Talk

Invited Tutorial

Regular Papers

System Descriptions

Rewriting Logic and Maude: Concepts and Applications*

José Meseguer

Computer Science Laboratory
SRI International, Menlo Park, CA 94025, USA

Abstract. For the most part, rewriting techniques have been developed and applied to support efficient equational reasoning and equational specification, verification, and programming. Therefore, a rewrite rule $t \longrightarrow t'$ has been usually interpreted as a *directed equation $t = t'$*. Rewriting logic is a substantial broadening of the semantics given to rewrite rules. The equational reading is abandoned, in favor of a more dynamic interpretation. There are now in fact two complementary readings of a rule $t \longrightarrow t'$, one computational, and another logical: (i) computationally, the rewrite rule $t \longrightarrow t'$ is interpreted as a *local transition* in a concurrent system; (ii) logically, the rewrite rule $t \longrightarrow t'$ is interpreted as an *inference rule*. The experience gained so far strongly suggest that rewriting is indeed a very flexible and general formalism for both computational and logical applications. This means that from the computational point of view rewriting logic is a very expressive *semantic framework*, in which many different models of concurrency, languages, and distributed systems can be specified and programmed; and that from a logical point of view is a general *logical framework* in which many different logics can be represented and implemented. This paper introduces the main concepts of rewriting logic and of the Maude rewriting logic language, and discusses a wide range of semantic framework and logical framework applications that have been developed in rewriting logic using Maude.

1 Introduction

For the most part, rewriting techniques have been developed and applied to support efficient equational reasoning and equational specification, verification, and programming. Therefore, a rewrite rule $t \longrightarrow t'$ has been usually interpreted as a *directed equation $t = t'$*. This equational semantics does of course suggest long-term research directions to advance the applicability of rewriting techniques to equational reasoning. For example, notions of confluence and termination are central for equational purposes, and completion techniques for different variants

---

* Supported by DARPA through Rome Laboratories Contract F30602-C-0312, by DARPA and NASA through Contract NAS2-98073, by Office of Naval Research Contract N00014-99-C-0198, and by National Science Foundation Grant CCR-9900334.

L. Bachmair (Ed.): RTA 2000, LNCS 1833, pp. 1–26, 2000.

of equational logic, for rewriting modulo axioms, and in support of theorem proving methods are important ongoing research topics.

Rewriting logic [61] is a substantial broadening of the semantics given to rewrite rules. The equational reading is abandoned, in favor of a more dynamic, state-changing, and in general irreversible interpretation. There are now in fact two complementary readings of a rule $t \longrightarrow t'$, one computational, and another logical:

- *computationally*, the rewrite rule $t \longrightarrow t'$ is interpreted as a *local transition* in a concurrent system; that is, t and t' describe patterns for *fragments* of the distributed state of a system, and the rule explains how a local concurrent transition can take place in such a system, changing the local state fragment from the pattern t to the pattern t'.
- *logically*, the rewrite rule $t \longrightarrow t'$ is interpreted as an *inference rule*, so that we can infer formulas of the form t' from formulas of the form t.

The computational and logical viewpoints are not exclusive: they complement each other. This can be illustrated with a simple Petri net example. Indeed, Petri nets [81] provide some of the simplest examples of systems exhibiting concurrent change, and therefore it is interesting to see how the computational and logical viewpoints appear in them as two sides of the same coin.

Usually, a Petri net is presented as a set of places, a disjoint set of transitions and a relation of causality between them that associates to each transition the set of resources consumed as well as produced by its firing. Ugo Montanari and I recast this idea in an algebraic framework in [68]. From this point of view, resources are represented as multisets of places, and therefore we have a binary operation (multiset union, denoted $\otimes$ here) that is associative, commutative, and has the empty multiset as an identity[1] but is not idempotent. Then, a Petri net is viewed as a graph whose arcs are the transitions and whose nodes are multisets over the set of places, usually called *markings*.

The following Petri net represents a machine to buy cakes and apples; a cake costs a dollar and an apple three quarters. Due to an unfortunate design, the machine only accepts dollars, and it returns a quarter when the user buys an apple; to alleviate in part this problem, the machine can change four quarters into a dollar.

As a graph, this net has the following arcs:

$$
\begin{aligned}
buy\text{-}c : \quad &\$ \longrightarrow c \\
buy\text{-}a : \quad &\$ \longrightarrow a \otimes q \\
change : \quad &q \otimes q \otimes q \otimes q \longrightarrow \$
\end{aligned}
$$

---

[1] From now on the associativity, commutativity, and identity axioms are denoted by the acronym ACI.

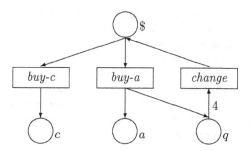

The expression of this Petri net in rewriting logic is now obvious. We can view each of the labeled arcs of the Petri net as a rewrite rule in a rewrite theory having a binary *associative, commutative* operator $\otimes$ (multiset union) with *identity* 1 so that rewriting happens modulo ACI, that is, is multiset rewriting. Then, the concurrent computations of the Petri net from a given initial marking in fact coincide with the ACI-rewriting computations that can be performed from that marking using the above four rules. This is the obvious computational interpretation.

But we can just as well adopt a logical interpretation in which the multiset union operator $\otimes$ can be viewed as a form or resource-conscious non-idempotent conjunction. Then, the state $a \otimes q \otimes q$ corresponds to having an apple *and* a quarter *and* a quarter, which is a strictly better situation than having an apple *and* a quarter (non-idempotence of $\otimes$). Several researchers realized independently that this ACI operation on multisets corresponds to the conjunctive connective $\otimes$ (*tensor*) in linear logic [3,45,56,57]. This complementary point of view sees a net as a theory in this fragment of linear logic.

For example, in order to get the tensor theory corresponding to our Petri net above, it is enough to change the arrows in the graph presentation into turnstiles, getting the following axioms:

$$buy\text{-}c : \quad \$ \vdash c$$
$$buy\text{-}a : \quad \$ \vdash a \otimes q$$
$$change : q \otimes q \otimes q \otimes q \vdash \$$$

Rewriting logic faithfully supports the computational and logical interpretations, in the sense that a marking M' is reachable from a marking M by a concurrent computation of the above Petri net iff there is an ACI-rewriting computation $M \longrightarrow M'$ iff the above tensor theory can derive the sequent $M \vdash M'$ in linear logic (see [56,61]).

The above example illustrates another important point, namely, that equational logic, although surpassed by the rewriting logic interpretation of rewrite rules is nevertheless not abandoned. In fact, a rewrite theory $\mathcal{R}$ is a 4-tuple $\mathcal{R} = (\Sigma, E, L, R)$, where (Σ, E) is the equational theory *modulo which* we rewrite, L is a set of labels, and R is a set of labeled rules. In the above example Σ consists of the binary operator $\otimes$ and the constants a, c, q, and \$, and E consists of the ACI axioms.

This means that, by identifying equational theories with rewrite theories such that the sets L and R are empty, equational logic can be naturally regarded as a *sublogic* of rewriting logic. In this way, the equational world is preserved intact within the broader semantic interpretation proposed by rewriting logic, both at the level of deduction and at the level of models. Note that rewriting techniques are applicable at two different levels of a rewrite theory $\mathcal{R}$, namely, at the level of its equations E—which in general may not only contain axioms like ACI, but may for example contain Church-Rosser equations modulo such axioms—and at the level of the rewrite rules R which in general may not be Church-Rosser and may not terminate.

The experience that we have gained so far strongly suggest that rewriting is indeed a very flexible and general formalism for both computational and logical applications. This means that from the computational point of view rewriting logic is a very expressive *semantic framework*, in which many different models of concurrency, languages, and distributed systems can be specified and programmed; and that from a logical point of view is a general *logical framework* in which many different logics can be represented and implemented [58].

The goal of this paper is to introduce the main concepts of rewriting logic and of the Maude rewriting logic language that we have implemented at SRI [19,20], and to give a flavor for the many semantic framework and logical framework applications that rewriting logic makes possible, and that a language implementation such as Maude can support by building tools and performing a variety of formal analyses on executable rewriting logic specifications. The paper does *not* intend to give a survey of research in rewriting logic, for which I refer the reader to the Workshop Proceedings [64,51] and to the survey [66]. In particular, I do not cover the research associated with two other important rewriting logic languages, namely, ELAN [52,10,9] and CafeOBJ [39].

The paper is organized as follows. Sections 2 and 3 introduce the main concepts of rewriting logic and Maude. Section 4 discusses some semantic framework applications, and Section 5 is dedicated to some logical framework applications. I finish with some concluding remarks.

2   Rewriting Logic

This section introduces the basic concepts of rewriting logic, including its inference rules, initial and free models, reflection, and executability issues.

2.1   Inference Rules and their Meaning

A *signature* in rewriting logic is an equational theory[2] (Σ, E), where Σ is an equational signature and E is a set of Σ-equations. Rewriting will operate on equivalence classes of terms modulo E.

---

[2] Rewriting logic is parameterized by the choice of its underlying equational logic, that can be unsorted, many-sorted, order-sorted, membership equational logic, and so on. To ease the exposition I give an *unsorted* presentation.

Given a signature (Σ, E), *sentences* of rewriting logic are sequents of the form

$$[t]_E \longrightarrow [t']_E,$$

where t and t' are Σ-terms possibly involving some variables, and $[t]_E$ denotes the equivalence class of the term t modulo the equations E. A *rewrite theory* $\mathcal{R}$ is a 4-tuple $\mathcal{R} = (\Sigma, E, L, R)$ where Σ is a ranked alphabet of function symbols, E is a set of Σ-equations, L is a set of *labels*, and R is a set of pairs $R \subseteq L \times T_{\Sigma, E}(X)^2$ whose first component is a label and whose second component is a pair of E-equivalence classes of terms, with $X = \{x_1, \ldots, x_n, \ldots\}$ a countably infinite set of variables. Elements of R are called *rewrite rules*.[3] We understand a rule $(r, ([t], [t']))$ as a labeled sequent and use for it the notation $r : [t] \longrightarrow [t']$. To indicate that $\{x_1, \ldots, x_n\}$ is the set of variables occurring in either t or t', we write $r : [t(x_1, \ldots, x_n)] \longrightarrow [t'(x_1, \ldots, x_n)]$, or in abbreviated notation $r : [t(\overline{x})] \longrightarrow [t'(\overline{x})]$.

Given a rewrite theory $\mathcal{R}$, we say that $\mathcal{R}$ *entails* a sentence $[t] \longrightarrow [t']$, or that $[t] \longrightarrow [t']$ is a *(concurrent)* $\mathcal{R}$-rewrite, and write $\mathcal{R} \vdash [t] \longrightarrow [t']$ if and only if $[t] \longrightarrow [t']$ can be obtained by finite application of the following *rules of deduction* (where we assume that all the terms are well formed and $t(\overline{w}/\overline{x})$ denotes the simultaneous substitution of w_i for x_i in t):

1. **Reflexivity.** For each $[t] \in T_{\Sigma, E}(X)$, $\dfrac{}{[t] \longrightarrow [t]}$.

2. **Congruence.** For each $f \in \Sigma_n$, $n \in \mathbb{N}$,

$$\frac{[t_1] \longrightarrow [t'_1] \quad \cdots \quad [t_n] \longrightarrow [t'_n]}{[f(t_1, \ldots, t_n)] \longrightarrow [f(t'_1, \ldots, t'_n)]}.$$

3. **Replacement.** For each rule $r : [t(x_1, \ldots, x_n)] \longrightarrow [t'(x_1, \ldots, x_n)]$ in R,

$$\frac{[w_1] \longrightarrow [w'_1] \quad \cdots \quad [w_n] \longrightarrow [w'_n]}{[t(\overline{w}/\overline{x})] \longrightarrow [t'(\overline{w'}/\overline{x})]}.$$

4. **Transitivity**

$$\frac{[t_1] \longrightarrow [t_2] \quad [t_2] \longrightarrow [t_3]}{[t_1] \longrightarrow [t_3]}.$$

Rewriting logic is a logic for reasoning correctly about *concurrent systems* having *states*, and evolving by means of *transitions*. The signature of a rewrite theory describes a particular structure for the states of a system—e.g., multiset, binary tree, etc.—so that its states can be distributed according to such a structure. The rewrite rules in the theory describe which *elementary local transitions*

---

[3] To simplify the exposition the rules of the logic are given for the case of *unconditional* rewrite rules. However, all the ideas presented here have been extended to conditional rules in [61] with very general rules of the form

$$r : [t] \longrightarrow [t'] \ \text{ if } \ [u_1] \longrightarrow [v_1] \wedge \ldots \wedge [u_k] \longrightarrow [v_k].$$

This increases considerably the expressive power of rewrite theories.

are possible in the distributed state by concurrent local transformations. The rules of rewriting logic allow us to reason correctly about which *general* concurrent transitions are possible in a system satisfying such a description. Thus, computationally, each rewriting step is a parallel local transition in a concurrent system.

Alternatively, however, we can adopt a logical viewpoint instead, and regard the rules of rewriting logic as *metarules* for correct deduction in a *logical system*. Logically, each rewriting step is a logical *entailment* in a formal system.

2.2   Initial and Free Models

Equational logic enjoys very good model-theoretic properties. In particular, the existence of initial and free models, and of free extensions relative to theory interpretations are very useful and are at the basis of the initial algebra semantics of equational specifications and programming languages. All such good model-theoretic properties are preserved when equational logic is generalized to rewriting logic. For this, the notion of model itself must of course be generalized.

I sketch the construction of initial and free models for a rewrite theory $\mathcal{R} = (\Sigma, E, L, R)$. Such models capture nicely the intuitive idea of a "rewrite system" in the sense that they are systems whose states are E-equivalence classes of terms, and whose transitions are concurrent rewritings using the rules in R. By adopting a logical instead of a computational perspective, we can alternatively view such models as "logical systems" in which formulas are validly rewritten to other formulas by concurrent rewritings which correspond to proofs for the logic in question. Such models have a natural *category* structure, with states (or formulas) as objects, transitions (or proofs) as morphisms, and sequential composition as morphism composition, and in them dynamic behavior exactly corresponds to deduction.

Given a rewrite theory $\mathcal{R} = (\Sigma, E, L, R)$, for which we assume that different labels in L name different rules in R, the model that we are seeking is a category $\mathcal{T}_{\mathcal{R}}(X)$ whose objects are equivalence classes of terms $[t] \in T_{\Sigma,E}(X)$ and whose morphisms are equivalence classes of "proof terms" representing proofs in rewriting deduction, i.e., concurrent $\mathcal{R}$-rewrites. The rules for generating such proof terms, with the specification of their respective domains and codomains, are given below; they just "decorate" with proof terms the rules 1–4 of rewriting logic. Note that we always use "diagrammatic" notation for morphism composition, i.e., $\alpha; \beta$ always means the composition of α *followed by* β.

1. **Identities.** For each $[t] \in T_{\Sigma,E}(X)$, $\dfrac{}{[t] : [t] \longrightarrow [t]}.$

2. **Σ-Structure.** For each $f \in \Sigma_n$, $n \in \mathbb{N}$,
$$\frac{\alpha_1 : [t_1] \longrightarrow [t_1'] \quad \ldots \quad \alpha_n : [t_n] \longrightarrow [t_n']}{f(\alpha_1, \ldots, \alpha_n) : [f(t_1, \ldots, t_n)] \longrightarrow [f(t_1', \ldots, t_n')]}.$$

3. **Replacement.** For each rewrite rule $r : [t(\overline{x}^n)] \longrightarrow [t'(\overline{x}^n)]$ in R,
$$\frac{\alpha_1 : [w_1] \longrightarrow [w_1'] \quad \ldots \quad \alpha_n : [w_n] \longrightarrow [w_n']}{r(\alpha_1, \ldots, \alpha_n) : [t(\overline{w}/\overline{x})] \longrightarrow [t'(\overline{w'}/\overline{x})]}.$$

4. **Composition** $\dfrac{\alpha : [t_1] \longrightarrow [t_2] \quad \beta : [t_2] \longrightarrow [t_3]}{\alpha; \beta : [t_1] \longrightarrow [t_3]}$.

Each of the above rules of generation defines a different operation taking certain proof terms as arguments and returning a resulting proof term. In other words, proof terms form an algebraic structure $\mathcal{P}_\mathcal{R}(X)$ consisting of a graph with nodes $T_{\Sigma,E}(X)$, with identity arrows, and with operations f (for each $f \in \Sigma$), r (for each rewrite rule), and $\_;\_$ (for composing arrows). Our desired model $\mathcal{T}_\mathcal{R}(X)$ is the quotient of $\mathcal{P}_\mathcal{R}(X)$ modulo the following equations:[4]

1. **Category**
   (a) *Associativity.* For all α, β, γ, $(\alpha; \beta); \gamma = \alpha; (\beta; \gamma)$.
   (b) *Identities.* For each $\alpha : [t] \longrightarrow [t']$, $\alpha; [t'] = \alpha$ and $[t]; \alpha = \alpha$.
2. **Functoriality of the Σ-Algebraic Structure.** For each $f \in \Sigma_n$,
   (a) *Preservation of composition.* For all $\alpha_1, \ldots, \alpha_n, \beta_1, \ldots, \beta_n$,

$$f(\alpha_1; \beta_1, \ldots, \alpha_n; \beta_n) = f(\alpha_1, \ldots, \alpha_n); f(\beta_1, \ldots, \beta_n).$$

   (b) *Preservation of identities.* $f([t_1], \ldots, [t_n]) = [f(t_1, \ldots, t_n)]$.
3. **Axioms in** E. For $t(x_1, \ldots, x_n) = t'(x_1, \ldots, x_n)$ an axiom in E, for all $\alpha_1, \ldots, \alpha_n$, $t(\alpha_1, \ldots, \alpha_n) = t'(\alpha_1, \ldots, \alpha_n)$.
4. **Exchange.** For each $r : [t(x_1, \ldots, x_n)] \longrightarrow [t'(x_1, \ldots, x_n)]$ in R,

$$\frac{\alpha_1 : [w_1] \longrightarrow [w_1'] \quad \ldots \quad \alpha_n : [w_n] \longrightarrow [w_n']}{r(\overline{\alpha}) = r(\overline{[w]}); t'(\overline{\alpha}) = t(\overline{\alpha}); r(\overline{[w']})}.$$

Note that the set X of variables is actually a parameter of these constructions, and we need not assume X to be fixed and countable. In particular, for $X = \emptyset$, we adopt the notation $\mathcal{T}_\mathcal{R}$. The equations in 1 make $\mathcal{T}_\mathcal{R}(X)$ a category, the equations in 2 make each $f \in \Sigma$ a functor, and 3 forces the axioms E. The exchange law states that any rewriting of the form $r(\overline{\alpha})$—which represents the *simultaneous* rewriting of the term at the top using rule r *and* "below," i.e., in the subterms matched by the variables, using the rewrites $\overline{\alpha}$—is equivalent to the sequential composition $r(\overline{[w]}); t'(\overline{\alpha})$, corresponding to first rewriting on top with r and then below on the subterms matched by the variables with $\overline{\alpha}$, and is also equivalent to the sequential composition $t(\overline{\alpha}); r(\overline{[w']})$ corresponding to first rewriting below with $\overline{\alpha}$ and then on top with r. Therefore, the exchange law states that rewriting at the top by means of rule r and rewriting "below" using $\overline{\alpha}$ are processes that are independent of each other and can be done either simultaneously or in any order.

Since each proof term is a description of a concurrent computation, what these equations provide is an equational theory of *true concurrency*, allowing us to characterize when two such descriptions specify the same abstract computation. From a logical viewpoint they provide a notion of *abstract proof*, where equivalent syntactic descriptions of the same proof object are identified.

---

[4] In the expressions appearing in the equations, when compositions of morphisms are involved, we always implicitly assume that the corresponding domains and codomains match.

The models $\mathcal{T}_{\mathcal{R}}$ and $\mathcal{T}_{\mathcal{R}}(X)$ are, respectively, the initial model and the free model on the set of generators X for a general category of models of a rewrite theory $\mathcal{R}$; each model interprets the rewrite rules in $\mathcal{R}$ as natural transformations in an underlying category endowed with algebraic structure [61].

2.3   Reflection

Intuitively, a logic is reflective if it can represent its metalevel at the object level in a sound and coherent way. More precisely, Manuel Clavel and I have shown that rewriting logic logic is reflective [23,18,25] in the sense that there is a finitely presented rewrite theory $\mathcal{U}$ that is *universal* in the sense that for any finitely presented rewrite theory $\mathcal{R}$ (including $\mathcal{U}$ itself) we have the following equivalence

$$\mathcal{R} \vdash t \rightarrow t' \;\Leftrightarrow\; \mathcal{U} \vdash \langle \overline{\mathcal{R}}, \overline{t} \rangle \rightarrow \langle \overline{\mathcal{R}}, \overline{t'} \rangle,$$

where $\overline{\mathcal{R}}$ and $\overline{t}$ are terms representing $\mathcal{R}$ and t as data elements of $\mathcal{U}$. Since $\mathcal{U}$ is representable in itself, we can achieve a "reflective tower" with an arbitrary number of levels of reflection, since we have

$$\mathcal{R} \vdash t \rightarrow t' \;\Leftrightarrow\; \mathcal{U} \vdash \langle \overline{\mathcal{R}}, \overline{t} \rangle \rightarrow \langle \overline{\mathcal{R}}, \overline{t'} \rangle \;\Leftrightarrow\; \mathcal{U} \vdash \langle \overline{\mathcal{U}}, \overline{\langle \overline{\mathcal{R}}, \overline{t} \rangle} \rangle \rightarrow \langle \overline{\mathcal{U}}, \overline{\langle \overline{\mathcal{R}}, \overline{t'} \rangle} \rangle \ldots$$

Reflection is a very powerful property. It is systematically exploited in the Maude rewriting logic language implementation [19], that provides key features of the universal theory $\mathcal{U}$ in a built-in module called META-LEVEL. In particular, META-LEVEL has sorts Term and Module, so that the representations $\overline{t}$ and $\overline{\mathcal{R}}$ of a term t and a module $\mathcal{R}$ have sorts Term and Module, respectively. META-LEVEL has also functions meta-reduce($\overline{\mathcal{R}}$, $\overline{t}$) and meta-apply($\overline{\mathcal{R}}$, $\overline{t}$, $\overline{l}$, $\overline{\sigma}$, n) which return, respectively, the representation of the reduced form of a term t using the equations in a module $\mathcal{R}$, and the (representation of the) result of applying a rule labeled l in the module $\mathcal{R}$ to a term t at the top with the $(n+1)$th match consistent with the partial substitution σ. As the universal theory $\mathcal{U}$ that it implements in a built-in fashion, META-LEVEL can also support a reflective tower with an arbitrary number of levels of reflection.

2.4   Executability Issues

How should rewrite theories be executed in practice? First of all, in a general rewrite theory $\mathcal{R} = (\Sigma, E, L, R)$ the equations E can be arbitrary, and therefore, E-equality may be undecidable. Faced with such a general problem, an equally general solution is to transform $\mathcal{R}$ into a rewrite theory $\mathcal{R}^{\sharp} = (\Sigma, \emptyset, L \cup L_E, R \cup E \cup E^{-1})$ in which we view the equations E as rules from left to right (E) and from right to left (E^{-1}), labeled by appropriate new labels L_E. In this way, we reduce the problem of rewriting modulo E to the problem of standard rewriting, since we have the equivalence

$$\mathcal{R} \vdash [t] \rightarrow [t'] \;\Leftrightarrow\; \mathcal{R}^{\sharp} \vdash t \rightarrow t'.$$

In actual specification and programming practice we can do much better than this, because the equational theory (Σ, E) is typically decidable. For computational applications this assumption is reasonable, because in the initial model $T_{\mathcal{R}}$ the algebra $T_{\Sigma,E}$ axiomatizes the *state space* of the concurrent system in question, which should be computable. For applications in which $\mathcal{R}$ is interpreted as a logic, the decidability of (Σ, E) is again the norm, since then $T_{\Sigma,E}$ is interpreted as the set of *formulas* for the logic, which should be computable even if deducibility of one formula from another may be undecidable.

An attractive and commonly occurring form for the decidable equational theory (Σ, E) is with $E = E' \cup A$, where A is a set of equational axioms for which we have a matching algorithm, and E' is a set of Church-Rosser and terminating equations *modulo* A. In these circumstances, a very attractive possibility is to transform $\mathcal{R} = (\Sigma, E' \cup A, L, R)$ into the theory $\mathcal{R}^{\dagger} = (\Sigma, A, L \cup L_{E'}, R \cup E')$. That is, we now view the equations E' as rules added to R, labeled with appropriate new labels $L_{E'}$. In this way, we reduce the problem, of rewriting modulo E to the much simpler problem of rewriting module A, for which, by assumption, we have a matching algorithm.

The question is, of course, under which conditions is this transformation complete. That is, under which conditions do we have an equivalence

$$\mathcal{R} \vdash [t]_E \to [t']_E \quad \Leftrightarrow \quad \mathcal{R}^{\dagger} \vdash [t]_A \to [t']_A.$$

The above equivalence can be guaranteed if $\mathcal{R}$ satisfies the following *weak coherence* condition, that generalizes the coherence condition originally proposed by Viry [90], and a similar condition in [62], Section 12.5.2.1, namely, whenever $\mathcal{R}^{\dagger} \vdash [t]_A \to [t']_A$, we then also have $\mathcal{R}^{\dagger} \vdash can_{E',A}(t) \to [t'']_A$ for some t'' such that $can_{E',A}(t') = can_{E',A}(t'')$, where $can_{E',A}(t)$ denotes the A-equivalence class of the canonical form of t when rewritten modulo A with the equations E'.

Weak coherence, and Viry's coherence, express of course notions of *relative confluence* between the equations and the rules. Methods for checking or for achieving coherence, that generalize to the rewriting logic context similar techniques for equational completion modulo axioms have been proposed by Viry in several papers [90,89].

In actual executable specification and programming practice, one tends to write specifications $\mathcal{R}$ that are weakly coherent, or at least weakly *ground* coherent. This happens because, given an equivalence classes $[t]_E$ the associated canonical form $can_{E',A}(t)$ typically is built up by constructors modulo A, and the patterns in the lefhand sides of the rules in R typically involve only such constructors. The operational semantics of Maude assumes that the specification $\mathcal{R}$ is weakly (ground) coherent, and therefore always reduces terms to canonical form with the equations modulo the given equational axioms A before any step of rewriting with the rules R.

Even if we have a rewrite theory $\mathcal{R}$ that is weakly (ground) coherent, executing such a theory is a nontrivial matter for the following reasons:

1. $\mathcal{R}$ need not be confluent;
2. $\mathcal{R}$ need not be terminating;
3. some rules in $\mathcal{R}$ may have extra variables in their righhand sides and/or in their conditions.

In fact, when we think of a rewrite theory as a means to specify a concurrent system, that can be highly nondeterministic, and that can be forever reacting to new events in its environment, assumptions such as confluence, which intuitively is a weak form of determinism, and termination cannot be expected to hold in general. Logical applications lead to similarly low expectations, since logical deduction can in general progress in many different directions.

Therefore, even though some form of *default execution*[5] may be quite useful in practice for some classes of applications, in general rewrite theories should be executed with a *strategy*, and such strategies may be quite different depending on the given application. Therefore, rewriting logic languages such as Maude and ELAN [52,10,9] support rewriting with strategies.

Reflection can be exploited to define *internal rewriting strategies* [23,24,18], that is, strategies to guide the rewriting process whose semantics can be defined inside the logic by rewrite rules at the metalevel. In fact, there is great freedom for defining many different strategy languages inside Maude. This can be done in a completely user-definable way, so that users are not limited by a fixed and closed strategy language. The idea is to use the operations `meta-reduce` and `meta-apply` as basic strategy expressions, and then to extend the module `META-LEVEL` by additional strategy expressions and corresponding semantic rules. A number of strategy languages that have been defined following this methodology can be found in [18,20].

3   Maude

Maude [19,20] is a language and system developed at SRI International whose modules are theories in rewriting logic. The most general Maude modules are called *system modules*. They have the syntax mod $\mathcal{R}$ endm with $\mathcal{R}$ the rewrite theory in question, expressed with a syntax quite close to the corresponding mathematical notation[6]. The equations E in the equational theory (Σ, E) underlying the rewrite theory $\mathcal{R} = (\Sigma, E, L, R)$ are presented as a union $E = E' \cup A$, with A a set of *equational axioms* introduced as *attributes* of certain operators in the signature Σ—for example, an operator + can be declared associative and commutative by keywords `assoc` and `comm`—and where E' is a set of equations that are assumed to be Church-Rosser and terminating *modulo* the axioms A. Furthermore, as already pointed out in Section 2.4, $\mathcal{R}$ is assumed to be weakly (ground) coherent. Maude supports rewriting modulo different combinations of

---

[5] Maude supports a fair top-down default execution of this kind in which the user may also specify a bound on the number of rewrites.

[6] See [19] for a detailed description of Maude's syntax, which is quite similar to that of OBJ3 [43].

such equational attributes A: operators can be declared with any combination of associative, commutative, and with left, right, or two-sided identity attributes.

Since we can view an equational theory as a rewrite theory whose set L and R of labels and rules are both empty, Maude contains a sublanguage of *functional modules* of the form fmod (Σ, E) endfm, where, as before, $E = E' \cup A$, with E' Church-Rosser and terminating modulo A. The equational logic of functional modules and of the equational part of system modules is membership equational logic [65], an expressive logic whose atomic formulas are equations $t = t'$ and membership predicates $t : s$ with s a sort. Membership equational logic is quite expressive. It supports sorts, subsorts, operator overloading, and equational partiality. But this expressiveness is achieved while being efficiently executable by rewriting and having suitable techniques for completion and for equational theorem proving [11].

Maude has a third class of modules, namely, *object-oriented modules* that specify concurrent object-oriented systems and have syntax omod $\mathcal{R}$ endom, with $\mathcal{R}$ a sugared form for a rewrite theory. That is, object-oriented modules are internally translated into ordinary system modules, but Maude provides a more convenient syntax for them, supporting concepts such as objects, messages, object classes, and multiple class inheritance.

Modules in Maude have an *initial semantics*. Therefore, a system module mod $\mathcal{R}$ endm specifies the initial model $\mathcal{T}_\mathcal{R}$ of the rewrite theory $\mathcal{R}$. Similarly, a functional module fmod (Σ, E) endfm specifies the initial algebra $T_{\Sigma,E}$ of the equational theory (Σ, E).

In addition, like in OBJ3, Maude supports a module algebra with parameterized modules, parameter theories (with "loose" semantics) views (that is, theory interpretations) and module expressions. In Maude, this module algebra, called Full Maude, is defined inside the language by reflection, and is easily extensible with new module composition operations [37,35]. Such a module algebra is one important concrete application of the general *metaprogramming* capabilities made possible by Maude's support of reflection through the META-LEVEL module. As already mentioned in Section 2.4, another important application of reflection is the capacity for defining internal strategy languages that can guide the execution of rewrite theories by means of rewrite rules at the metalevel.

Even though the Maude system is an interpreter, its use of advanced semi-compilation techniques that compile each rewrite rule into matching and replacement automata [38] make it a high-performance system that can reach up to 1.66 million rewrites per second in the free theory case ($A = \emptyset$) and from 130,000 to one million rewrites per second in the associative-commutative case on a 500 MHz Alpha for some applications. In addition, a Maude compiler currently under development can reach up 13 million rewrites per second on the same hardware for some applications.

This means that Maude can be used effectively not only for executable specification purposes, but also for declarative programming. In addition, the design of a mobile language extension of Maude called Mobile Maude is currently under development [36].

Executables for the Maude interpreter, a language manual, a tutorial, examples and case studies, and relevant papers can be found in http://maude.csl.sri.com.

4   Semantic Framework Applications

Semantic framework applications include both concurrency models and formal specification, analysis, and programming of communication systems.

4.1   Concurrency Models

From its beginning, one of the key applications of rewriting logic has been as a semantic framework that can unify a wide range of models of concurrency. The key idea is that rewriting logic is not yet another concurrency *model*; it is instead a *logic* in which different *theories*, specifying different concurrency models, can be defined. The initial models of such rewrite theories then provide "true concurrency" semantic *models* for each desired concurrency style. In fact, many different concurrency models have been formalized in this way within rewriting logic. Detailed surveys for many of these formalizations can be found in [61,63,66]. I list below some of the models that have been studied, giving appropriate references for the corresponding formalizations:

- Concurrent Objects [62]
- Actors [86,85,87]
- Petri nets [61,82]
- CSP (see Section 5.2)
- CCS [67,59]
- Parallel functional programming and the lambda calculus [61,53]
- The π-Calculus [91]
- The UNITY language [61]
- Dataflow [63]
- Concurrent Graph Rewriting [63]
- Neural Networks [63]
- The Chemical Abstract Machine [61]
- Real-time and hybrid systems [75,76]
- Tile models [69,15]

One important feature of these formalizations is that their initial models—which provide a general "true concurrency" semantics—or models closely related to such initial models, are often isomorphic to well-known true concurrency models. This has been shown for several variants of standard rewriting ($E = \emptyset$) [28,53], for the lambda calculus [53], for Petri nets [61,29], for CCS [17], and for concurrent objects [70].

4.2   Specifying, Analyzing, and Programming Communication Systems

I review here the experience gained so far in specifying in Maude communication systems such as communication protocols, cryptographic protocols, active networks algorithms, composable communication services, and distributed software architectures. I also explain how such specifications can be subjected to a flexible range of formal analysis techniques. In addition, Maude can be used not only to specify communication systems, but also to program them in the Mobile Maude language currently under development.

The general idea is to have a series of *increasingly stronger formal methods*, to which a system specification is subjected. Only after less costly and "lighter" methods have been used, leading to a better understanding and to important improvements and corrections of the original specification, is it meaningful and worthwhile to invest effort on "heavier" and costlier methods. Maude and its theorem proving tools [21] can be used to support the following, increasingly stronger methods:

1. *Formal specification.* This process results in a first *formal model* of the system, in which many ambiguities and hidden assumptions present in an informal specification are clarified. A rewriting logic specification provides a formal model in exactly this sense.
2. *Execution of the specification.* Executable rewriting logic specifications can be used directly for simulation and debugging purposes, leading to increasingly better designs. Maude's default interpreter can be used for this purpose.
3. *Model-checking analysis.* Errors in highly distributed and nondeterministic systems not revealed by a particular execution can be found by a model-checking analysis that considers all behaviors of a system from an initial state, up to some level or condition. Maude's metalevel strategies can be used to model check a specification this way.
4. *Narrowing analysis.* By using symbolic expressions with logical variables, one can carry out a symbolic model-checking analysis in which all behaviors not from a single initial state, but from the possibly infinite set of states described by a symbolic expression are analyzed. A planned unification mechanism will allow Maude to perform this narrowing analysis in an efficient way.
5. *Formal Proof.* For highly critical properties it is also possible to carry out a formal proof of correctness, which can be assisted by formal tools such as those in Maude's formal environment [21].

We are still in an early phase in the task of applying rewriting logic to communication systems. However, in addition to the work on foundations, on models of concurrent computation, some recent research by different authors focusing specifically on this area seems quite promising. The paper [32] surveys some of these advances, which can be summarized s follows:

– The paper [34] reports on joint work by researchers at Stanford and SRI with the group led by J.J. García-Luna at the Computer Communications

Research Group at University of California Santa Cruz in which we used
Maude very early in the design of a new reliable broadcast protocol for Active
Networks. In this work, we have developed precise executable specifications of
the new protocol and, by analyzing it through execution and model-checking
techniques, we have found many deadlocks and inconsistencies, and have
clarified incomplete or unspecified assumptions about its behavior.

- Maude has also been applied to the specification and analysis of cryptographic protocols [30], showing how reflective model-checking techniques can
  be used to discover attacks.
- The positive experience with security protocols has led to the adoption of
  Maude by J. Millen and G. Denker at SRI as the basis for giving a formal
  semantics to their new secure protocol specification language CAPSL and as
  the meta-tool used to endow CAPSL with an execution and formal analysis
  environment [33].
- The paper [92] reports joint work with Y. Wang and C. Gunter at the University of Pennsylvania in using Maude to formally specify and analyze a
  PLAN [47] active network algorithm.
- The paper [31] presents an executable specification of a general middle-
  ware architecture for composable distributed communication services such
  as fault-tolerance, security, and so on, that can be composed and can be dy-
  namically added to selected subsets of a distributed communications system.
- In [19] (Appendix E) a substantial case study showing how Maude can
  be used to execute very high level software designs, namely architectural
  descriptions, is presented. It focuses on a difficult case, namely, *heteroge-
  neous* architectures illustrated by a command and control example featuring
  dataflow, message passing, and implicit invocation sub-architectures. Using
  Maude, each of the different subarchitectures can not only be executed, but
  they can also be interoperated in the execution of the resulting overall system.
- As part of a project to represent the Wright architecture description language
  [1] in Maude, Nodelman an Talcott have developed a representation of CSP
  in Maude. This is compatible with existing tools for analyzing CSP specifications, complements them by providing a rich execution environment and
  the ability to analyze non-finite state specifications, and provides a means of
  combining CSP specifications with other notations for specifying concurrent
  systems.
- Najm and Stefani have used rewriting logic to specify computational models
  for open distributed systems [73].
- Talcott has used rewriting logic to define a very general model of open distributed components [85].
- Pita and Martí-Oliet have used the reflective features of Maude to specify
  the management process of broadband telecommunication networks [78,79].
- Nakajima has used rewriting logic to give semantics to the calculus of mobile
  ambients and to specify a Java/ORB implementation of a network management system [74].

– Wirsing and Knapp have defined a systematic translation from object-oriented design notations for distributed object systems to Maude specifications
[93].

As already mentioned, Maude can be used not only for specifying communication systems, but also for programming them. At SRI, F. Durán, S. Eker, P.
Lincoln and I are currently advancing the design of Mobile Maude [36]. This is
an extension of Maude supporting mobile computation that uses reflection in a
systematic way to obtain a simple and general declarative mobile language design. The two key notions are *processes* and *mobile objects*. Processes are located
computational environments where mobile objects can reside. Mobile objects
can move between different process in different locations, and can communicate asynchronously with each other by means of messages. Each mobile object
contains its own code—that is a rewrite theory $\mathcal{R}$—metarepresented as a term
$\overline{\mathcal{R}}$. In this way, reflection endows mobile objects with powerful "higher-order"
capabilities whithin a simple first-order framework.

We expect that Mobile Maude will have good support for *secure* mobile computation for two reasons. Firstly, mobile objects will communicate with each
other and will move from one location to another using state-of-the-art encryp-
tion mechanisms. Secondly, because of the logical basis of Mobile Maude, we
expect to be able to prove critical properties of applications developed in it with
much less effort than what it would be required if the same applications were
developed in a conventional language such as Java.

5   Logical Framework Applications

When we look at rewriting logic from the logical point of view, it becomes a
logical framework in which many other logics can be naturally represented [58].
Furthermore, reflection gives it particularly powerful representational powers, so
that Maude can be used as a formal meta-tool to build many other formal tools
[22].

5.1   A Reflective Logical Framework

The basic reason why rewriting logic can easily represent many other logics is
that the syntax of a logic can typically be defined as an algebraic data type
of formulas, satisfying perhaps some equations, such as associative and commutativity of logical operators like conjunction and disjunctions, or equations
for explicit substitution to equationally axiomatize quantifiers and other binding operators. That is, formulas can typically be expressed as elements of the
initial algebra of a suitable equational theory. Then, the typical inference rules
of a logic are nothing but rewrite rules that rewrite formulas, or proof-theoretic
structures such as sequents, in the deduction process; if an inference rule has side
conditions, then the corresponding rewrite rule is a conditional rule. Therefore,
we can typically represent inference in a logic—or in a theory within a logic in

the case when different signatures are possible—by means of a rewrite theory. Furthermore, such a representation is usually very natural, because it mirrors very closely both the syntax and the inference rules of the original logic.

The representation of a logic $\mathcal{L}$ into rewriting logic can therefore be understood as a *mapping*

$$\Psi : \mathcal{L} \longrightarrow RW\,Logic.$$

This suggest using the theory of general logics [60] to define the space of logics as a *category*, in which the objects are the different logics, and the morphisms are the different mappings translating one logic into another. In general, we can axiomatize a translation Θ from a logic $\mathcal{L}$ to a logic $\mathcal{L}'$ as a morphism

$$\Theta : \mathcal{L} \longrightarrow \mathcal{L}'$$

in the category of logics. A *logical framework* is then a logic $\mathcal{F}$ such that a very wide class of logics can be mapped to it by maps of logics

$$\Psi : \mathcal{L} \longrightarrow \mathcal{F}$$

called *representation maps*, that have particularly good properties such as conservativity[7]. By choosing $\mathcal{F} = RWLogic$ we explore the use of rewriting logic as a logical framework.

One reason why reflection makes rewriting logic particularly powerful as a logical framework is that *maps between logics* can be reified and executed within rewriting logic. We can do so by extending the universal theory $\mathcal{U}$ with equational abstract data type definitions for the data type of theories $Module_{\mathcal{L}}$ for each logic $\mathcal{L}$ of interest. Then, a map $Theta : \mathcal{L} \longrightarrow \mathcal{L}'$ can be reified as an equationally-defined function

$$\overline{\Phi} : Module_{\mathcal{L}} \longrightarrow Module_{\mathcal{L}'}.$$

Similarly, a representation map $\Psi : \mathcal{L} \longrightarrow RW\,Logic$ can be reified by a function

$$\overline{\Psi} : Module_{\mathcal{L}} \longrightarrow Module.$$

If the maps Φ and Ψ are computable, then, by a metatheorem of Bergstra and Tucker [7] it is possible to define the functions $\overline{\Phi}$ and $\overline{\Psi}$ by means of corresponding finite sets of Church-Rosser and terminating equations. That is, such functions can be effectively defined and executed within rewriting logic.

In summary, mappings between logics, including maps representing other logics in rewriting logic, can be internalized and executed within rewriting logic, as indicated in the picture below.

There is yet another reason why rewriting logic reflection is very important for logical framework applications. By reflection rewriting logic can not only be used as a logical framework in which the deduction of a logic $\mathcal{L}$ can be faithfully

---

[7] A map of logics is *conservative* [60] if the translation of a sentence is a theorem if and only if the sentence was a theorem in the original logic. Conservative maps are sometimes said to be *adequate* and *faithful* by other autors.

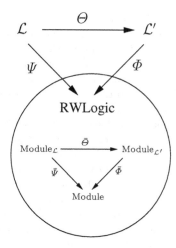

simulated, but also as a *meta-logical framework* in which we can reason about the metalogical properties of a logic $\mathcal{L}$. David Basin, Manuel Clavel and I have begun studying the use of the Maude inductive theorem prover (see Section 5.2) enriched with reflective reasoning principles to prove such metalogical properties [5,6].

5.2 Formal Meta-tool Applications

All this suggest using Maude as a *formal meta-tool* [22] to build other formal tools such as theorem provers or translators between logics. Furthermore, such tools can be built from executable specifications of the inference systems and the mappings themselves. The reflective features of Maude, besides making this possible, also allow building entire *environments* for such formal tools. This can be achieved by using the general meta-parsing and meta-pretty printing functions of the META-LEVEL module, and the LOOP-MODE module, that provides a general read-eval-print loop that can be customized with appropriate rewrite rules to define the interaction with the user for each tool. Our experience suggest that it is much easier to build and maintain formal tools this way than with conventional implementation techniques. Also, because of the high performance of the Maude engine, the tools generated this way often have quite reasonable performance. Of course, special-purpose algorithms can be needed for performance reasons as components of a specific tool, but this is not excluded by our general methodology.

The paper [22] gives a detailed account of different formal tools, including theorem proving tool, execution and analysis environments for different formalisms, a module algebra, and logic traslators, that have been defined in Maude by different authors using the above reflective methodology. I summarize below our experience so far:

An Inductive Theorem Prover. Using the reflective features of Maude, we have built an inductive theorem prover for equational logic specifications [21] that can be used to prove inductive properties of both CafeOBJ specifications [39] and of functional modules in Maude. As already mentioned, this tool can be extended with reflective reasoning principles to reason about the metalogical properties of a logic represented in rewriting logic or, more generally, to prove metalevel properties [6].

A Church-Rosser Checker. We have also built a Church-Rosser checker tool [21] that analyzes equational specifications to check whether they satisfy the Church-Rosser property. This tool can be used to analyze order-sorted [42] equational specifications in CafeOBJ and in Maude. The tool outputs a collection of proof obligations that can be used to either modify the specification or to prove them. Extensions of this tool to perform equational completion and to check coherence of rewrite theories are currently under development.

Full Maude. Maude has been extended with special syntax for object-oriented specifications, and with a rich *module algebra* of parameterized modules and module composition in the Clear/OBJ style [16,43] giving rise to the Full Maude language. All of Full Maude has been formally specified in Maude using reflection [37,35]. This formal specification—about 7,000 lines—is in fact its implementation, which is part of the Maude distribution. Our experience in this regard is very encouraging in several respects. Firstly, because of how quickly it was possible to develop Full Maude. Secondly, because of how easy it will be to maintain it, modify it, and extend it with new features and new module operations. Thirdly, because of the competitive performance with which it can carry out complex module composition and module transformation operations, that makes the interaction with Full Maude quite reasonable.

A Proof Assistant for OCC. Coquand and Huet's calculus of constructions [27] CC, provides higher-order (dependent) types, but it is based on a fixed notion of computation, namely β-reduction, which is quite restrictive in practice. This situation has been addressed by addition of inductive definitions [77][54] and algebraic extensions in the style of abstract data type systems [8]. Also, the idea of overcoming these limitations using some combination of membership equational logic with the calculus of constructions has been suggested as a long-term goal in [50]. Using the general results on the mapping of pure type systems to rewriting logic (see the translation $PTS \rightarrow RWLogic$ below) Mark-Oliver Stehr is currently investigating, and has built a proof assitant for, the *open calculus of constructions* (OCC) an equational variant of the calculus of constructions with an open computational system and a flexible universe hierarchy.

Real-Time Maude. Based on a notion of *real-time rewrite theory* that can naturally represent many existing models of real-time and hybrid systems, and that has a straightforward translation into an ordinary rewrite theory [76], Peter Ölveczky and I are developing an execution and analysis environment for specifications of real-time and hybrid systems called Real-Time Maude. This tool translates real-time rewrite theories into Maude modules and can execute and analyze such theories by means of a library of strategies that can be easily extended by the user to perform other kinds of formal analysis.

Maude Action Tool. Action semantics [71] is a formal framework for specifying the semantics of programming languages in a modular and readable way. Modular structural operational semantics (MSOS) is also a modular formalism for SOS definitions [72] that in particular can give an operational semantics to action notation preserving its modularity features. Christiano Braga, Hermann Haeusler, Peter Mosses, and I are currently developing a tool in Maude for the execution and analysis of programming language definitions written either in action notation or in MSOS notation [12].

A CCS Execution and Verification Environment. Using Maude, Alberto Verdejo and Narciso Martí-Oliet have built a flexible execution environment for CSS based on CCS' operational semantics, or on extensions of such a semantics to traces of actions or to the weak transition relation. In this environment, they can perform a variety of formal analyses using strategies, and they can model check a CCS process with respect to a modal formula in the Hennessy-Milner logic [88].

HOL → Nuprl. The HOL theorem proving system [44] has a rich library of theories that can save a lot of effort by not having to specify from scratch many commonly encountered theories. Howe [49] defined a model-theoretic map from the HOL logic into the logic of Nuprl [26], and implemented such a map to make possible the translation from HOL theories to Nuprl theories. However, the translation itself was carried out by conventional means, and therefore was not in a form suitable for metalogical analysis. Mark-Oliver Stehr and I have recently formally defined in Maude an *executable formal specification* of a proof-theoretic mapping that translates HOL theories into Nuprl theories. Large HOL libraries have already been translated into Nuprl this way. Furthermore, in collaboration with Pavel Naumov, an *abstract version of this mapping* has been proved correct in the categorical framework of general logics and the mapping itself has been used to translate in a systematic way HOL proofs into Nuprl proofs [84].

LinLogic → RWLogic. Narciso Martí-Oliet and I defined two simple mappings from linear logic [41] to rewriting logic: one for its propositional fragment, and another for first-order linear logic [58]. In addition, they explained how—using the fact that rewriting logic is reflective—these mappings could be specified and executed in Maude, thus endowing linear logic with an executable environment. Based on these ideas, Manuel Clavel and Narciso Martí-Oliet have specified in Maude the mapping from propositional linear logic to rewriting logic [18].

Wright → CSP → RWLogic. Architectural description languages (ADLs) can be useful in the early phases of software design, maintenance, and evolution. Furthermore, if architectural descriptions can be subjected to formal analysis, design flaws and inconsistencies can be detected quite early in the design process. The Wright language [2] is an ADL with the attractive feature of having a formal semantics based on CSP [48]. Uri Nodelman, and Carolyn Talcott have recently developed in Maude a prototype executable environment for Wright using two mappings. The first mapping gives an executable formal specification of the CSP semantics of Wright, that is, it associates to each Wright architectural description a CSP process. The second mapping gives an executable rewriting logic semantics to CSP itself. The composition of both mappings provides a prototype executable environment for Wright, which can be used—in conjunction with appropriate rewrite strategies—to both animate Wright architectural descriptions, and to submit such descriptions to different forms of formal analysis.

PTS → RWLogic. *Pure type systems* (PTS) [4] generalize the λ-cube [4], which already contains important higher-order systems, like the simply typed and the (higher-order) polymorphic lambda calculi, a system λP close to the logical framework LF [46], and their combination, the calculus of constructions CC [27]. In [83] Mark-Oliver Stehr and I show how the definition of PTS systems can be easily formalized in membership equational logic and define *uniform pure type systems* (UPTS) a more concrete variant of PTS systems that do not abstract from the treatment of names, but use a uniform notion of names based on CINNI, a new first-order calculus of names and substitutions. UPTS systems solve the problem of closure under α-conversion [80][55] in a very elegant way. Furthermore, [83] descibes how meta-operational aspects of UPTS systems, like type checking and type inference, can be formalized in rewriting logic.

Tile Logic → RWLogic. Tile logic is a flexible formalism for the specification of synchronous concurrent systems [40]. Robero Bruni, Ugo Montanari, and I have defined a mapping from Tile Logic to Rewriting Logic [14,15] that relates the semantic models of both formalisms. This mapping has then been used as a basis for executing tile logic specifications in Maude using appropriate strategies [13].

6   Conclusions

I have introduced the main concepts of rewriting logic and Maude, and have given many examples of how they can be used in a wide range of semantic framework and logical framework applications. I hope to have given enough evidence to suggest that rewriting techniques can be fruitfully extended an applied beyond the equational logic world in the broader semantic context of rewriting logic. There are indeed many theoretical and practical questions awaiting to be investigated.

Acknowledgments

I wish to thank the organizers of RTA 2000 for kindly suggesting the possibility of giving this talk. The work on rewriting logic and Maude is joint work with all my colleagues in the Maude team and in other research teams in the US, Europe, and Japan. It is therefore a pleasure for me to warmly thank all these colleagues, many of whom are mentioned by name in the references, for their many important contributions. As already mentioned, this is not a survey paper; therefore, for other important contributions not mentioned here I refer the reader to [66].

References

1. R. Allen and D. Garlan. Formalizing architectural connection. In *Proceedings 16th International Conference on Software Engineering*, 1994.
2. R. Allen and D. Garlan. A formal basis for architectural connection. *ACM Trans. Soft. Eng. and Meth.*, July 1997.
3. A. Asperti. A logic for concurrency. Unpublished manuscript, November 1987.
4. H. P. Barendregt. Lambda-calculi with types. In S. Abramsky, D. M. Gabbay, and T. Maibaum, editors, *Background: Computational Structures*, volume 2 of *Handbook of Logic in Computer Science*. Oxford: Clarendon Press, 1992.
5. D. Basin, M. Clavel, and J. Meseguer. Reflective metalogical frameworks. In Proc. LFM'99, (Paris, France, September 1999) http://www.cs.bell-labs.com/~felty/LFM99/.
6. D. Basin, M. Clavel, and J. Meseguer. Rewriting logic as a metalogical framework. http://www.informatik.uni-freiburg.de/~basin/pubs/pubs.html.
7. J. Bergstra and J. Tucker. Characterization of computable data types by means of a finite equational specification method. In J. W. de Bakker and J. van Leeuwen, editors, *Automata, Languages and Programming, Seventh Colloquium*, pages 76–90. Springer-Verlag, 1980. LNCS, Volume 81.
8. F. Blanqui, J. Jouannaud, and M. Okada. The calculus of algebraic constructions. In *Proc. RTA'99: Rewriting Techniques and Applications*, Lecture Notes in Computer Science. Springer-Verlag, 1999.
9. P. Borovanský, C. Kirchner, and H. Kirchner. Controlling rewriting by rewriting. In J. Meseguer, editor, *Proc. First Intl. Workshop on Rewriting Logic and its Applications*, volume 4 of *Electronic Notes in Theoretical Computer Science*. Elsevier, 1996. http://www.elsevier.nl/cas/tree/store/tcs/free/noncas/pc/volume4.htm.
10. P. Borovanský, C. Kirchner, H. Kirchner, P.-E. Moreau, and M. Vittek. ELAN: A logical framework based on computational systems. In J. Meseguer, editor, *Proc. First Intl. Workshop on Rewriting Logic and its Applications*, volume 4 of *Electronic Notes in Theoretical Computer Science*. Elsevier, 1996. http://www.elsevier.nl/cas/tree/store/tcs/free/noncas/pc/volume4.htm.
11. A. Bouhoula, J.-P. Jouannaud, and J. Meseguer. Specification and proof in membership equational logic. *Theoretical Computer Science*, 236:35–132, 2000.
12. C. Braga, H. Haeusler, J. Meseguer, and P. Mosses. Maude Action Tool: using reflection to map action semantics to rewriting logic. In *Proceedings of AMAST'2000*. Springer LNCS, 2000. To appear.
13. R. Bruni, J. Meseguer, and U. Montanari. Internal strategies in a rewriting implementation of tile systems. *Proc. 2nd Intl. Workshop on Rewriting Logic and its Applications*, ENTCS, North Holland, 1998.
14. R. Bruni, J. Meseguer, and U. Montanari. Process and term tile logic. Technical Report SRI-CSL-98-06, SRI International, July 1998.
15. R. Bruni, J. Meseguer, and U. Montanari. Executable tile specifications for process calculi. In *Proc. of FASE'99, 2nd Intl. Conf. on Fundamental Approaches to Software Engineering*, volume 1577 of *Lecture Notes in Computer Science*, pages 60–76. Springer-Verlag, 1992.

16. R. Burstall and J. A. Goguen. The semantics of Clear, a specification language. In D. Bjorner, editor, *Proceedings of the 1979 Copenhagen Winter School on Abstract Software Specification*, pages 292–332. Springer LNCS 86, 1980.

17. G. Carabetta, P. Degano, and F. Gadducci. CCS semantics via proved transition systems and rewriting logic. *Proc. 2nd Intl. Workshop on Rewriting Logic and its Applications*, ENTCS, North Holland, 1998.

18. M. Clavel. Reflection in general logics and in rewriting logic, with applications to the Maude language. Ph.D. Thesis, University of Navarre, 1998.

19. M. Clavel, F. Durán, S. Eker, P. Lincoln, N. Martí-Oliet, J. Meseguer, and J. Quesada. Maude: specification and programming in rewriting logic. SRI International, January 1999, http://maude.csl.sri.com.

20. M. Clavel, F. Durán, S. Eker, P. Lincoln, N. Martí-Oliet, J. Meseguer, and J. Quesada. A tutorial on Maude. SRI International, March 2000, http://maude.csl.sri.com.

21. M. Clavel, F. Durán, S. Eker, and J. Meseguer. Building equational proving tools by reflection in rewriting logic. In *Proc. of the CafeOBJ Symposium '98, Numazu, Japan*. CafeOBJ Project, April 1998. http://maude.csl.sri.com.

22. M. Clavel, F. Durán, S. Eker, J. Meseguer, and M.-O. Stehr. Maude as a formal meta-tool. In J. Wing and J. Woodcock, editors, *FM'99 — Formal Methods*, volume 1709 of *Lecture Notes in Computer Science*, pages 1684–1703. Springer-Verlag, 1999.

23. M. Clavel and J. Meseguer. Reflection and strategies in rewriting logic. In J. Meseguer, editor, *Proc. First Intl. Workshop on Rewriting Logic and its Applications*, volume 4 of *Electronic Notes in Theoretical Computer Science*. Elsevier, 1996. http://www.elsevier.nl/cas/tree/store/tcs/free/noncas/pc/volume4.htm.

24. M. Clavel and J. Meseguer. Internal strategies in a reflective logic. In B. Gramlich and H. Kirchner, editors, *Proceedings of the CADE-14 Workshop on Strategies in Automated Deduction (Townsville, Australia, July 1997)*, pages 1–12, 1997.

25. M. Clavel and J. Meseguer. Reflection in conditional rewriting logic. Submitted for publication, 2000.

26. R. Constable. *Implementing Mathematics with the Nuprl Proof Development System*. Prentice Hall, 1987.

27. T. Coquand and G. Huet. The calculus of constructions. *Information and Computation*, 76(2/3):95–120, 1988.

28. A. Corradini, F. Gadducci, and U. Montanari. Relating two categorical models of term rewriting. In J. Hsiang, editor, *Proc. Rewriting Techniques and Applications, Kaiserslautern*, pages 225–240, 1995.

29. P. Degano, J. Meseguer, and U. Montanari. Axiomatizing the algebra of net computations and processes. *Acta Informatica*, 33:641–667, 1996.

30. G. Denker, J. Meseguer, and C. Talcott. Protocol Specification and Analysis in Maude. In N. Heintze and J. Wing, editors, *Proc. of Workshop on Formal Methods and Security Protocols, 25 June 1998, Indianapolis, Indiana*, 1998. http://www.cs.bell-labs.com/who/nch/fmsp/index.html.

31. G. Denker, J. Meseguer, and C. Talcott. Rewriting Semantics of Distributed Meta Objects and Composable Communication Services, 1999. working draft.

32. G. Denker, J. Meseguer, and C. Talcott. Formal specification and analysis of active networks and communication protocols: the Maude experience. In *Proc. DARPA Information Survivability Conference and Exposition DICEX 2000, Vol. 1, Hilton Head, South Carolina, January 2000*, pages 251–265. IEEE, 2000.

33. G. Denker and J. Millen. CAPSL Intermediate Language. In N. Heintze and E. Clarke, editors, *Workshop on Formal Methods and Security Protocols (FMSP'99), July 5, 1999, Trento, Italy (part of FLOC'99)*, 1999. http://cm.bell-labs.com/cm/cs/who/nch/fmsp99/.
34. Denker, G. and Garcia-Luna-Aceves, J.J. and Meseguer, J. and Ölveczky, P. and Raju, J. and Smith, B. and Talcott, C. Specification and Analysis of a Reliable Broadcasting Protocol in Maude. In B. Hajek and R. Sreenivas, editors, *Proc. 37th Allerton Conference on Communication, Control and Computation*, 1999. url-http://www.comm.csl.uiuc.edu/allerton.
35. F. Durán. A reflective module algebra with applications to the Maude language. Ph.D. Thesis, University of Málaga, 1999.
36. F. Durán, S. Eker, P. Lincoln, and J. Meseguer. Principles of Mobile Maude. SRI International, March 2000, http://maude.csl.sri.com.
37. F. Durán and J. Meseguer. An extensible module algebra for Maude. *Proc. 2nd Intl. Workshop on Rewriting Logic and its Applications*, ENTCS, North Holland, 1998.
38. S. Eker. Fast matching in combination of regular equational theories. In J. Meseguer, editor, *Proc. First Intl. Workshop on Rewriting Logic and its Applications*, volume 4 of *Electronic Notes in Theoretical Computer Science*. Elsevier, 1996. http://www.elsevier.nl/cas/tree/store/tcs/free/noncas/pc/volume4.htm.
39. K. Futatsugi and R. Diaconescu. CafeOBJ report. AMAST Series, World Scientific, 1998.
40. F. Gadducci and U. Montanari. The tile model. In G. Plotkin, C. Stirling and M. Tofte, eds., *Proof, Language and Interaction: Essays in Honour of Robin Milner*, MIT Press. Also, TR-96-27, C.S. Dept., Univ. of Pisa, 1996.
41. J.-Y. Girard. Linear Logic. *Theoretical Computer Science*, 50:1–102, 1987.
42. J. Goguen and J. Meseguer. Order-sorted algebra I: Equational deduction for multiple inheritance, overloading, exceptions and partial operations. *Theoretical Computer Science*, 105:217–273, 1992.
43. J. Goguen, T. Winkler, J. Meseguer, K. Futatsugi, and J.-P. Jouannaud. Introducing OBJ. In *Software Engineering with OBJ: Algebraic Specification in Action*. Kluwer, 2000.
44. M. Gordon. *Introduction to HOL: A Theorem Proving Environment*. Cambridge University Press, 1993.
45. C. Gunter and V. Gehlot. Nets as tensor theories. Technical Report MS-CIS-89-68, Dept. of Computer and Information Science, University of Pennsylvania, 1989.
46. R. Harper, F. Honsell, and G. Plotkin. A framework for defining logics. *Journal of the Association Computing Machinery*, 40(1):143–184, 1993.
47. M. Hicks, P. Kakkar, J. T. Moore, C. A. Gunter, and S. Nettles. PLAN: A Packet Language for Active Networks. In *Proceedings of the Third ACM SIGPLAN International Conference on Functional Programming Languages*, pages 86–93. ACM, 1998.
48. C. Hoare. *Communicating Sequential Processes*. Prentice Hall, 1985.
49. D. J. Howe. Semantic foundations for embedding HOL in Nuprl. In M. Wirsing and M. Nivat, editors, *Algebraic Methodology and Software Technology*, volume 1101 of *Lecture Notes in Computer Science*, pages 85–101, Berlin, 1996. Springer-Verlag.
50. J. P. Jouannaud. Membership equational logic, calculus of inductive constructions, and rewrite logic. In *2nd Workshop on Rewrite Logic and Applications*, 1998.
51. C. Kirchner and H. Kirchner. (eds.). *Proc. 2nd Intl. Workshop on Rewriting Logic and its Applications*, ENTCS, North Holland, 1998.

52. C. Kirchner, H. Kirchner, and M. Vittek. Designing constraint logic programming languages using computational systems. In V. Saraswat and P. van Hentenryck, editors, *Principles and Practice of Constraint Programming: The Newport Papers*, pages 133–160. MIT Press, 1995.

53. C. Laneve and U. Montanari. Axiomatizing permutation equivalence. *Mathematical Structures in Computer Science*, 6:219–249, 1996.

54. Z. Luo. *Computation and Reasoning: A Type Theory for Computer Science.* International Series of Monographs on Computer Science. Oxford University Press, 1994.

55. L. Magnussen. *The Implementation of ALF – a Proof Editor based on Martin-Löf's Monomorphic Type Theory with Explicit Substitutions.* PhD thesis, University of Göteborg, Dept. of Computer Science, 1994.

56. N. Martí-Oliet and J. Meseguer. From Petri nets to linear logic. In D. P. et al., editor, *Category Theory and Computer Science*, pages 313–340. Springer LNCS 389, 1989. Final version in *Mathematical Structures in Computer Science*, 1:69–101, 1991.

57. N. Martí-Oliet and J. Meseguer. From Petri nets to linear logic through categories: a survey. *Intl. J. of Foundations of Comp. Sci.*, 2(4):297–399, 1991.

58. N. Martí-Oliet and J. Meseguer. Rewriting logic as a logical and semantic framework. Technical Report SRI-CSL-93-05, SRI International, Computer Science Laboratory, August 1993. To appear in D. Gabbay, ed., *Handbook of Philosophical Logic*, Kluwer Academic Publishers.

59. N. Martí-Oliet and J. Meseguer. Rewriting logic as a logical and semantic framework. In J. Meseguer, editor, *Proc. First Intl. Workshop on Rewriting Logic and its Applications*, volume 4 of *Electronic Notes in Theoretical Computer Science*. Elsevier, 1996. http://www.elsevier.nl/cas/tree/store/tcs/free/noncas/pc/volume4.htm.

60. J. Meseguer. General logics. In H.-D. E. et al., editor, *Logic Colloquium'87*, pages 275–329. North-Holland, 1989.

61. J. Meseguer. Conditional rewriting logic as a unified model of concurrency. *Theoretical Computer Science*, 96(1):73–155, 1992.

62. J. Meseguer. A logical theory of concurrent objects and its realization in the Maude language. In G. Agha, P. Wegner, and A. Yonezawa, editors, *Research Directions in Concurrent Object-Oriented Programming*, pages 314–390. MIT Press, 1993.

63. J. Meseguer. Rewriting logic as a semantic framework for concurrency: a progress report. In *Proc. CONCUR'96, Pisa, August 1996*, pages 331–372. Springer LNCS 1119, 1996.

64. J. Meseguer. (ed.). *Proc. First Intl. Workshop on Rewriting Logic and its Applications*, ENTCS, North Holland, 1996.

65. J. Meseguer. Membership algebra as a semantic framework for equational specification. In F. Parisi-Presicce, ed., *Proc. WADT'97*, 18–61, Springer LNCS 1376, 1998.

66. J. Meseguer. Research directions in rewriting logic. In U. Berger and H. Schwichtenberg, editors, *Computational Logic, NATO Advanced Study Institute, Marktoberdorf, Germany, July 29 – August 6, 1997.* Springer-Verlag, 1999.

67. J. Meseguer, K. Futatsugi, and T. Winkler. Using rewriting logic to specify, program, integrate, and reuse open concurrent systems of cooperating agents. In *Proceedings of the 1992 International Symposium on New Models for Software Architecture, Tokyo, Japan, November 1992*, pages 61–106. Research Institute of Software Engineering, 1992.

68. J. Meseguer and U. Montanari. Petri nets are monoids. *Information and Computation*, 88:105–155, 1990.
69. J. Meseguer and U. Montanari. Mapping tile logic into rewriting logic. in F. Parisi-Presicce, ed., Proc. WADT'97, Springer LNCS 1376, 1998.
70. J. Meseguer and C. Talcott. A partial order event model for concurrent objects. In *Proc. CONCUR'96, Eindhoven, The Netherlands, August 1999*, pages 415–430. Springer LNCS 1664, 1999.
71. P. Mosses. *Action Semantics*. Cambridge University Press, 1992.
72. P. Mosses. Foundations of modular SOS. In *Proceedings of MFCS'99, 24th International Symposium on Mathematical Foundations of Computer Science*, pages 70–80. Springer LNCS 1672, 1999.
73. E. Najm and J.-B. Stefani. Computational models for open distributed systems. In H. Bowman and J. Derrick, editors, *Formal Methods for Open Object-based Distributed Systems, Vol. 2*, pages 157–176. Chapman & Hall, 1997.
74. S. Nakajima. Encoding mobility in CafeOBJ: an exercise of describing mobile code-based software architecture. In *Proc. of the CafeOBJ Symposium '98, Numazu, Japan.* CafeOBJ Project, April 1998.
75. P. C. Ölveczky and J. Meseguer. Specifying real-time systems in rewriting logic. In J. Meseguer, editor, *Proc. First Intl. Workshop on Rewriting Logic and its Applications*, volume 4 of *Electronic Notes in Theoretical Computer Science*. Elsevier, 1996. http://www.elsevier.nl/cas/tree/store/tcs/free/noncas/pc/volume4.htm.
76. P. C. Ölveczky and J. Meseguer. Specification of real-time and hybrid systems in rewriting logic. Submitted for publication, 1999.
77. C. Paulin-Mohring. Inductive Definitions in the system Coq – Rules and Properties. In M. Bezem and J. . F. Groote, editors, *Typed Lambda Calculi and Applications, International Conference on Typed Lambda Calculi and Applications, TLCA 93*, volume 664 of *Lecture Notes in Computer Science*. Springer Varlag, 1993.
78. I. Pita and N. Martí-Oliet. A Maude specification of an object oriented database model for telecommunication networks. In J. Meseguer, editor, *Proc. First Intl. Workshop on Rewriting Logic and its Applications*, volume 4 of *Electronic Notes in Theoretical Computer Science*. Elsevier, 1996. http://www.elsevier.nl/cas/tree/store/tcs/free/noncas/pc/volume4.htm.
79. I. Pita and N. Martí-Oliet. Using reflection to specify transaction sequences in rewriting logic. In J. Fiadeiro, editor, *Recent Trends in Algebraic Development Techniques*, volume 1589 of *Lecture Notes in Computer Science*, pages 261–276. Springer-Verlag, 1999.
80. R. Pollack. Closure under alpha-conversion. In H. Barendregt and T. Nipkow, editors, *Types for Proofs and Programs: International Workshop TYPES'93, Nijmegen, May 1993, Selected Papers.*, volume 806 of *Lecture Notes in Computer Science*, pages 313–332. Springer-Verlag, 1993.
81. W. Reisig. *Petri Nets*. Springer-Verlag, 1985.
82. M.-O. Stehr. A rewriting semantics for algebraic nets. In C. Girault and R. Valk, editors, *Petri Nets for System Engineering – A Guide to Modelling, Verification, and Applications*. Springer-Verlag. To appear.
83. M.-O. Stehr and J. Meseguer. Pure type systems in rewriting logic. In *Proc. of LFM'99: Workshop on Logical Frameworks and Meta-languages, Paris, France, September 28, 1999*.
84. M.-O. Stehr, P. Naumov, and J. Meseguer. A proof-theoretic approach to the HOL-Nuprl connection with applications to proof translation. Manuscript, SRI International, February 2000. http://www.csl.sri.com/~stehr/fi_eng.html,

85. C. L. Talcott. An actor rewrite theory. In J. Meseguer, editor, *Proc. First Intl. Workshop on Rewriting Logic and its Applications*, volume 4 of *Electronic Notes in Theoretical Computer Science*. Elsevier, 1996.
http://www.elsevier.nl/cas/tree/store/tcs/free/noncas/pc/volume4.htm.

86. C. L. Talcott. Interaction semantics for components of distributed systems. In E. Najm and J.-B. Stefani, editors, *Formal Methods for Open Object-based Distributed Systems*, pages 154–169. Chapman & Hall, 1997.

87. C. L. Talcott. Actor theories in rewriting logic, 1999. Submitted for publication.

88. A. Verdejo and N. Martí-Oliet. Executing and verifying CCS in Maude. Technical Report 99-00, Dto. Sistemas Informáticos y Programación, Universidad Complutense, Madrid; also, http://maude.csl.sri.com.

89. P. Viry. Rewriting modulo a rewrite system. TR-95-20, C.S. Department, University of Pisa, 1996.

90. P. Viry. Rewriting: An effective model of concurrency. In C. Halatsis et al., editors, *PARLE'94, Proc. Sixth Int. Conf. on Parallel Architectures and Languages Europe, Athens, Greece, July 1994*, volume 817 of *LNCS*, pages 648–660. Springer-Verlag, 1994.

91. P. Viry. Input/output for ELAN. In J. Meseguer, editor, *Proc. First Intl. Workshop on Rewriting Logic and its Applications*, volume 4 of *Electronic Notes in Theoretical Computer Science*. Elsevier, 1996.
http://www.elsevier.nl/cas/tree/store/tcs/free/noncas/pc/volume4.htm.

92. B.-Y. Wang, J. Meseguer, and C. A. Gunter. Specification and formal analysis of a PLAN algorithm in Maude. To appear in Proc. DSVV 2000.

93. M. Wirsing and A. Knapp. A formal approach to object-oriented software engineering. In J. Meseguer, editor, *Proc. First Intl. Workshop on Rewriting Logic and its Applications*, volume 4 of *Electronic Notes in Theoretical Computer Science*. Elsevier, 1996.
http://www.elsevier.nl/cas/tree/store/tcs/free/noncas/pc/volume4.htm.

Tree Automata and Term Rewrite Systems
(Extended Abstract)

Sophie Tison

LIFL, Bât M3, Université Lille 1
F59655 Villeneuve d'Ascq cedex, France
`tison@lifl.fr`

Abstract. This tutorial is devoted to tree automata. We will present some of the most fruitful applications of tree automata in rewriting theory and we will give an outline of the current state of research on tree automata. We give here just a sketch of the presentation. The reader can also refer to the on-line book "Tree Automata and Their Applications" [CDG+97].

1 Introduction

Tree Automata theory and term rewriting theory are strongly connected [Dau94, Ott99]. On one side, tree automata can be viewed as a subclass of ground term rewrite systems [Bra69a, Bra69b]. On the other hand, tree automata have been used successfully as decision tools in rewriting theory. In this tutorial, we will present some of the most fruitful applications of tree automata in rewriting theory and we point out some promising research directions in this area. For definitions and properties of tree automata the reader can refer to [GS96, CDG+97]. Most of the results we mention here and more references can be found in [CDG+97].

2 Classical Tree Automata and Rewrite Systems

If you want to use tree automata in rewriting theory, the ideal situation occurs when the reducibility relation is recognizable: a binary relation is said recognizable Iff the set of its encodings is a recognizable tree language; a couple is just encoded by overlapping its two terms: e.g. the [f(a,b), f(f(a,a,),a)] will be encoded into [f,f] ([a,f]([⊥,a],[⊥,a]),[b,a]). E.g., reducibility relations are recognizable for ground rewrite systems (more generally for linear term rewriting system such that left and right members of the rules do not share variables). Now, let us consider the following logical theory: the set of formulas is the set of all first-order formulas using no function symbols and a single binary predicate symbol, the predicate symbol is interpreted as the reducibility relation associated with a given rewrite system. When the reducibility relation is recognizable, you get easily the decidability of the theory, thanks to the good closure and decision properties of tree automata; this implies decidability of any property expressible

L. Bachmair (Ed.): RTA 2000, LNCS 1833, pp. 27–30, 2000.

in this theory, like confluence. Furthermore, under some counditions, you can enrich the theory for expressing termination properties [DT90].

Clearly, recognizability of the reducibility relation is a very strong property restricted to limited subclasses of rewrite systems. Now, you can just require that the reducibility relation preserves recognizability, i.e. that the set of descendants (resp. the ancestors) of a recognizable tree language is recognizable. (Let us note that this is not the case even for linear systems). For example, reachability can then be easily reduced to membership to recognizable tree language. If preservation of regularity is undecidable [GT95], some conditions ensure it and it leads to decidability of reachability and joinability for some subclasses of term rewriting systems. ([Sal88, CDGV94, Jac96, FJSV98, NT99]).

You can also require recognizability of the language of ground normal forms: it provides for example a very simple procedure for testing the ground reducibility of a term. Clearly, the set of ground normal forms is a regular tree language for left-linear rewrite systems. Moreover, recognizability of the set of normal forms has been proven decidable [VG92, Kou92]. Finally, recognizability of the set of normalizable terms (it's clearly ensured when the set of ground normal forms is recognizable and the inverse reducibilty relation preseves recognizability) ensures decidability of the sequentiality of the system, when left-linear [Com95].

All the previous approaches provide good decision procedures but only for very restricted classes of rewrite systems. If you are interested in one particular rewrite system, you can also try to prove "experimentally" its good behavior w.r.t. to recognizability. E.g. J. Waldmann has proven by computer the recognizability of the set of normalizing S-terms [Wal98]. Some pointers to software for manipulating tree regular expressions and automata can be found in [CDG+97]. But a question rises: How far can we go in using tree automata when describing properties of term rewriting systems? Can we find new "interesting" classes of t.r.s. with good properties w.r.t. recognizability?

3   How to Go beyond the Limits of Usual Tree Automata?

A way of going beyond the limits of the previous approaches is to consider approximation of rewrite systems. For example, Comon and Jacquemard study by these means reduction strategies and sequentiality [Com95, Jac96]. More recently, T. Genet and F. Klay compute regular over-approximations of the set of the descendants [Gen98] and use them for the verification of cryptographic protocols [GK00].

But to go beyond the limits of tree automata, you can also use extensions of tree automata. The idea is roughly to enrich the notion of tree automata for dealing with the non-linearity while keeping good closure and decision properties. Several classes have been defined in this view and have been sucessfully applied in term rewriting theory. For example, the reduction automata provide decision procedures for emptiness and finiteness of the language of ground normal forms for every term rewriting system and they give a new procedure for testing ground reducibility [Pla85, DCC95, CJ97, CJ94]. Tree automata with tests between brothers which have good decision properties allow also to get some new results for non-linear rewrite systems [CSTT99, STT99]. Let us finally cite also the powerful notion of tree t-uple synchronized grammars which has been used in unification theory end rewriting theory [LR98, LR99].

Of course, you can combine these two last approaches, e.g. by approximating the set of the descendants by extended recognizable tree languages. This opens new prospects and will require design of software for dealing with extended tree automata.

References

[Bra69a]   W. S. Brainerd. Semi-thue systems and representations of trees. In *Proc. 10th IEEE Symposium on Foundations of Computer Science (FOCS)*, pages 240–244, 1969.

[Bra69b]   W. S. Brainerd. Tree generating regular systems. *Information and Control*, 14(2):217–231, February 1969.

[CDG+97]   H. Comon, M. Dauchet, R. Gilleron, F. Jacquemard, D. Lugiez, S. Tison, and M. Tommasi. Tree automata techniques and applications. Available on: http://l3ux02.univ-lille3.fr/tata, 1997.

[CDGV94]   J.L. Coquide, M. Dauchet, R. Gilleron, and S. Vagvolgyi. Bottom-up tree pushdown automata : Classification and connection with rewrite systems. *Theorical Computer Science*, 127:69–98, 1994.

[CJ94]   H. Comon and F. Jacquemard. Ground reducibility and automata with disequality constraints. In Patrice Enjalbert, editor, *11th Annual Symposium on Theoretical Aspects of Computer Science*, volume 775 of *LNCS*, pages 151–162, 1994.

[CJ97]   H. Comon and F. Jacquemard. Ground reducibility is EXPTIME-complete. In *Proceedings, 12th Annual IEEE Symposium on Logic in Computer Science*, pages 26–34. IEEE Computer Society Press, 1997.

[Com95]   H. Comon. Sequentiality, second-order monadic logic and tree automata. In *Proceedings, Tenth Annual IEEE Symposium on Logic in Computer Science*. IEEE Computer Society Press, 26–29 June 1995.

[CSTT99]   A.-C. Caron, F. Seynhaeve, S. Tison, and M. Tommasi. Deciding the satisfiability of quantifier free formulae on one-step rewriting. In M. Rusinowitch F. Narendran, editor, *10th International Conference on Rewriting Techniques and Applications*, volume 1631 of *LNCS*, Trento, Italy, 1999. Springer Verlag.

[Dau94]   M. Dauchet. Rewriting and tree automata. In H. Comon and J.-P. Jouannaud, editors, *Proc. Spring School on Theoretical Computer Science: Rewriting*, LNCS, Odeillo, France, 1994. Springer Verlag.

[DCC95]   M. Dauchet, A.-C. Caron, and J.-L. Coquidé. Reduction properties and automata with constraints. *Journal of Symbolic Computation*, 20:215–233, 1995.

[DT90]   M. Dauchet and S. Tison. The theory of ground rewrite systems is decidable. In *Proceedings, Fifth Annual IEEE Symposium on Logic in Computer Science*, pages 242–248. IEEE Computer Society Press, 4–7 June 1990.

[FJSV98]   A. Flp, E. Jurvanen, M. Steinby, and S. Vagvlgy. On one-pass term rewriting. In L. Brim, J. Gruska, and J. Zlatusaksv, editors, *Proceedings of Mathematical Foundations of Computer Science*, volume 1450 of *LNCS*, pages 248–256. Springer Verlag, 1998.

[Gen98]   T. Genet. Decidable approximations of sets of descendants and sets of normal forms. In T. Nipkow, editor, *9th International Conference on Rewriting Techniques and Applications*, volume 1379 of *LNCS*, pages 151–165, Tsukuba, Japan, 1998. Springer Verlag.

[GK00]   T. Genet and F. Klay. Rewriting for cryptographic protocol verification. *Technical Report, INRIA, 2000, to appear in CADE2000*, 2000.

[GS96]   F. Gécseg and M. Steinby. Tree languages. In G. Rozenberg and A. Salomaa, editors, *Handbook of Formal Languages*, volume 3, pages 1–68. Springer Verlag, 1996.

[GT95]   R. Gilleron and S. Tison. Regular tree languages and rewrite systems. *Fundamenta Informaticae*, 24:157–176, 1995.

[Jac96]   F. Jacquemard. Decidable approximations of term rewriting systems. In H. Ganzinger, editor, *Proceedings. Seventh International Conference on Rewriting Techniques and Applications*, volume 1103 of *LNCS*, 1996.

[Kou92]   E. Kounalis. Testing for the ground (co)-reducibility in term rewriting systems. *Theorical Computer Science*, 106(1):87–117, 1992.

[LR98]   S. Limet and P. Réty. Solving Disequations modulo some Class of Rewrite System. In T. Nipkow, editor, *9th International Conference on Rewriting Techniques and Applications*, volume 1379 of *LNCS*, pages 121–135, Tsukuba, Japan, 1998. Springer Verlag.

[LR99]   S. Limet and P. Réty. A new result about the decidability of the existential one-step rewriting theory. In M. Rusinowitch F. Narendran, editor, *10th International Conference on Rewriting Techniques and Applications*, volume 1631 of *LNCS*, Trento, Italy, 1999. Springer Verlag.

[NT99]   T. Nagaya and Y. Toyama. Decidability for left-linear growing term rewriting systems. In M. Rusinowitch F. Narendran, editor, *10th International Conference on Rewriting Techniques and Applications*, volume 1631 of *LNCS*, Trento, Italy, 1999. Springer Verlag.

[Ott99]   F. Otto. On the connections between rewriting and formal languauge theory. In M. Rusinowitch F. Narendran, editor, *10th International Conference on Rewriting Techniques and Applications*, volume 1631 of *LNCS*, pages 332–355, Trento, Italy, 1999. Springer Verlag.

[Pla85]   D.A. Plaisted. Semantic confluence tests and completion method. *Information and Control*, 65:182–215, 1985.

[Sal88]   K. Salomaa. Deterministic tree pushdown automata and monadic tree rewriting systems. *Journal of Comput. and Syst. Sci.*, 37:367–394, 1988.

[STT99]   F. Seynhaeve, S. Tison, and M. Tommasi. Homomorphisms and concurrent term rewriting. In G. Ciobanu and G. Paun, editors, *Proceedings of the twelfth International Conference on Fundamentals of Computation theory*, number 1684 in Lecture Notes in Computer Science, Iasi, Romania, 1999.

[VG92]   S. Vágvölgyi and R. Gilleron. For a rewrite system it is decidable whether the set of irreducible ground terms is recognizable. *Bulletin of the European Association of Theoretical Computer Science*, 48:197–209, October 1992.

[Wal98]   J. Waldmann. Normalization of s-terms is decidable. In T. Nipkow, editor, *9th International Conference on Rewriting Techniques and Applications*, volume 1379 of *LNCS*, pages 138–150, Tsukuba, Japan, 1998. Springer Verlag.

Absolute Explicit Unification

Nikolaj Bjørner[1]* and César Muñoz[2]**

[1] Kestrel Institute
nikolaj@kestrel.edu
[2] Institute for Computer Applications in Science and Engineering (ICASE)
munoz@icase.edu

Abstract. This paper presents a system for explicit substitutions in Pure Type Systems (PTS). The system allows to solve type checking, type inhabitation, higher-order unification, and type inference for PTS using purely first-order machinery. A novel feature of our system is that it combines substitutions and variable declarations. This allows as a side-effect to type check let-bindings. Our treatment of meta-variables is also explicit, such that instantiations of meta-variables is internalized in the calculus. This produces a confluent λ-calculus with distinguished holes and explicit substitutions that is insensitive to α-conversion, and allows directly embedding the system into rewriting logic.

1 Introduction

Explicit substitutions provide a convenient framework for encoding higher-order typed λ-calculus using first-order machinery. In particular, this allows to integrate higher-order unification with first-order provers, rewriting logic, and to delay evaluation and resolve scoping when type checking dependent-typed terms. On the other hand, several problems related to type theory, such as type checking (with definitions), type inference, checking equality of well typed terms, proof-term refinement, and the inhabitation problem, can be solved using the same machinery, once it is properly developed. We therefore here combine explicit substitutions, variable declarations as explicit substitutions, and explicit instantiation of meta-variables using first-order rewrite rules. The combination is formulated for pure type systems, and applies therefore for arbitrary type systems as those of the λ-cube. A higher-order unification procedure for systems in the λ-cube is a particular payoff.

It is well-known that definitions, i.e., let-in expressions, are problematic in dependent-type systems [24, 4]. Two approaches have been used to extend the λ-calculus with definitions. Severi and Poll [24] consider definitions as terms and extend the reduction relation to unfold definitions during the typing process. Bloo *et al* [4] do not extend the syntax of terms (although they use a different

---

* This research is supported by NASA under award No NAG 2-1227, and DARPA under Contract No. F30602-96-C-0282.
** This research was supported by the National Aeronautics and Space Administration under NASA Contract NAS1-97046.

L. Bachmair (Ed.): RTA 2000, LNCS 1833, pp. 31–46, 2000.

notation for terms called *item notation*), but they consider definitions as part of the contexts. Combining the two approaches, Bloo [3] proposes a calculus of explicit substitutions where substitutions are also part of the contexts. In Bloo's system, α-conversion remains as an implicit rule and non-explicit substitutions are still required to unfold definitions on types. The unfolding of definitions on types is inconvenient when they contain meta-variables. In this paper, we propose a system where context are special case of explicit substitutions, we can safely extend our calculus with definitions and meta-variables, and yet be completely insensitive to α-conversion.

Finally, our extended calculus of explicit substitutions, where instantiations are also explicit, realizes the spirit of delayed evaluation to also cover instantiation of meta-variables. Indeed, our calculus can be considered a λ-calculus of contexts[1] i.e., a λ-calculus with place-holders where the mechanism of filling in holes is explicit.

The rest of this paper is organized as follows. We continue with a short review of *pure type systems* and *explicit substitutions*. More detailed descriptions can be found in [2] and [1]. Section 2 introduces the system $\text{PTS}_{\mathcal{L}}$, which is a system of pure types with explicit substitutions. Section 3 presents two novel aspects of $\text{PTS}_{\mathcal{L}}$ with respect to other proposals: definitions and explicit instantiations. Section 4 summarizes a meta-theoretical investigation, and Section 5 applies the system to higher-order unification and related typing problems.

Pure Type Systems: *Pure Type Systems* [14, 2] is a formalism to describe a family of type systems which generalizes the cube of typed lambda calculi. A *Pure Type System*, PTS for short, is a typed λ-calculus given by a triple $(\mathcal{S}, \mathcal{A}, \mathcal{R})$, where $\mathcal{S}$ is a base set of *sorts*, $\mathcal{A} \subseteq \mathcal{S} \times \mathcal{S}$ is a set of *axioms*, and $\mathcal{R} \subseteq \mathcal{S} \times \mathcal{S} \times \mathcal{S}$ is a set of *rules*. A sort s_1 is a *top sort* if it does not exist a sort s_2 such that $(s_1, s_2) \in \mathcal{A}$.

The set of PTS *(pseudo-)terms* is formed by variables $x, y, \ldots$, applications: $(M\ N)$, abstractions: $\lambda x{:}A.M$, products: $\Pi x{:}A.B$, and sorts $s \in \mathcal{S}$. Abstractions and products are binding structures. As usual, in higher-order formal systems, terms are considered equivalents modulo α-conversion, i.e., renaming of bound variables. Notice that there is no syntactical distinction between terms denoting objects and terms denoting types. For sake of readability, we use the uppercase Roman letters $A, B, \ldots$ to range over terms denoting types. The notation $A \to B$ is used for $\Pi x{:}A.B$ when x does not appear free in B.

A PTS type judgment has the form $\Gamma \vdash M : A$ where Γ is a *typing context*, i.e., a list $x_1 : A_1, \ldots, x_n : A_n$ of type assignments for variables, where $n \geq 0$ and $x_i \neq x_j$ for $i \neq j$. We use the Greek letters Γ, Δ to range over typing contexts. A variable x is *fresh* (in Γ) if $x \neq x_i$ for $1 \leq i \leq n$. The typing rules of the PTS defined by $(\mathcal{S}, \mathcal{A}, \mathcal{R})$ are shown in Figure 1. In rules (Start) and (Weak), we assume that x is fresh in Γ.

---

[1] In a typed calculus, the word *context* has two meanings: variable declaration and expression with a distinguished hole. When confusion may arise, we will write *typing context* in the former case and *expression context* in the latter one.

$$\frac{(s_1, s_2) \in \mathcal{A}}{\vdash s_1 : s_2} \text{ (Sort)} \qquad \frac{\Gamma \vdash A : s}{\Gamma, x : A \vdash x : A} \text{ (Start)}$$

$$\frac{\Gamma \vdash A : s \quad \Gamma \vdash M : B}{\Gamma, x : A \vdash M : B} \text{ (Weak)} \qquad \frac{\Gamma \vdash A : s_1 \quad \Gamma, x : A \vdash B : s_2 \quad (s_1, s_2, s_3) \in \mathcal{R}}{\Gamma \vdash \Pi x{:}A.B : s_3} \text{ (Prod)}$$

$$\frac{\Gamma, x : A \vdash M : B \quad \Gamma \vdash \Pi x{:}A.B : s}{\Gamma \vdash \lambda x{:}A.M : \Pi x{:}A.B} \text{ (Abs)} \qquad \frac{\Gamma \vdash M : \Pi x{:}A.B \quad \Gamma \vdash N : A}{\Gamma \vdash (M\ N) : B[N/x]} \text{ (Appl)} \qquad \frac{\Gamma \vdash M : A \quad \Gamma \vdash B : s \quad A \equiv B}{\Gamma \vdash M : B} \text{ (Conv)}$$

Fig. 1. PTS typing rules

We here consider the extension of PTS with the η-equality, i.e., the relation $\equiv$ in rule (Conv) is the congruence relation induced by the rewrites β: $(\lambda x{:}A.M\ N) \longrightarrow M[N/x]$ and η: $\lambda y{:}A.(M\ y) \longrightarrow M$ where y is not free in M. Recall that $M[N/x]$ denotes atomic substitution of the free occurrences of x in M by N, with renaming of bound variables in M when necessary.

A PTS is *functional* if (1) $(s, s_1), (s, s_2) \in \mathcal{A}$ implies $s_1 = s_2$ and (2) $(s, s', s_1), (s, s', s_2) \in \mathcal{R}$ implies $s_1 = s_2$. The *cube of type systems* (Barendregt's cube) [2] are the PTS such that $\mathcal{S} = \{\star, \square\}$, $\mathcal{A} = \{(\star, \square)\}$, $(\star, \star, \star) \in \mathcal{R}$, and for (s_1, s_2, s_3) in $\mathcal{R}$, it holds $s_2 = s_3$. Well-known type systems of Barendregt's cube: the simply-typed λ-calculus, system F, LF, and the calculus of constructions, are all functional. For example, the simply-typed λ-calculus has $\mathcal{R} = \{(\star, \star, \star)\}$.

Explicit Substitutions: The $\lambda\sigma$-calculus [1] is a first-order rewrite system with two sorts of expressions: *terms* and *substitutions*. By using substitutions as first-class objects and de Bruijn indices notation for variables, the $\lambda\sigma$-calculus allows a first-order encoding of the λ-calculus. In consequence, technical nuisances due to higher-order aspects of the λ-calculus can be minimized or eliminated (e.g., α-conversion) in explicit substitution calculi.

The rewrite system of $\lambda\sigma$ includes a surjective-pairing rule (SCons): $\underline{1}[S] \cdot (\uparrow \circ S) \longrightarrow S$. Rule (SCons) is responsible of confluence and typing problems in $\lambda\sigma$ [9, 20]. These problems are overcame in a variant of $\lambda\sigma$, called $\lambda_{\mathcal{L}}$ [20]. The $\lambda_{\mathcal{L}}$-calculus has the same general features as $\lambda\sigma$, i.e., simple, finite, and first-order presentation, but it does not contain rule (SCons).

Expressions of the untyped version of $\lambda_{\mathcal{L}}$ are *terms*: $\underline{1}$, applications $(M\ N)$, abstractions λM, and closures $M[S]$; and *explicit substitutions*: $\uparrow^n$, $M \cdot S$, and $S \circ T$; where M, N range over terms, S, T range over substitutions, and n ranges over natural numbers constructed with 0 and $n + 1$. The $\lambda_{\mathcal{L}}$-calculus is given in Figure 2. Free and bound variables are represented by de Bruijn indices. They are encoded by means of the constant $\underline{1}$ and the substitution $\uparrow^n$. We overload the notation $\underline{n}$ to represent the $\lambda_{\mathcal{L}}$-term corresponding to the index n, i.e., $\underline{n+1} = \underline{1}[\uparrow^n]$. The occurrence of an index i in a term M is *free* when that occurrence is bound by j λ-abstractions and $j < i$. By convenience, we write *free indices* to mean free occurrences of indices in a given term.

An explicit substitution denotes a mapping from indices to terms. Thus, $\uparrow^n$ maps each index i to the term $\underline{i+n}$, $S \circ T$ is the composition of the mapping denoted by T with the mapping denoted by S (notice that the composition of substitution follows a reverse order with respect to the usual notation of function composition), and finally, $M \cdot S$ maps the index 1 to the term M, and recursively, the index $i + 1$ to the term mapped by the substitution S on the index i.

The $\lambda_{\mathcal{L}}$-calculus, just as $\lambda\sigma$, uses the composition operation to achieve confluence on the calculus of substitutions (the calculus without (Beta)). [2]

$(\lambda M \ N)$	$\longrightarrow$	$M[N \cdot \uparrow^0]$	(Beta)	$(M \cdot S) \circ T$	$\longrightarrow$ $M[T] \cdot (S \circ T)$	(Map)
$(\lambda M)[S]$	$\longrightarrow$	$\lambda M[\underline{1} \cdot (S \circ \uparrow^1)]$	(Lambda)	$\uparrow^0 \circ S$	$\longrightarrow$ S	(IdS)
$(M \ N)[S]$	$\longrightarrow$	$(M[S] \ N[S])$	(Application)	$\uparrow^{n+1} \circ (M \cdot S)$	$\longrightarrow$ $\uparrow^n \circ S$	(ShiftCons)
$M[S][T]$	$\longrightarrow$	$M[S \circ T]$	(Clos)	$\uparrow^{n+1} \circ \uparrow^m$	$\longrightarrow$ $\uparrow^n \circ \uparrow^{m+1}$	(ShiftShift)
$\underline{1}[M \cdot S]$	$\longrightarrow$	M	(VarCons)	$\underline{1} \cdot \uparrow^1$	$\longrightarrow$ $\uparrow^0$	(Shift0)
$M[\uparrow^0]$	$\longrightarrow$	M	(Id)	$\underline{1}[\uparrow^n] \cdot \uparrow^{n+1}$	$\longrightarrow$ $\uparrow^n$	(ShiftS)

Fig. 2. The $\lambda_{\mathcal{L}}$-rewrite system

2 Pure Type Systems with Explicit Substitutions

In this section, we present an explicit substitution λ-calculus for PTS, namely PTS$_{\mathcal{L}}$. The main features of PTS$_{\mathcal{L}}$ are: a first-order setting insensitive to α-conversion, typing contexts as explicit substitutions, and support for expression contexts. As previously pointed out, higher-order aspects of the λ-calculus, including α-conversion, may be handled in a first-order setting via explicit substitutions *and* de Bruijn indices. We use the $\lambda_{\mathcal{L}}$-calculus as the base calculus of PTS$_{\mathcal{L}}$.

The PTS$_{\mathcal{L}}$-System: As in the case of PTS, a PTS$_{\mathcal{L}}$ is defined by a triple $(\mathcal{S}, \mathcal{A}, \mathcal{R})$ of sorts, axioms, and rules. The grammar of well-formed PTS$_{\mathcal{L}}$ *(pseudo-)expressions* extends the one of $\lambda_{\mathcal{L}}$ with sorts s, meta-variables X, type annotated abstractions $\lambda_A.M$, products $\Pi_A.B$, and type annotated substitutions $M \cdot_A S$. In PTS$_{\mathcal{L}}$, meta-variables, as well as substitutions, are first-class objects. However, only meta-variables on the sort of terms are allowed. We assume a set $\mathcal{V}$ of meta-variables. This set will be precisely defined in Section 3.

An expression is *ground* if it does not contain meta-variables. A ground expression is also *pure* if it does not contain other explicit substitutions than those appearing in the terms denoting de Bruijn indices (i.e., in terms of the form $\underline{i}$).

We define the $\lambda\Pi_{\mathcal{L}}$-rewrite system as the extension to $\lambda_{\mathcal{L}}$ with products and sorts given by Figure 3. The system $\Pi_{\mathcal{L}}$ is obtained by dropping rule (Beta) from $\lambda\Pi_{\mathcal{L}}$.

$(\lambda_A.M \ N)$	$\longrightarrow$	$M[N \cdot_A \uparrow^0]$	$(M \cdot_A S) \circ T$	$\longrightarrow$	$M[T] \cdot_A (S \circ T)$
$(\lambda_A.M)[S]$	$\longrightarrow$	$\lambda_{A[S]}.M[\underline{1} \cdot_A (S \circ \uparrow^1)]$	$\uparrow^0 \circ S$	$\longrightarrow$	S
$(\Pi_A.B)[S]$	$\longrightarrow$	$\Pi_{A[S]}.B[\underline{1} \cdot_A (S \circ \uparrow^1)]$	$\uparrow^{n+1} \circ (M \cdot_A S)$	$\longrightarrow$	$\uparrow^n \circ S$
$(M \ N)[S]$	$\longrightarrow$	$(M[S] \ N[S])$	$\uparrow^{n+1} \circ \uparrow^m$	$\longrightarrow$	$\uparrow^n \circ \uparrow^{m+1}$
$M[S][T]$	$\longrightarrow$	$M[S \circ T]$	$\underline{1} \cdot_A \uparrow^1$	$\longrightarrow$	$\uparrow^0$
$\underline{1}[M \cdot_A S]$	$\longrightarrow$	M	$\underline{1}[\uparrow^n] \cdot_A \uparrow^{n+1}$	$\longrightarrow$	$\uparrow^n$
$M[\uparrow^0]$	$\longrightarrow$	M	$s[S]$	$\longrightarrow$	s

Fig. 3. The $\lambda\Pi_{\mathcal{L}}$-rewrite system

---

[2] The $\lambda\sigma$-calculus is not confluent on general open expressions [9]. However, it is confluent on semi-open expressions [23]. The $\lambda\sigma_{\Uparrow}$-calculus, a variant of $\lambda\sigma$, achieves confluence on general open expressions [9].

Lemma 1. *The $\Pi_{\mathcal{L}}$-calculus is terminating.*

Proof. See [20]. The proof, due to Zantema, uses the semantic labeling technique [27]. □

The set of $\Pi_{\mathcal{L}}$-normal forms of an expression x (term or substitution) is denoted by $(x){\downarrow}_{\Pi_{\mathcal{L}}}$. The equivalence relation $\equiv_{\lambda\Pi_{\mathcal{L}}}$ (resp. $\equiv_{\Pi_{\mathcal{L}}}$) is defined as the congruence relation induced by the rewrite system $\lambda\Pi_{\mathcal{L}}$ (resp. $\Pi_{\mathcal{L}}$).

Typing Contexts: Typing contexts in PTS$_{\mathcal{L}}$ contain more information than in PTS. In order to combine declaration and definition of variables within a typing context, we associate a type *and a term* to the each variable declaration. Indeed, a *typing context* in PTS$_{\mathcal{L}}$ is an explicit substitution having the form $M_1 \cdot_{A_1} \cdots M_n \cdot_{A_n} \uparrow^n$, where, for $0 < i \leq n$, M_i is either $\underline{i}$ or is equal to a term $N_i[\uparrow^i]$. When M_i is $\underline{i}$, we say that index i is *declared*; otherwise, we say that the index i is *defined*. Furthermore, M_i and A_i are called the *term definition* and *type declaration* of the index i, respectively. Note that not every explicit substitution denotes a typing context. We use the Greek letters Γ and Δ to range over explicit substitutions denoting typing contexts.

Free indices in term definitions and in type declarations obey different conventions. Let Γ be $M_1 \cdot_{A_1} \cdots M_n \cdot_{A_n} \uparrow^n$. Free indices in a term definition M_i are absolute, i.e., they refer to the whole context Γ. Cyclic *definitions* are avoided by construction. Notice that we require M_i to be equal to $N_i[\uparrow^i]$, for some term N_i. In that case, for all free indices j of M_i, it holds that $j > i$. On the other hand, free indices in a type declaration A_i are relative, i.e., they refer to the portion of the context where the index i is declared or defined. Therefore, by using this convention, cyclic *declarations* are impossible.

Although a different convention for the free indices in a typing context is still possible, we prefer the one sketched above since it allows an elegant encoding of contexts as explicit substitutions. The intention is to identify the evaluation of a term M in a context Γ as the $\Pi_{\mathcal{L}}$-reduction of $M[\Gamma]$. In particular, notice that a typing context without definitions has the form $\underline{1} \cdot_{A_1} \cdots \underline{n} \cdot_{A_n} \uparrow^n$. This substitution $\Pi_{\mathcal{L}}$-reduces to $\uparrow^0$. Therefore, as expected, the evaluation of a term M in a context Γ which does not contain definitions results in M.

The fact that indices are either *defined* or *declared* is rather a convenient way to explain typing contexts. The type system does not really distinguish both classes of indices.

Meta-variables and Constraints: *Meta-variables* and *constraints* are used to deal with higher-order unification problems. Informally, meta-variables stand for instantiation variables and constraints are term equalities to be solved.

Meta-variables are first-class objects in PTS$_{\mathcal{L}}$. Just as variables, they have to be declared in order to keep track of possible dependencies between terms and types. A *meta-variable declaration* has the form $X{:}_{\Gamma}A$, where Γ and A are, respectively, a context and a type assigned to the meta-variable X. Indices in A are relative to Γ.

A *constraint* $M \simeq_\Gamma N$ relates two terms M, N and a context Γ. Indices in M and N are relative to Γ. Similarly to meta-variables, constraints respect a typing discipline: terms M and N have the same type in Γ.

Definition 1 (Constrained Signatures). *A (constrained) signature is a list containing meta-variable declarations and constraint declarations. An empty signature is denoted by ϵ. Furthermore, if Σ is a constrained signature, $X\!:_\Gamma A.\ \Sigma$ and $M \simeq_\Gamma N.\ \Sigma$ are well-formed constrained signatures.*

We overload the notation $\Sigma_1.\ \Sigma_2$ to write the concatenation of the signatures Σ_1 and Σ_2. A meta-variable X is *fresh* with respect to a signature Σ, denoted $X \notin \Sigma$, if there are not A and Γ such that $X\!:_\Gamma A$ is in Σ.

Equality: In order to deal with constraints, PTS$_\mathcal{L}$ needs a finer notion of convertibility than that for PTS (Section 1).

Definition 2 (Equivalence Modulo Constraints). *Let Σ be a signature; we define the relation $\equiv_\Sigma$ as the smallest congruence such that (1) if $M \equiv_{\lambda \Pi_\mathcal{L}} N$, then $M \equiv_\Sigma N$, (2) if $M \simeq_\Gamma N \in \Sigma$, then $M[\Gamma] \equiv_\Sigma N[\Gamma]$, and (3) if $M \equiv_\Sigma N[\uparrow^1]$, then $\lambda_A.(M\ \underline{1}) \equiv_\Sigma N$.*

The last case of the definition above handles η-conversions. In this way, we avoid to consider an η-rule explicitly in the $\lambda \Pi_\mathcal{L}$-calculus.

We extend $\equiv_\Sigma$ to relate typing contexts as follows: (1) $\uparrow^n \equiv_\Sigma \uparrow^n$ and (2) $M \cdot_A \Gamma \equiv_\Sigma N \cdot_B \Delta$, if $M \equiv_\Sigma N$, $A[\Gamma] \equiv_\Sigma B[\Delta]$, and $\Gamma \equiv_\Sigma \Delta$.

PTS$_\mathcal{L}$ Typing Rules: In PTS, typing rules for validity of typing contexts are implicit. However, in that case, structural rules (Start) and (Weak) are necessary to create an initial context and to add new variable declarations to it. The notation $\vDash \Sigma; \Gamma$ captures that Γ is valid in the valid signature Σ. We write $\Sigma; \Gamma \vDash M : A$ to state that the term M has type A in $\Sigma; \Gamma$. For explicit substitutions S we write $\Sigma; \Gamma \vDash S \triangleright \Delta$ to state that S has type Δ in $\Sigma; \Gamma$. The type system for PTS$_\mathcal{L}$ is given in Figure 4. The judgments omits Σ, when it is ϵ, and Γ, when it is $\uparrow^0$. We reserve $\vdash$ for judgments where Σ does not contain constraints, otherwise we use $\vDash$. A signature Σ is *valid* if $\vDash \Sigma$ holds.

We use the following functions on contexts:

- $\Gamma \circ \uparrow^i = \Delta$, where Δ is the normal form of $\Gamma \circ \uparrow^i$ with respect to the rewrite system composed by the rules (Map), (Ids), (ShiftCons), and (ShiftShift). We also have to consider the case when i is negative: if $\Gamma \equiv_{\Pi_\mathcal{L}} \Delta \circ \uparrow^j$ and $j > i$, we consider $\Gamma \circ \uparrow^{-i}$ equal to $\Delta \circ \uparrow^{j-i}$.
- To add and remove elements from a context we use the shorthands $push(M, A, \Gamma) = M \cdot_A (\Gamma \circ \uparrow^1)$, $top(M \cdot_A \Gamma) = A[\Gamma]$, and $pop(M \cdot_A \Gamma) = \Gamma \circ \uparrow^{-1}$.
- The *order* of a substitution S indicates how many term definitions will be either consumed, if the number is negative, or produced, if the number is positive, when S is applied to a context Γ. Formally, $order(\uparrow^n) = -n$, $order(M \cdot_A S) = 1 + order(S)$, $order(S \circ T) = order(S) + order(T)$.

$$\frac{}{\;\vdash\!\!\sim \epsilon;\,\uparrow^0}\ \text{(Empty)} \qquad\qquad \frac{\Sigma;\Gamma\vdash\!\!\sim M_1 : A \quad \Sigma;\Gamma\vdash\!\!\sim M_2 : A}{\vdash\!\!\sim M_1 \simeq_\Gamma M_2.\,\Sigma;\Gamma}\ \text{(Constraint)}$$

$$\frac{\Sigma;\Gamma\vdash\!\!\sim A : s}{\vdash\!\!\sim \Sigma; push(\underline{1},A,\Gamma)}\ \text{(Var-Decl)} \qquad \frac{\Sigma;\Gamma\vdash\!\!\sim M : A \quad \Sigma;\Gamma\vdash\!\!\sim A : s}{\vdash\!\!\sim \Sigma; push(M[\Gamma\circ\uparrow^1],A,\Gamma)}\ \text{(Let-Decl)}$$

$$\frac{\vdash\!\!\sim \Sigma;\Gamma \quad X\notin\Sigma}{\vdash\!\!\sim X :_\Gamma s.\ \Sigma;\Gamma}\ \text{(Meta-Var-Decl}_1) \qquad \frac{\Sigma;\Gamma\vdash\!\!\sim A : s \quad X\notin\Sigma}{\vdash\!\!\sim X :_\Gamma A[\Gamma].\ \Sigma;\Gamma}\ \text{(Meta-Var-Decl}_2)$$

$$\frac{\vdash\!\!\sim \Sigma;\Gamma \quad (s_1,s_2)\in\mathcal{A}}{\Sigma;\Gamma\vdash\!\!\sim s_1 : s_2}\ \text{(Sort)} \qquad\qquad \frac{\vdash\!\!\sim \Sigma;\Gamma}{\Sigma;\Gamma\vdash\!\!\sim \underline{1} : top(\Gamma)}\ \text{(Var)}$$

$$\frac{\begin{array}{c}\Sigma;\Gamma\vdash\!\!\sim A : s_1 \\ \Sigma; push(\underline{1},A,\Gamma)\vdash\!\!\sim B : s_2 \\ (s_1,s_2,s_3)\in\mathcal{R}\end{array}}{\Sigma;\Gamma\vdash\!\!\sim \Pi_A.B : s_3}\ \text{(Prod)} \qquad \frac{\begin{array}{c}\Sigma; push(\underline{1},A,\Gamma)\vdash\!\!\sim M : B \\ \Sigma;\Gamma\vdash\!\!\sim \Pi_A.B : s\end{array}}{\Sigma;\Gamma\vdash\!\!\sim \lambda_A.M : \Pi_{A[\Gamma]}.B}\ \text{(Abs)}$$

$$\frac{\begin{array}{c}\Sigma;\Gamma\vdash\!\!\sim M : \Pi_A.B \\ \Sigma;\Gamma\vdash\!\!\sim N : A\end{array}}{\Sigma;\Gamma\vdash\!\!\sim (M\ N) : B[N[\Gamma]\cdot_A\uparrow^0]}\ \text{(Appl)}$$

$$\frac{\begin{array}{c}\vdash\!\!\sim \Sigma;\Gamma \quad X:_\Delta A\in\Sigma \\ \Sigma;\Delta\vdash\!\!\sim A : s \quad \lfloor\Delta\rfloor_0 \equiv_\Sigma \lfloor\Gamma\rfloor_0\end{array}}{\Sigma;\Gamma\vdash\!\!\sim X : A[\Gamma]}\ \text{(Meta-Var)} \qquad \frac{\begin{array}{c}\vdash\!\!\sim \Sigma;\Gamma \quad X:_\Delta s\in\Sigma \\ s \text{ is a top sort} \quad \lfloor\Delta\rfloor_0 \equiv_\Sigma \lfloor\Gamma\rfloor_0\end{array}}{\Sigma;\Gamma\vdash\!\!\sim X : s}\ \text{(Meta-Var-Sort)}$$

$$\frac{\begin{array}{c}\Sigma;\Gamma\vdash\!\!\sim S\rhd\Delta \quad \Sigma;\Delta\vdash\!\!\sim M : A \\ \Sigma;\Delta\vdash\!\!\sim A : s \quad n = order(S)\end{array}}{\Sigma;\Gamma\vdash\!\!\sim M[S] : A[\uparrow^{-n}]}\ \text{(Clos)} \qquad \frac{\begin{array}{c}\Sigma;\Gamma\vdash\!\!\sim S\rhd\Delta \\ \Sigma;\Delta\vdash\!\!\sim M : s \quad s \text{ is a top sort}\end{array}}{\Sigma;\Gamma\vdash\!\!\sim M[S] : s}\ \text{(Clos-Sort)}$$

$$\frac{\vdash\!\!\sim \Sigma;\Gamma}{\Sigma;\Gamma\vdash\!\!\sim \uparrow^0\rhd\Gamma}\ \text{(Id)} \qquad \frac{\vdash\!\!\sim \Sigma;\Gamma \quad \Sigma; pop(\Gamma)\vdash\!\!\sim \uparrow^n\rhd\Delta}{\Sigma;\Gamma\vdash\!\!\sim \uparrow^{n+1}\rhd\Delta}\ \text{(Shift)}$$

$$\frac{\Sigma;\Gamma\vdash\!\!\sim S\rhd\Delta_1 \quad \Sigma;\Delta_1\vdash\!\!\sim T\rhd\Delta_2}{\Sigma;\Gamma\vdash\!\!\sim T\circ S\rhd\Delta_2}\ \text{(Comp)} \qquad \frac{\begin{array}{c}\Sigma;\Gamma\vdash\!\!\sim M : A[S] \quad \Sigma;\Gamma\vdash\!\!\sim S\rhd\Delta \\ \Sigma;\Delta\vdash\!\!\sim A : s \quad n = order(S)\end{array}}{\Sigma;\Gamma\vdash\!\!\sim M\cdot_A S\rhd push(M[\Gamma\circ\uparrow^{n+1}],A,\Delta)}\ \text{(Cons)}$$

$$\frac{\begin{array}{c}\Sigma;\Gamma\vdash\!\!\sim M : A \\ \Sigma;\Gamma\vdash\!\!\sim B : s \\ A\equiv_\Sigma B[\Gamma]\equiv_{\lambda\Pi_\mathcal{L}} B\end{array}}{\Sigma;\Gamma\vdash\!\!\sim M : B}\ \text{(Conv)} \qquad \frac{\begin{array}{c}\Sigma;\Gamma\vdash\!\!\sim S\rhd\Delta_1 \\ \vdash\!\!\sim \Sigma;\Delta_2 \\ \Delta_1 \equiv_\Sigma \Delta_2\end{array}}{\Sigma;\Gamma\vdash\!\!\sim S\rhd\Delta_2}\ \text{(Conv-Subs)}$$

Fig. 4. The $\lambda\Pi_\mathcal{L}$-type system

- Given a context Γ, the operation $\lfloor\Gamma\rfloor_0$ computes a new context where all the definitions in Γ have been transformed into declarations, as follows $\lfloor\uparrow^m\rfloor_n = \uparrow^n$, $\lfloor M\cdot_A\Gamma\rfloor_n = \underline{n+1}\cdot_A\lfloor\Gamma\rfloor_{n+1}$.

Relating PTS$_\mathcal{L}$ to PTS: Since PTS$_\mathcal{L}$ allows arbitrary constraints between terms, strong normalization, as well as other usual typing properties, can be easily violated in arbitrary PTS$_\mathcal{L}$. For instance, the term $(\lambda x{:}A.(x\ x)\ \lambda x{:}A.(x\ x))$ can be typed in any PTS$_\mathcal{L}$ containing the constraint $A\simeq_\Gamma A\to A$. However, as we will see below, PTS$_\mathcal{L}$ is a conservative extension of PTS. Furthermore, when only pure expressions are considered, i.e., signatures are empty (and then we use $\vdash$ rather than $\vdash\!\!\sim$), PTS$_\mathcal{L}$ types as many terms as PTS (modulo $\Pi_\mathcal{L}$-reductions).

We say that a typing context is *pure* if it is the identity substitution or it has the form $\underline{1}\cdot_{A_1}\ldots\underline{n}\cdot_{A_n}\uparrow^n$ and A_i is pure for $1\le i\le n$. Notice that pure contexts reduce to $\uparrow^0$ via application of rules (Shift0) and (Shift S).

Theorem 1 (Conservative Extension). *Let M, A be pure terms and Γ be a pure context. Then, $\Gamma \vdash M : A$ in a PTS, if and only if $\Gamma \vdash M : A$ in $PTS_{\mathcal{L}}$ (we assume de Bruijn indices translation. For details of this translation see [9]).*

Proof. Both cases require typing properties for PTS and $PTS_{\mathcal{L}}$. We refer to [14] for the proofs of properties in PTS and to Section 4 for a summary of properties in $PTS_{\mathcal{L}}$. The case PTS $\Rightarrow PTS_{\mathcal{L}}$, proceeds by induction on the typing derivation. Note that if $M \equiv_{\beta\eta} N$ then $M \equiv_{\Sigma} N$. For the case $PTS_{\mathcal{L}} \Rightarrow$ PTS, we first prove by induction on the typing derivation that for arbitrary terms N, B, and pure context Γ, $\Gamma \vdash N : B$ in $PTS_{\mathcal{L}}$ implies $\Gamma \vdash (N)\downarrow_{\Pi_{\mathcal{L}}} : (B)\downarrow_{\Pi_{\mathcal{L}}}$.   □

3   Definitions and Explicit Instantiations

In this section, we address two novel aspects of $\lambda \Pi_{\mathcal{L}}$: the ability to encode let-in expressions and the support for expression contexts.

Definitions: Let-in expressions are a convenient way to support definitions in λ-calculus. In the simply-typed λ-calculus an expression as let $x := N :$ A in M can be encoded as $(\lambda x{:}A.M \ N)$. In this case, definition unfolding is performed by the β-reduction mechanism. The behavior of definitions in a dependent-type system cannot be straightforwardly encoded as a β-reduction. Consider, for example, the expression let $x := 0 : nat$ in $(m \ (l \ x))$ in a context $\Gamma = m : (A \ 0) \to nat. \ 0 : nat. \ l : (\Pi n{:}nat.(A \ n)). \ A : nat \to Type. \ nat : Type.$[3] Although this term is unfolded into $(m \ (l \ 0))$ in the same way that the term $((\lambda x{:}nat.(m \ (l \ x))) \ 0)$ β-reduces to $(m \ (l \ 0))$, the term $((\lambda x{:}nat.(m \ (l \ x))) \ 0)$ cannot be typed in Γ. This is because the information that the variable x will be substituted by 0 in $(m \ (l \ x))$ is not taken into account by the application rule. Indeed, the type of $(l \ x)$ is $(A \ x)$, not $(A \ 0)$ as expected by m, and, therefore, the term $\lambda x{:}nat.(m \ (l \ x))$ is ill-typed in Γ. Solutions to the above problem require either to consider definitions as first-class terms (not just a macro expansion to a β-redex) [24] or to use a different notation and typing rules for applications [4]. In [3], Bloo proposes a calculus of explicit substitutions for PTS with contexts extended with term definitions.

As a side effect to combine definitions and declarations in contexts, let-in expressions can be encoded as explicit substitutions. In $PTS_{\mathcal{L}}$, as well as in Bloo's system, let $x := N : A$ in M is just a shorthand for the term $M[x := N : A]$ (or $M[N \cdot_A \uparrow^0]$ in our nameless notation). On the other hand, the typing rule for applications remains unmodified.

In contrast to Bloo's approach, we use a uniform notation for typing contexts and explicit substitutions. Furthermore, we internalize completely the substitution mechanism within the theory. In particular, we do not require any implicit substitution mechanism. Implicit substitutions are problematic when meta-variables are allowed. Notice that meta-variables, in a dependent-type theory, may also appear in typing contexts.

---

[3] For readability, we use named variables when discussing examples. Nevertheless, as we have said, $PTS_{\mathcal{L}}$ uses a de Bruijn nameless notation of variables.

Explicit Instantiations: In term rewriting, an *expression context* is an expression with distinguished holes. Filling a hole in a expression context with an arbitrary term is a first-order substitution that does not care of renaming variables. It is well-known that expression contexts in λ-calculus, and in general in higher-order rewrite systems, raise technical problems. For instance, contexts are not compatible with α and β-conversion. Calculus of contexts for simply-typed λ-calculus have been studied in for instance [5]. In previous approaches, either β-redexes cannot be reduced in context terms or new binding structures together with delicate type systems have been required to handle holes and explicit filling of holes.

Explicit substitutions overcome these difficulties in λ-calculus, and thus, they allow us to formulate $\mathrm{PTS}_{\mathcal{L}}$ as λ-calculus of contexts where meta-variables denote holes. To complete the framework, we must provide an explicit mechanism to perform instantiations.

Up till now, meta-variables were given in an abstract set $\mathcal{V}$. In order to remain in a first-order setting, we index meta-variables with positive natural numbers. Therefore, we write X_p for a meta-variable indexed by the positive natural number p. Expressions of a $\mathrm{PTS}_{\mathcal{L}}$-system with explicit instantiations include $M\{\theta\}$ and $S\{\theta\}$ as first-class terms and substitutions where θ is a set of instantiations. We use list of terms to represent a set of instantiations.

An instantiation θ is *well-formed* if it has the form $M_1 \cdot \ldots M_n \cdot m$, where $n \leq m$ and each term M_i is either X_i or a term which does not contain the meta-variable X_i. In the latter case, M_i is called the *instantiation term* of X_i in θ. The *grade* of θ, also written $grade(\theta)$, is given by $m - n$.

The structure of explicit instantiation is analogous to the structure of explicit substitutions. Indeed, an instantiation $M \cdot \theta$ denotes the replacement of meta-variables X_1 by M, X_2 by the head of θ and so forth. The lookup mechanism of meta-variables in an explicit instantiation is also analogous to the lookup of variables in an explicit substitution. The index p of the meta-variable X_p is consumed at the same time that the instantiation θ is traversed. The grade of θ helps to reconstruct the original index of the meta-variable in the cases of failed look-ups, i.e., the meta-variable X_p does not appear in θ.

In contrast to explicit substitutions, instantiations do not care about recalculation of free-indices in expressions. The calculus of explicit instantiations, called $\mathcal{X}_{\{\mathcal{L}\}}$, is depicted in Figure 5.

$$
\begin{array}{ll}
s\{\theta\} \longrightarrow s & (N \cdot_A S)\{\theta\} \longrightarrow N\{\theta\} \cdot_{A\{\theta\}} S\{\theta\} \\
\underline{1}\{\theta\} \longrightarrow \underline{1} & (S \circ T)\{\theta\} \longrightarrow S\{\theta\} \circ T\{\theta\} \\
(\Pi_A.B)\{\theta\} \longrightarrow \Pi_{A\{\theta\}}.B\{\theta\} & X_p\{0\} \longrightarrow X_p \\
(\lambda_A.N)\{\theta\} \longrightarrow \lambda_{A\{\theta\}}.N\{\theta\} & X_p\{n+1\} \longrightarrow X_{p+1}\{n\} \\
(N_1\ N_2)\{\theta\} \longrightarrow (N_1\{\theta\}\ N_2\{\theta\}) & X_1\{M \cdot \theta\} \longrightarrow M \\
(N[S])\{\theta\} \longrightarrow N\{\theta\}[S\{\theta\}] & X_{p+1}\{M \cdot \theta\} \longrightarrow X_p\{\theta\} \\
\uparrow^n\{\theta\} \longrightarrow \uparrow^n &
\end{array}
$$

Fig. 5. The $\mathcal{X}_{\{\mathcal{L}\}}$-rewrite system

Lemma 2. *The $\mathcal{X}_{\{\mathcal{L}\}}$-rewrite system is confluent and terminating.*

Proof. To show that $\mathcal{X}_{\{c\}}$ is terminating, we use a measure that penalizes top-level application of instantiations and the height of trees representing instantiations. Furthermore, $\mathcal{X}_{\{c\}}$ is orthogonal, then it is confluent. □

Typing rules for explicit instantiations are shown in Figure 6. We extend the relation $\equiv_\Sigma$ to consider the case of $\mathcal{X}_{\{c\}}$-reductions as follows: if $M \xrightarrow{\mathcal{X}_{\{c\}}} N$, then $M \equiv_\Sigma N$. We also have a new kind of typing judgment: $\Sigma; n \mathrel{|\!\!\sim} \theta \Rightarrow \Sigma'$. It has a double purpose. In first place, it enforces the invariant $n = grade(\theta)$. Secondly, it states that for all instantiation term M_i in θ, M_i is well-typed with respect to the meta-variable X_i. In that case, a constraint $X_i \simeq_\Gamma M_i$ is accumulated in a new signature Σ'.

$$\frac{\Sigma;0 \mathrel{|\!\!\sim} \theta \Rightarrow \Sigma' \quad \Sigma;\Gamma \mathrel{|\!\!\sim} M : A}{\Sigma'.\ \Sigma;\Gamma \mathrel{|\!\!\sim} M\{\theta\} : A} \text{ (Inst-Term)} \qquad \frac{\Sigma;0 \mathrel{|\!\!\sim} \theta \Rightarrow \Sigma' \quad \Sigma;\Gamma \mathrel{|\!\!\sim} S \triangleright \Delta}{\Sigma'.\ \Sigma;\Gamma \mathrel{|\!\!\sim} S\{\theta\} \triangleright \Delta} \text{ (Inst-Subs)}$$

$$\frac{\Sigma; grade(\theta) \mathrel{|\!\!\sim} \theta \Rightarrow \Sigma' \quad \Sigma;\Gamma \mathrel{|\!\!\sim} X_{grade(\theta)+p} : A}{\Sigma'.\ \Sigma;\Gamma \mathrel{|\!\!\sim} X_p\{\theta\} : A} \text{ (Lookup)} \qquad \frac{\mathrel{|\!\!\sim} \Sigma}{\Sigma;n \mathrel{|\!\!\sim} n \Rightarrow \epsilon} \text{ (Empty-Inst)}$$

$$\frac{\Sigma;n+1 \mathrel{|\!\!\sim} \theta \Rightarrow \Sigma'}{\Sigma;n \mathrel{|\!\!\sim} X_{n+1} \cdot \theta \Rightarrow \Sigma'} \text{ (List-Inst}_1) \qquad \frac{\Sigma;n+1 \mathrel{|\!\!\sim} \theta \Rightarrow \Sigma' \quad \Sigma = \Sigma_1.\ X_{n+1}:_\Gamma A.\ \Sigma_2 \quad \Sigma_2;\Gamma \mathrel{|\!\!\sim} M : A}{\Sigma;n \mathrel{|\!\!\sim} M \cdot \theta \Rightarrow X_{n+1} \simeq_\Gamma M.\ \Sigma'} \text{ (List-Inst}_2)$$

Fig. 6. Typing rules for explicit instantiations

We denote by $x\{\theta\}_\downarrow$, the $\mathcal{X}_{\{c\}}$-normal form of an expression $x\{\theta\}$ where x is either a term or a substitution. We extend this notation to contexts and signatures as follows: $\epsilon\{\theta\}_\downarrow = \epsilon$, $(A.\Gamma)\{\theta\}_\downarrow = A\{\theta\}_\downarrow.\Gamma\{\theta\}_\downarrow$, $(X:_\Gamma A.\ \Sigma)\{\theta\}_\downarrow = X:_{\Gamma\{\theta\}_\downarrow} A\{\theta\}_\downarrow.\ \Sigma\{\theta\}_\downarrow$, and $(M \simeq_\Gamma N.\ \Sigma)\{\theta\}_\downarrow = M\{\theta\}_\downarrow \simeq_{\Gamma\{\theta\}_\downarrow} N\{\theta\}_\downarrow.\ \Sigma\{\theta\}_\downarrow$.

PTS$_\mathcal{L}$ allows to represent open terms in PTS, i.e., terms with meta-variables. The *instantiation* mechanism of meta-variables fills the place-holders in open terms. This way, an open term in PTS$_\mathcal{L}$ gradually becomes a pure term in PTS.

Lemma 3 (Instantiation). *Assume* $\Sigma;0 \mathrel{|\!\!\sim} \theta \Rightarrow \Sigma'$. *(1) If* $\mathrel{|\!\!\sim} \Sigma;\Delta$, *then* $\mathrel{|\!\!\sim} \Sigma\{\theta\}_\downarrow;\Delta\{\theta\}_\downarrow$, *(2) if* $\Sigma;\Delta \mathrel{|\!\!\sim} N : B$, *then* $\Sigma\{\theta\}_\downarrow;\Delta\{\theta\}_\downarrow \mathrel{|\!\!\sim} N\{\theta\}_\downarrow : B\{\theta\}_\downarrow$, *and (3) if* $\Sigma;\Delta_1 \mathrel{|\!\!\sim} S \triangleright \Delta_2$, *then* $\Sigma\{\theta\}_\downarrow;\Delta_1\{\theta\}_\downarrow \mathrel{|\!\!\sim} S\{\theta\}_\downarrow \triangleright \Delta_2\{\theta\}_\downarrow$.

Proof. First, we prove by simultaneous induction on terms and substitutions that $\mathcal{X}_{\{c\}}$-reductions preserve typing, i.e. if $\Sigma;\Gamma \mathrel{|\!\!\sim} M : A$ and $M \xrightarrow{\mathcal{X}_{\{c\}}} N$, then $\Sigma;\Gamma \mathrel{|\!\!\sim} N : A$, and if $\Sigma;\Gamma \mathrel{|\!\!\sim} S \triangleright \Delta$ and $S \xrightarrow{\mathcal{X}_{\{c\}}} T$, then $\Sigma;\Gamma \mathrel{|\!\!\sim} S \triangleright \Delta$. Hence, we can prove that (1) $\mathrel{|\!\!\sim} \Sigma'.\ \Sigma\{\theta\}_\downarrow;\Delta\{\theta\}_\downarrow$, (2) $\Sigma'.\ \Sigma\{\theta\}_\downarrow;\Delta\{\theta\}_\downarrow \mathrel{|\!\!\sim} N\{\theta\}_\downarrow : B\{\theta\}_\downarrow$, and (3) $\Sigma'.\ \Sigma\{\theta\}_\downarrow;\Delta_1\{\theta\}_\downarrow \mathrel{|\!\!\sim} S\{\theta\}_\downarrow \triangleright \Delta_2\{\theta\}_\downarrow$. Finally, we use a weakening property on signatures which states that constraints of the form $X_p \simeq_\Gamma M$ can be safely removed from a signature when the meta-variable X_p does not appear as a sub-term in the remaining signature. □

4   Meta-theory

We here summarize a meta-theoretical study for dependent-type versions for $\lambda \Pi_{\mathcal{L}}$. It extends [20] by considering explicit definitions and instantiations. Proofs are available at [21].

First of all, we prove some properties about contexts. The first explains the meaning of the typing rules for substitutions. The second states that terms in valid contexts always are in normal form with respect to term definitions. The third property states that the order of a valid contexts is 0. The last one states that contexts type themselves.

Lemma 4 (Properties about Valid Contexts). *(1) If $\Sigma; \Gamma \mathrel{\mkern-5mu\vert\mkern-5mu\sim} S \vartriangleright \Delta$, then $S \circ \Gamma \equiv_{\Sigma} \Delta \circ \uparrow^{-order(S)}$. (2) If $\mathrel{\mkern-5mu\vert\mkern-5mu\sim} \Sigma; \Gamma$, then $\Gamma \circ \Gamma \equiv_{\Pi_{\mathcal{L}}} \Gamma$. (3) If $\mathrel{\mkern-5mu\vert\mkern-5mu\sim} \Sigma; \Gamma$, then $order(\Gamma) = 0$. (4) If $\mathrel{\mkern-5mu\vert\mkern-5mu\sim} \Sigma; \Gamma$, then $\Sigma; \Gamma \mathrel{\mkern-5mu\vert\mkern-5mu\sim} \Gamma \vartriangleright \Gamma$.*

Proof. All the proofs are by induction on the typing derivations[21].   □

We cannot go further with usual properties as *subject reduction, confluence,* or *normalization.* As we have mentioned before, when arbitrary constraints are allowed, these properties can be easily violated. To continue our meta-theoretical study we make some assumptions. First of all, we consider only functional $\text{PTS}_{\mathcal{L}}$. Hence, we include all the type-systems of the Barendregt's cube. Furthermore, we consider signatures without constraints. Then we use $\vdash$ rather that $\mathrel{\mkern-5mu\vert\mkern-5mu\sim}$ to denote typing judgments.

Even with the assumptions above, the system $\lambda \Pi_{\mathcal{L}}$ is not confluent on pseudo-expressions, i.e., on expressions that are not well-typed. To handle this problem, we follow the development in [20]. The approach is originally due to Geuvers [13]. It exploits the fact that the rewrite-system $\lambda \Pi_{\mathcal{L}}$ without type annotations is confluent. We prove the following key lemma.

Lemma 5 (Geuvers' Lemma). *Given a signature Σ without constraints, (1) if $\Pi_{A_1}.B_1 \equiv_{\Sigma} \Pi_{A_2}.B_2$, then $A_1 \equiv_{\Sigma} A_2$ and $B_1 \equiv_{\Sigma} B_2$, and (2) if $(c\ M_1 \dots M_n) \equiv_{\Sigma} N$, then $N \xrightarrow{\lambda \Pi_{\mathcal{L}}} (c\ N_1 \dots N_n)$ where $M_i \equiv_{\Sigma} N_i$.*

Proof. First, we prove that $\Pi_{\mathcal{L}}$ is confluent on expressions without type annotations on substitutions. Notice that η is not a rewrite-rule and that only meta-variables on the sort of terms are allowed. Then, we show that $\Pi_{\mathcal{L}}$ commutes with the parallelization of (Beta) [20]. The conclusion follows from a positive use of the counter-example for the confluence of the system with type annotations [13, 20].   □

Geuvers' lemma is enough to prove the following properties. Most of them require simultaneous induction on terms and substitutions.

Sort Soundness: If $\Sigma; \Gamma \vdash M : A$, then either $A = s$, where s is a top sort, or $\Sigma; \Gamma \vdash A : s$, where $s \in \mathcal{S}$.

Type Uniqueness: For a functional $\text{PTS}_{\mathcal{L}}$, if $\Sigma; \Gamma \vdash M : A$ and $\Sigma; \Gamma \vdash M : B$, then $A \equiv_{\Sigma} B$.

Subject Reduction: For a functional PTS$_\mathcal{L}$, if $M \xrightarrow{\lambda\Pi_\mathcal{L}^*} N$ and $\Sigma; \Gamma \vdash M : A$, then $\Sigma; \Gamma \vdash N : A$.

Church-Rosser: For a functional weakly normalizing PTS$_\mathcal{L}$, if $M_1 \equiv_{\lambda\Pi_\mathcal{L}} M_2$, $\Sigma; \Gamma \vdash M_1 : A$, and $\Sigma; \Gamma \vdash M_2 : A$, then M_1 and M_2 are $\lambda\Pi_\mathcal{L}$-joinable, i.e., there exists M such that $M_1 \xrightarrow{\lambda\Pi_\mathcal{L}^*} M$ and $M_2 \xrightarrow{\lambda\Pi_\mathcal{L}^*} M$.

It is well-known that explicit substitutions calculi based on $\lambda\sigma$ do not satisfy the strong-normalization property, not even in the simply-typed theory [19]. In [20], it is shown that $\lambda\Pi_\mathcal{L}$ without constraints and without η-equality is weakly normalizing on the Calculus of Constructions.

5 Applications

The PTS$_\mathcal{L}$-type system cannot be applied bottom up because some of the rules require premises that are not present in the conclusion, and others require to check equality under $\equiv_\Sigma$. On the other hand, it is possible to divide type checking into three sub-tasks that can be treated separately: (1) type inference, (2) checking (ground) equality, and (3) unification. This section therefore develops the relevant machinery for accomplishing these tasks.

Type Inference: We obtain a type inference system from $\equiv_\Sigma$ by converting the judgments $\Sigma; \Gamma \mathrel{\vdash\!\!\!\sim} M : A$ and $\Sigma; \Gamma \mathrel{\vdash\!\!\!\sim} S \rhd \Delta$ into $\Sigma; \Gamma \mathrel{\vdash\!\!\!\sim}_\mathcal{M} M \implies A, \Sigma'$, respectively, $\Sigma; \Gamma \mathrel{\vdash\!\!\!\sim}_\mathcal{M} S \implies \Delta, \Sigma'$, where the type A, context Δ, and auxiliary constraints Σ' are produced by applying the type inference rules. For lack of space we will not give the full inference system here, but illustrate some of the more interesting cases: the rule to infer the type of an application. The full system appears in [21].

$$\frac{\begin{array}{c} \Sigma; \Gamma \mathrel{\vdash\!\!\!\sim}_\mathcal{M} M \implies A_1, \Sigma_1 \quad \Sigma_1; \Gamma \mathrel{\vdash\!\!\!\sim}_\mathcal{M} N \implies A_2, \Sigma_2 \\ \Sigma_2; \Gamma \mathrel{\vdash\!\!\!\sim}_\mathcal{M} A_1 : \mathcal{S} \implies s_3, \Sigma_3 \quad \Sigma_3; \Gamma \mathrel{\vdash\!\!\!\sim}_\mathcal{M} A_2 : \mathcal{S} \implies s_1, \Sigma_4 \\ H \notin \Sigma_4 \quad (s_1, s_2, s_3) \in \mathcal{R} \quad \Delta = push(1, A_2, \Gamma) \\ \Sigma_5 = A_1 \simeq_\Gamma \Pi_{A_2}.H. \ H :_\Delta s_2. \ \Sigma_4 \end{array}}{\Sigma; \Gamma \mathrel{\vdash\!\!\!\sim}_\mathcal{M} (M \ N) \implies H[N \cdot_{A_2} \uparrow^0], \Sigma_5}$$

These rules can be encoded directly with Braga's conditional rewrite rule extension to Maude [7].

Checking Equality: The rewrite-system $\lambda\Pi_\mathcal{L}$ provides a way to check equality under β-reduction, but not η equality. One could add an η-rule as a conditional rewrite rule $\lambda_A.(M \ \underline{1}) \longrightarrow N$ if $M \equiv_{\lambda\Pi_\mathcal{L}} N[\uparrow^1]$, and check equality of terms by first applying reduction to normal form. A more economical approach is to interleave weak head normal form conversion with rules that decompose equalities into smaller equalities. By weak head normal form we here take the minimal reduction under $\lambda\Pi_\mathcal{L}$-system that allows to decompose an equality using the other rules in Fig. 7. The cases not mentioned there are assumed as failures.

Our rules are directly structural on the shape of terms. We should note that the $\lambda - N$ rule realizes η-equality, and that the $\lambda - \lambda$ rule does not check

whnf	$M \simeq_\Gamma N$	$\Longrightarrow$	$\mathrm{whnf}(M) \simeq_\Gamma \mathrm{whnf}(N)$
$\lambda - \lambda$	$\lambda_A.M \simeq_\Gamma \lambda_B.N$	$\Longrightarrow$	$M \simeq_{push(\underline{1},A,\Gamma)} N$
$\Pi - \Pi$	$\Pi_{A_1}.B_1 \simeq_\Gamma \Pi_{A_2}.B_2$	$\Longrightarrow$	$A_1 \simeq_\Gamma A_2.\ B_1 \simeq_{push(\underline{1},A_1,\Gamma)} B_2$
$\lambda - N$	$\lambda_A.M \simeq_\Gamma N$	$\Longrightarrow$	$M \simeq_{push(\underline{1},A,\Gamma)} (N[\uparrow^1]\ \underline{1})$
Subst	$M \simeq_\Gamma N$	$\Longrightarrow$	$M[\Gamma] \simeq_\Gamma N[\Gamma]$
$\mathcal{S} - \mathcal{S}$	$s \simeq_\Gamma s$	$\Longrightarrow$	ϵ
App-App	$(\underline{n}\ M_1\ \dots\ M_k) \simeq_\Gamma (\underline{n}\ N_1\ \dots\ N_k) \Longrightarrow$		$M_1 \simeq_\Gamma N_1.\dots M_k \simeq_\Gamma N_k$

Fig. 7. Rules for checking term equality

equality of types because we assume that type constraints on the equated terms are checked by other constraints. The alternative notion of *algorithmic equality* studied by Harper and Pfenning [15] recurses on the structure of types in the context of LF. It is so far unclear how this idea adapts to type systems with impredicative type polymorphism.

Unification and Inhabitation: With the available machinery, the inhabitation and unification problems for pure type systems is now a relatively straightforward adaption from [20, 10], and for the case of higher-order pattern unification from [11]. The full paper [21] contains a detailed treatment, but here we highlight the main ideas. As usual in higher-order unification, we distinguish between the cases of rigid-rigid, flex-rigid, and flex-flex pairs.

Rigid-Rigid. The rules from Fig. 7 can be directly adapted to decompose rigid-rigid pairs.

Flex-Rigid. To solve flex-rigid pairs we need in general to generate terms with de Brujin indices applied to meta-variables. To do this in a way that preserves types we here introduce an auxiliary operation χ_i that produces an application to i fresh meta-variables and calculates its type.

$$
\frac{\Sigma;\Gamma \mathrel{\vdash\mkern-9mu\sim}_{\chi^i} M \implies N : A, \Sigma_1 \quad \Sigma_1;\Gamma \mathrel{\vdash\mkern-9mu\sim} A : \mathcal{S} \implies s_2, \Sigma_2}{}
$$

$$
\frac{X_{i+1}, Y_1, Y_2 \notin \Sigma_2 \quad s_1 \in \mathcal{S} \quad (s_1, s_2) \in \mathcal{A} \quad \Delta = push(\underline{1}, Y_1, \Gamma)}{}
$$

$$
\frac{\Sigma;\Gamma \mathrel{\vdash\mkern-9mu\sim} M \implies A, \Sigma_1 \quad X_{i+1}: {}_\Gamma Y_1.\ Y_2: {}_\Delta s_2.\ Y_1: {}_\Gamma s_1.\ \Sigma_2;\Gamma \mathrel{\vdash\mkern-9mu\sim} \{A \simeq \Pi_{Y_1}.Y_2\} \implies \Sigma_3}{\Sigma;\Gamma \mathrel{\vdash\mkern-9mu\sim}_{\chi^0} M \implies M : A, \Sigma_1 \qquad \Sigma;\Gamma \mathrel{\vdash\mkern-9mu\sim}_{\chi^{i+1}} M \implies (N\ X_{i+1}) : Y_2[X_{i+1} \cdot_{Y_1} \uparrow^0], \Sigma_3}
$$

A useful property about these judgments is that if $\Sigma;\Gamma \mathrel{\vdash\mkern-9mu\sim}_{\chi^i} M$ fails (produces an unsatisfiable constraint), then $\Sigma;\Gamma \mathrel{\vdash\mkern-9mu\sim}_{\chi^j} M$ fails for every $j \geq i$. This restricts unbounded branching in the cases where we do not make direct use of impredicative type abstraction. For example, in LF one computes the necessary i by inspecting the number of Π nested in A.

To prepare elimination of a meta-variable in a flex-rigid pair by instantiation, we furthermore must order meta-variable declarations in signatures according to their dependencies. For this purpose we introduce an operation to permute signatures. Permutations respect type derivation in $PTS_\mathcal{L}$.

As the final preparation to the flex-rigid rules, we replace occurrences of $(X[S]\ M)$ in Σ by $((\lambda_{A_1}.Y)[S]\ M)$, for a fresh meta-variable $Y:_{push(1,A_1,\Gamma)}A_2$, when $X:_{\Gamma}\Pi_{A_1}.A_2 \in \Sigma$. This produces flex-rigid pairs, that we can finally reduce as follows:

$$\Sigma \implies \Sigma_3\{N/X\}_{\downarrow} \quad \text{if} \quad \begin{cases} (X[M_1 \cdot_{A_1} \ldots M_p \cdot_{A_p} \uparrow^n] \simeq_{\Gamma} (\underline{m}\ N_1\ \ldots\ N_q)) \in \Sigma, \quad X:_{\Delta}B \in \Sigma \\ h \in \{1,\ldots,p\} \cup (\text{if } m > n \text{ then } \{m-n+p\} \text{ else } \emptyset) \\ i \geq 0 \quad \Sigma; \Gamma \vdash_{\chi^i}\underline{h} \implies N:A, \Sigma_1 \\ \Sigma_1; \Gamma \vdash \{A \simeq B\} \implies \Sigma_2 \quad permute(\Sigma_2) \implies \Sigma_4. \Delta:_{X}B. \Sigma_3 \end{cases}$$

Flex-Flex. When all rigid-rigid and flex-rigid pairs are solved, we are left with flex-flex pairs of the form $X[M_1 \cdot_{A_1} \ldots M_p \cdot_{A_p} \uparrow^n] \simeq_{\Gamma} Y[N_1 \cdot_{B_1} \ldots N_q \cdot_{B_q} \uparrow^m]$. For an arbitrary PTS these equations are solved by enumerating inhabitants for X and Y and checking equality using the other rules. The main tool for enumerating the inhabitants is χ_i as exemplified for the calculus of constructions in [20]. This is unlike the situation when one of the substitutions is a *pattern* substitution [11], or if the types of X and Y are *simple* [16].

6   Related Work and Conclusion

Dependent-type versions of $\lambda\Pi_{\mathcal{L}}$ have been studied in [20]. The definition of PTS$_{\mathcal{L}}$ that we present here generalizes those versions in several ways. First of all, we consider $\beta\eta$-equality instead of just β-equality. Secondly, we also identify contexts with a particular kind of explicit substitutions. Thus, let-expressions can be easily encoded in PTS$_{\mathcal{L}}$. Finally, we advocate for explicit instantiations as a way to obtain a calculus of contexts.

Pure type systems and explicit substitutions have been studied in [25, 3]. A polymorphic calculus with explicit substitutions has been proposed in [6]. In contrast to our work, in those approaches explicit substitutions are not first-class objects and atomic implicit substitutions are required by the type systems. In [18], Magnusson uses meta-variables and explicit substitutions to represent incomplete proof-terms in the Martin Löf Type Theory, but a complete meta-theoretical study of the system is missing. More recently, Strecker has developed in [26] a complete meta-theory for a variant of the Extended Calculus of Constructions [17] that supports meta-variables. Strecker's system provides an explicit notation for substitutions. However, it cannot be really considered as an explicit substitution *calculus*. Indeed, in Strecker's approach, the substitution mechanism is implemented by means of an external substitution operation, and the explicit notation is just used to suspend the application of external substitutions to meta-variables. That is, explicit substitutions are only allowed when they are attached to meta-variables. The PTS$_{\mathcal{L}}$-system completely internalizes the substitution mechanism into the theory. This theoretical treatment of substitutions allows a finer and granular control on the application of substitutions. In particular, Strecker's definition of β corresponds to a strategy in $\lambda\Pi_{\mathcal{L}}$ where a (Beta)-rule if followed by $\Pi_{\mathcal{L}}$-normalization. Unification algorithms for dependent types have been studied previously in [22, 12], and for the case of explicit substitutions in [10]. In comparison to Agda [8] our calculus erases the distinction between standard term and term with an explicit substitution.

Acknowledgments

The authors would like to thank C. Braga for his help with Maude and to the anonymous referees for useful remarks.

References

[1] M. Abadi, L. Cardelli, P.-L. Curien, and J.-J. Lévy. Explicit substitution. *Journal of Functional Programming*, 1(4):375–416, 1991.

[2] H. Barendregt. Lambda calculi with types. In Gabbai, Abramski, and Maibaum, editors, *Handbook of Logic in Comp. Sci.* Oxford Univ. Press, 1992.

[3] R. Bloo. *Preservation of Termination for Explicit Substitution.* PhD thesis, Eindhoven University of Technology, 1997.

[4] R. Bloo, F. Kamareddine, and R. Nederpelt. The Barendregt cube with definitions and generalised reduction. *Inf. and Computation*, 126(2):123–143, 1 May 1996.

[5] M. Bognar and Roel de Vrijer. The context calculus λ-c. In *LFM'99: Proceedings of Workshop on Logical Frameworks and Meta-languages*, 1999.

[6] E. Bonelli. The polymorphic λ-calculus with explicit substitutions. In *Proc. of 2nd Intl. Workshop on Explicit Substitutions*, 1999.

[7] C. Braga, H. Haeusler, J. Meseguer, and P. Mosses. Maude action tool: Using reflection to map action semantics to rewriting logic. Submitted, 2000.

[8] C. Coquand and T. Coquand. Structured Type Theory. In *Proc. of Workshop on Logical Frameworks and Meta-Languages*, September 1999.

[9] P.-L. Curien, T. Hardin, and J.-J. Lévy. Confluence properties of weak and strong calculi of explicit substitutions. *Journal of the ACM*, 43(2):362–397, March 1996.

[10] G. Dowek, T. Hardin, and C. Kirchner. Higher-order unification via explicit substitutions. In *Proc. of LICS*, pages 366–374, 1995.

[11] G. Dowek, T. Hardin, C. Kirchner, and F. Pfenning. Unification via explicit substitutions: The case of higher-order patterns. In *Proc. of LICS*, 1996.

[12] C. Elliott. Higher-order unification with dependent types. In N. Dershowitz, editor, *Proc. of RTA*, volume 355 of *LNCS*, pages 121–136. Spr.-Verlag, 1989.

[13] H. Geuvers. The Church-Rosser property for $\beta\eta$-reduction in typed λ-calculi. In *Proc. of LICS*, pages 453–460, 1992.

[14] H. Geuvers. *Logics and Type Systems.* PhD thesis, Kath. Univ. Nijmegen, 1993.

[15] R. Harper and F. Pfenning. On Equivalence and Canonical Forms in the LF Type Theory. In *LFM'99: Proc. of Workshop on Logical Frameworks and Meta-languages*, 1999.

[16] G. Huet. A unification algorithm for typed λ-calculus. *Th. C. S.*, 1(1):27–57, 1975.

[17] Z. Luo. *An Extended Calculus of Constructions.* Phd thesis CST-65-90, Lab. for Foundations of C. S., Dept. of C. S., Univ. of Edinburgh, July 1990.

[18] L. Magnusson. *The Implementation of ALF—A Proof Editor Based on Martin-Löf's Monomorphic Type Theory with Explicit Substitution.* PhD thesis, Chalmers and Göteborg Univ., 1995.

[19] P. A. Mellies. Typed λ-calculi with explicit substitutions may not terminate. *LNCS*, 902:328–338, 1995.

[20] C. Muñoz. *Un calcul de substitutions pour la représentation de preuves partielles en théorie de types*. Thèse de doctorat, Université Paris VII, 1997.

[21] C. Muñoz and N. Bjørner. Absolute explicit unification. Manuscript. Available from http://www.icase.edu/~munoz/, July 2000.

[22] F. Pfenning. Unification and anti-unification in the calculus of constructions. In *Proc. of LICS*, pages 74–85, 1991.

[23] A. Ríos. *Contributions à l'étude de λ-calculs avec des substitutions explicites*. Thèse de doctorat, Université Paris VII, 1993.

[24] P. Severi and E. Poll. Pure type systems with defn's. *LNCS*, 813:316–330, 1994.

[25] M.-O. Stehr and J. Meseguer. Pure type systems in rewriting logic. Proceedings of LFM'99: Workshop on Logical Frameworks and Meta-languages, sept 1999.

[26] M. Strecker. *Construction and Deduction in Type Theories*. PhD thesis, Univ. Ulm, 1999.

[27] H. Zantema. Termination of term rewriting by semantic labelling. *Fundamenta Informaticae*, 24:89–105, 1995.

Termination and Confluence
of Higher-Order Rewrite Systems

Frédéric Blanqui

LRI, Université de Paris-Sud
Bât. 490, 91405 Orsay, France
Tel: (33) 1.69.15.42.35,   Fax: (33) 1.69.15.65.86
Frederic.Blanqui@lri.fr
http://www.lri.fr/~blanqui/

Abstract. In the last twenty years, several approaches to higher-order
rewriting have been proposed, among which Klop's Combinatory Rewrite
Systems (CRSs), Nipkow's Higher-order Rewrite Systems (HRSs) and
Jouannaud and Okada's higher-order algebraic specification languages,
of which only the last one considers typed terms. The later approach has
been extended by Jouannaud, Okada and the present author into Inductive Data Type Systems (IDTSs). In this paper, we extend IDTSs with
the CRS higher-order pattern-matching mechanism, resulting in simply-
typed CRSs. Then, we show how the termination criterion developed
for IDTSs with first-order pattern-matching, called the General Schema,
can be extended so as to prove the strong normalization of IDTSs with
higher-order pattern-matching. Next, we compare the unified approach
with HRSs. We first prove that the extended General Schema can also
be applied to HRSs. Second, we show how Nipkow's higher-order critical
pair analysis technique for proving local confluence can be applied to
IDTSs.
Appendices A, B and C (proofs) are available from the web page.

1 Introduction

In 1980, after a work by Aczel [1], Klop introduced the Combinatory Rewrite
Systems (CRSs) [15, 16], to generalize both first-order term rewriting and rewrite
systems with bound variables like Church's λ-calculus.

In 1991, after Miller's decidability result of the pattern unification problem
[20], Nipkow introduced Higher-order Rewrite Systems (HRSs) [23] (called Pattern Rewrite Systems (PRSs) in [18]), to investigate the metatheory of logic
programming languages and theorem provers like λProlog [21] or Isabelle [25].
In particular, he extended to the higher-order case the decidability result of
Knuth and Bendix about local confluence of first-order term rewrite systems.

At the same time, after the works of Breazu-Tannen [6], Breazu-Tannen
and Gallier [7] and Okada [24] on the combination of Church's simply-typed
λ-calculus with first-order term rewriting, Jouannaud and Okada introduced
higher-order algebraic specification languages [11, 12] to provide a computational

L. Bachmair (Ed.): RTA 2000, LNCS 1833, pp. 47–61, 2000.

model for typed functional languages extended with first-order and higher-order rewrite definitions. Later, together with the present author, they extended these languages with (strictly positive) inductive types, leading to Inductive Data Type Systems (IDTSs) [5]. This approach has also been adapted to richer type disciplines like Coquand and Huet's Calculus of Constructions [2, 4], in order to extend the equality used in proof assistants based on the Curry-De Bruijn-Howard isomorphism like Coq [10] or Lego [17].

Although CRSs and HRSs seem quite different, they have been precisely compared by van Oostrom and van Raamsdonk [31], and shown to have the same expressive power, CRSs using a more lazy evaluation strategy than HRSs. On the other hand, although IDTSs seem very close in spirit to CRSs, the relation between both systems has not been clearly stated yet.

Other approaches have been proposed like Wolfram's Higher-Order Term Rewriting Systems (HOTRSs) [33], Khasidashvili's Expression Reduction Systems (ERSs) [14], Takahashi's Conditional Lambda-Calculus (CLC) [27], ... (see [29]). To tame this proliferation, van Oostrom and van Raamsdonk introduced Higher-Order Rewriting Systems (HORSs) [29, 32] in which the matching procedure is a parameter called "substitution calculus". It appears that most of the known approaches can be obtained by using an appropriate substitution calculus. Van Oostrom proved important confluence results for HORSs whose substitution calculus fulfill some conditions, hence factorizing the existing proofs for the different approaches.

Many results have been obtained so far about the confluence of CRSs and HRSs. On the other hand, for IDTSs, termination was the target of research efforts. A powerful and decidable termination criterion has been developed by Jouannaud, Okada and the present author, called the General Schema [5].

So, one may wonder whether the General Schema may be applied to HRSs, and whether Nipkow's higher-order critical pair analysis technique for proving local confluence of HRSs may be applied to IDTSs.

This paper answers positively both questions. However, we do not consider the *critical interpretation* introduced in [5] for dealing with function definitions over strictly positive inductive types (like Brouwer's ordinals or process algebra). In Section 3, we show how IDTSs relate to CRSs and extend IDTSs with the CRS higher-order pattern-matching mechanism, resulting in simply-typed CRSs. In Section 4, we adapt the General Schema to this new calculus and prove in Section 5 that the rewrite systems that follow this schema are strongly normalizing (every reduction sequence is finite). In Section 6, we show that it can be applied to HRSs. In Section 7, we show that Nipkow's higher-order critical pair analysis technique can be applied to IDTSs.

For proving the termination of a HRS, other criteria are available. Van de Pol extended to the higher-order case the use of strictly monotone interpretations [28]. This approach is of course very powerful but it cannot be automated. In [13], Jouannaud and Rubio defined an extension to the higher-order case of Dershowitz' Recursive Path Ordering (HORPO) exploiting the notion of computable closure introduced in [5] by Jouannaud, Okada and the present author for defining the General Schema. Roughly speaking, the General Schema may be seen as a non-recursive version of HORPO. However, HORPO has not yet been adapted to higher-order pattern-matching.

2   Preliminaries

We assume that the reader is familiar with simply-typed λ-calculus [3]. The set $T(\mathcal{B})$ of *types* $s, t, \ldots$ generated from a set $\mathcal{B}$ of *base types* $\mathbf{s}, \mathbf{t}, \ldots$ (in bold font) is the smallest set built from $\mathcal{B}$ and the function type constructor $\rightarrow$. We denote by $FV(u)$ the set of free variables of a term u, $u\downarrow_\beta$ (resp. $u\uparrow^\eta$) the β-normal form of u (resp. the η-long form of u).

We use a postfix notation for the application of substitutions, $\{x_1 \mapsto u_1, \ldots, x_n \mapsto u_n\}$ for denoting the substitution θ such that $x_i\theta = u_i$ for each $i \in \{1, \ldots, n\}$, and $\theta \uplus \{x \mapsto u\}$ when $x \notin dom(\theta)$, for denoting the substitution θ' such that $x\theta' = u$ and $y\theta' = y\theta$ if $y \neq x$. The domain of a substitution θ is the set $dom(\theta)$ of variables x such that $x\theta \neq x$. Its codomain is the set $cod(\theta) = \{x\theta \mid x \in dom(\theta)\}$.

Whenever we consider abstraction operators, like $\lambda_{-}._{-}$ in λ-calculus, we work modulo α-conversion, *i.e.* modulo renaming of bound variables. Hence, we can always assume that, in a term, the bound variables are pairwise distinct and distinct from the free variables. In addition, to avoid variable capture when applying a substitution θ to a term u, we can assume that the free variables of the terms of the codomain of θ are distinct from the bound variables of u.

We use words over positive numbers for denoting positions in a term. With a symbol f of fixed arity, say n, the positions of the arguments of f are the numbers $i \in \{1, \ldots, n\}$. We will denote by $Pos(u)$ the set of positions in a term u. The subterm at position p is denoted by $u|_p$. Its replacement by another term v is denoted by $u[v]_p$.

For the sake of simplicity, we will often use vector notations for denoting comma- or space-separated sequences of objects. For example, $\{\boldsymbol{x} \mapsto \boldsymbol{u}\}$ will denote $\{x_1 \mapsto u_1, \ldots, x_n \mapsto u_n\}$, $n = |\boldsymbol{u}|$ being the length of $\boldsymbol{u}$. Moreover, some functions will be naturally extended to sequences of objects. For example, $FV(\boldsymbol{u})$ will denote $\bigcup_{1\leq i\leq n} FV(u_i)$ and $\boldsymbol{u}\theta$ the sequence $u_1\theta \ldots u_n\theta$.

3   Extending IDTSs with Higher-Order Pattern-Matching *à la* CRS

In a Combinatory Rewrite System (CRS) [16], the terms are built from variables $x, y, \ldots$ function symbols $f, g, \ldots$ of fixed arity and an abstraction operator $[\_]\_$ such that, in $[x]u$, the variable x is bound in u. On the other hand, left-hand and right-hand sides of rules are not only built from variables, function symbols and the abstraction operator like terms, but also from metavariables $Z, Z', \ldots$ of fixed arity. In the left-hand sides of rules, the metavariables must be applied to distinct bound variables (a condition similar to the one for patterns *à la*

Miller [18]). By convention, a term $Z(x_{i_1}, \ldots, x_{i_k})$ headed by $[x_1], \ldots, [x_n]$ can be replaced only by a term u such that $FV(u) \cap \{x_1, \ldots, x_n\} \subseteq \{x_{i_1}, \ldots, x_{i_k}\}$.

For example, in a left-hand side of the form $f([x][y]Z(x))$, the metaterm $Z(x)$ stands for a term in which y cannot occur free, that is, the metaterm $[x][y]Z(x)$ stands for a function of two variables x and y not depending on y.

The λ-calculus itself may be seen as a CRS with the symbol @ of arity 2 for the application, the CRS abstraction operator $[\_]\_$ standing for λ, and the rule

$$@([x]Z(x), Z') \to Z(Z')$$

for the β-rewrite relation. Indeed, by definition of the CRS substitution mechanism, if $Z(x)$ stands for some term u and Z' for some other term v, then $Z(Z')$ stands for $u\{x \mapsto v\}$.

In [5], Inductive Data Type Systems (IDTSs) are defined as extensions of the simply-typed λ-calculus with function symbols of fixed arity defined by rewrite rules. So, an IDTS may be seen as the sub-CRS of well-typed terms, in which the free variables occuring in rewrite rules are metavariables of arity 0, and only β really uses the CRS substitution mechanism.

As a consequence, restricting matching to first-order matching clearly leads to non-confluence. For example, the rule

$$D(\lambda x.sin(F\ x)) \to \lambda x.(D(F)\ x) \times cos(F\ x)$$

defining a formal differential operator D over a function of the form $sin \circ F$, cannot rewrite a term of the form $D(\lambda x.sin(x))$ since x is not of the form $(u\ x)$.

On the other hand, in the CRS approach, thanks to the notions of metavariable and substitution, D may be properly defined with the rule

$$D([x]sin(F(x))) \to [x] @(D([y]F(y)), x) \times cos(F(x))$$

where F is a metavariable of arity 1.

This leads us to extend IDTSs with the CRS notions of metavariable and substitution, hence resulting in simply-typed CRSs.

Definition 1 (IDTS - New Definition). *An IDTS-alphabet $\mathcal{A}$ is a 4-tuple $(\mathcal{B}, \mathcal{X}, \mathcal{F}, \mathcal{Z})$ where:*
- *$\mathcal{B}$ is a set of base types,*
- *$\mathcal{X}$ is a family $(X_t)_{t \in T(\mathcal{B})}$ of sets of variables,*
- *$\mathcal{F}$ is a family $(F_{s_1, \ldots, s_n, s})_{n \geq 0, s_1, \ldots, s_n, s \in T(\mathcal{B})}$ of sets of function symbols,*
- *$\mathcal{Z}$ is a family $(Z_{s_1, \ldots, s_n, s})_{n \geq 0, s_1, \ldots, s_n, s \in T(\mathcal{B})}$ of sets of metavariables,*
such that all the sets are pairwise disjoint.

The set of IDTS-metaterms over $\mathcal{A}$ is $\mathcal{I}(\mathcal{A}) = \bigcup_{t \in T(\mathcal{B})} \mathcal{I}_t$ where $\mathcal{I}_t$ are the smallest sets such that:
(1) $X_t \subseteq \mathcal{I}_t$,
(2) if $x \in X_s$ and $u \in \mathcal{I}_t$, then $[x]u \in \mathcal{I}_{s \to t}$,
(3) if $f \in F_{s_1, \ldots, s_n, s}$, $u_1 \in \mathcal{I}_{s_1}, \ldots, u_n \in \mathcal{I}_{s_n}$, then $f(u_1, \ldots, u_n) \in \mathcal{I}_s$.
(4) if $Z \in Z_{s_1, \ldots, s_n, s}$, $u_1 \in \mathcal{I}_{s_1}, \ldots, u_n \in \mathcal{I}_{s_n}$, then $Z(u_1, \ldots, u_n) \in \mathcal{I}_s$.

We say that a metaterm u is of type $t \in T(\mathcal{B})$ *if* $u \in \mathcal{I}_t$. *The set of metavariables occuring in a metaterm u is denoted by* $Var(u)$. *A* term *is a metaterm with no metavariable.*

A metaterm l is an IDTS-pattern *if every metavariable occuring in l is applied to a sequence of distinct bound variables.*

An IDTS-rewrite rule *is a pair* $l \rightarrow r$ *of metaterms such that:*

(1) l is an IDTS-pattern,

(2) l is headed by a function symbol,

(3) $Var(r) \subseteq Var(l)$,

(4) r has the same type as l,

(5) l and r are closed $(FV(l) = FV(r) = \emptyset)$.

An n-ary substitute of type $s_1 \rightarrow \ldots \rightarrow s_n \rightarrow s$ *is an expression of the form* $\underline{\lambda}(\boldsymbol{x}).u$ *where* $\boldsymbol{x}$ *are distinct variables of respective types* $s_1, \ldots, s_n$ *and u is a term of type s. An* IDTS-valuation σ *is a type-preserving map associating an n-ary substitute to each metavariable of arity n. Its (postfix) application to a metaterm returns a term defined as follows:*

$- x\sigma = x$

$- ([x]u)\sigma = [x]u\sigma \quad (x \notin FV(cod(\sigma)))$

$- f(\boldsymbol{u})\sigma = f(\boldsymbol{u}\sigma)$

$- Z(\boldsymbol{u})\sigma = v\{\boldsymbol{x} \mapsto \boldsymbol{u}\sigma\} \quad \text{if} \quad \sigma(Z) = \underline{\lambda}(\boldsymbol{x}).v$

An IDTS $\mathcal{I}$ *is a pair* $(\mathcal{A}, \mathcal{R})$ *where* $\mathcal{A}$ *is an IDTS-alphabet and* $\mathcal{R}$ *is a set of IDTS-rewrite rules over* $\mathcal{A}$. *Its corresponding rewrite relation* $\rightarrow_{\mathcal{I}}$ *is the subterm compatible closure of the relation containing every pair* $l\sigma \rightarrow r\sigma$ *such that* $l \rightarrow r \in \mathcal{R}$ *and* σ *is an IDTS-valuation over* $\mathcal{A}$.

The following class of IDTSs will interest us especially:

Definition 2 (β-IDTS). *An IDTS* $(\mathcal{A}, \mathcal{R})$ *where* $\mathcal{A} = (\mathcal{B}, \mathcal{X}, \mathcal{F}, \mathcal{Z})$ *is a* β-IDTS *if, for every pair* $s, t \in T(\mathcal{B})$, *there is:*

(1) a function symbol $@_{s,t} \in \mathcal{F}_{s \rightarrow t, s, t}$,

(2) a rule $\beta_{s,t} = @([x]Z(x), Z') \rightarrow Z(Z') \in \mathcal{R}$,

and no other rule has a left-hand side headed by @.

Given an IDTS $\mathcal{I}$, *we can always add new symbols and new rules so as to obtain a* β-IDTS. *We will denote by* $\beta\mathcal{I}$ *this* β-extension *of* $\mathcal{I}$.

For short, we will denote $@(\ldots @(@(v, u_1), u_2), \ldots, u_n)$ by $@(v, \boldsymbol{u})$.

The strong normalization of $\beta\mathcal{I}$ trivially implies the strong normalization of $\mathcal{I}$. However, the study of $\beta\mathcal{I}$ seems a necessary step because the application symbol @ together with the rule β are the essence of the substitution mechanism. Should we replace in the right-hand sides of the rules every metaterm of the form $Z(\boldsymbol{u})$ by $@([\boldsymbol{x}]Z(\boldsymbol{x}), \boldsymbol{u})$, the system would lead to the same normal forms.

In Appendix A, we list some results about the relations between $\mathcal{I}$ and $\beta\mathcal{I}$.

4   Definition of the General Schema

All along this section and the following one, we fix a given β-IDTS $\mathcal{I} = (\mathcal{A}, \mathcal{R})$. Firstly, we adapt the definition of the General Schema given in [5] to take into account the notion of metavariable. Then, we prove that if the rules of $\mathcal{R}$ follow this schema, then $\rightarrow_{\mathcal{I}}$ is strongly normalizing.

The General Schema is a syntactic criterion which ensures the strong normalization of IDTSs. It has been designed so as to allow a strong normalization proof by the technique of *computability predicates* introduced by Tait for proving the normalization of the simply-typed λ-calculus [26, 9]. Hereafter, we only give basic definitions. The reader will find more details in [5].

Given a rule with left-hand side $f(l)$, we inductively define a set of admissible right-hand sides that we call the *computable closure* of l, starting from the *accessible* metavariables of l. The main problem will be to prove that the computable closure is indeed a set of "computable" terms whenever the terms in l are "computable". This is the objective of Lemma 13 below. The notion of computable closure has been first introduced by Jouannaud, Okada and the present author in [5, 4] for defining the General Schema, but it has been also used by Jouannaud and Rubio in [13] for strengthening their Higher-Order Recursive Path Ordering.

For each base type $\mathbf{s}$, we assume given a set $C_{\mathbf{s}} \subseteq \bigcup_{p \geq 0, s_1, \ldots, s_p \in T(\mathcal{B})} F_{s_1, \ldots, s_p, \mathbf{s}}$ whose elements are called the *constructors* of $\mathbf{s}$. When a function symbol is a constructor, we may denote it by the lower case letters $c, d, \ldots$

This induces the following relation on base types: $\mathbf{t}$ *depends on* $\mathbf{s}$ if there is a constructor $c \in C_{\mathbf{t}}$ such that $\mathbf{s}$ occurs in the type of one of the arguments of c. Its reflexive and transitive closure $\leq_{\mathcal{B}}$ is a quasi-ordering whose associated equivalence relation (resp. strict ordering) will be denoted by $=_{\mathcal{B}}$ (resp. $<_{\mathcal{B}}$).

We say that a constructor $c \in C_{\mathbf{s}}$ is *positive* if every base type $\mathbf{t} =_{\mathcal{B}} \mathbf{s}$ occurs only at positive positions (wrt. the type constructor $\rightarrow$) into the types of the arguments of c. c is *basic* if it is positive and has no functional arguments. A type is *positive* (resp. *basic*) if all its constructors are positive (resp. basic).

Definition 3 (Accessible Subterms). *The set $Acc(v)$ of accessible subterms of a metaterm v is the smallest set such that:*

(1) $v \in Acc(v)$
(2) if $[x]u \in Acc(v)$ then $u \in Acc(v)$
(3) if $c(\boldsymbol{u}) \in Acc(v)$ then each $u_i \in Acc(v)$
(4) if $f(\boldsymbol{u}) \in Acc(v)$ and u_i is of basic type then $u_i \in Acc(v)$
(5) if $@(u, x) \in Acc(v)$, $x \notin FV(u) \cup FV(v)$ then $u \in Acc(v)$
(6) if $@(x, \boldsymbol{u}) \in Acc(v)$, $x \notin FV(\boldsymbol{u}) \cup FV(v)$ then each $u_i \in Acc(v)$.

By abuse of notation, we will say that a metavariable Z is accessible in v if there are distinct bound variables $\boldsymbol{x}$ such that $Z(\boldsymbol{x}) \in Acc(v)$.

For example, F is accessible in $v = [x]sin(F(x))$ since $sin(F(x))$ is accessible in v by (2), and thus, $F(x)$ is accessible in v by (3).

Compared to [5], we express the accessibility with respect to a fixed v. This has no consequence on the definition of computable closure since, among the accessible subterms, only the free variables (here, the metavariables) are taken into account. Accessibility enjoys the following property:

Property 4. *If $u \in Acc(v)$ then $u\sigma \in Acc(v\sigma)$.*

For proving termination, we are led to compare the arguments of a function symbol with the arguments of the recursive calls generated by its reductions. To this end, each function symbol $f \in \mathcal{F}$ is equipped with a *status* $stat_f$ which specifies how to make the comparison as a simple combination of multiset and lexicographic comparisons. Then, an ordering on terms $\leq$ is easily extended to an ordering on sequences of terms $\leq_{stat_f}$. The reader will find precise definitions in [5]. To fix an idea, one can assume that $\leq_{stat_f}$ is the lexicographic extension $\leq_{lex}$ or the multiset extension $\leq_{mul}$ of $\leq$. We will denote by $<^{>}_{stat_f}$ (resp. $<^{\sim}_{stat_f}$) the strict ordering (resp. equivalence relation) associated to $\leq_{stat_f}$. $<^{>}_{stat_f}$ is well-founded if the strict ordering associated to $\leq$ is well-founded.

$\mathcal{R}$ induces the following relation on function symbols: g *depends on* f if there is a rewrite rule defining g (*i.e.* whose left-hand side is headed by g) in the right-hand side of which f occurs. Its reflexive and transitive closure is a quasi-ordering denoted by $\leq_{\mathcal{F}}$ whose associated equivalence relation (resp. strict ordering) will be denoted by $=_{\mathcal{F}}$ (resp. $<_{\mathcal{F}}$).

Finally, we will do the following

Assumptions (A)

(1) Every constructor is positive.
(2) No left-hand side of rule is headed by a constructor.
(3) Both $>_{\mathcal{B}}$ and $>_{\mathcal{F}}$ are well-founded.
(4) $stat_f = stat_g$ whenever $f =_{\mathcal{F}} g$.

The first assumption comes from the fact that, from non-positive inductive types, it is possible to build non-terminating terms [19]. The second assumption ensures that if a constructor-headed term is computable, then its arguments are computable too. The third assumption ensures that types and function definitions are not cyclic. The fourth assumption says that the arguments of equivalent symbols must be compared in the same way.

For comparing the arguments, the subterm ordering $\trianglelefteq$ used in [5] is not satisfactory anymore because of the metavariables which must be applied to some arguments. For example, $[x]F(x)$ is not a subterm of $[x]sin(F(x))$. This can be repaired by using the following ordering.

Definition 5 (Covered-Subterm Ordering). *We say that a metaterm u is a covered-subterm of a metaterm v, written $u \mathrel{\widehat{\trianglelefteq}} v$, if there are two positions $p \in Pos(v)$ and $q \in Pos(v|_p)$ such that (see the figure):*
  – $u = v[v|_{pq}]_p$,
  – $\forall r < p$, $v|_r$ is headed by an abstraction,
  – $\forall r < q$, $v|_{pr}$ is headed by a function symbol (which can be a constructor).

Property 6.

(1) $\widehat{\rhd}$ is stable by valuation: if $u \mathrel{\widehat{\rhd}} v$ and σ is a valuation, then $u\sigma \mathrel{\widehat{\rhd}} v\sigma$.
(2) $\widehat{\rhd}$ is stable by substitution: if $u \mathrel{\widehat{\rhd}} v$ and θ is a substitution, then $u\theta \mathrel{\widehat{\rhd}} v\theta$.
(3) $\widehat{\rhd}$ commutes with $\to$: if $u \mathrel{\widehat{\rhd}} v$ and $v \to w$ then there is a term v' such that $u \to v'$ and $v' \mathrel{\widehat{\rhd}} w$.

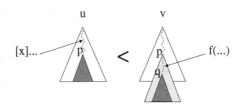

Finally, we come to the definition of computable closure.

Definition 7 (Computable Closure). *Given a function symbol $f \in F_{s_1,\ldots,s_n,s}$, the computable closure $CC_f(l)$ of a metaterm $f(l)$ is the least set CC such that:*

(1) if $Z \in Z_{t_1,\ldots,t_p,t}$ is accessible in l and u are p metaterms of CC of respective types $t_1,\ldots,t_p$, then $Z(u) \in CC$;
(2) if $x \in X_t$ then $x \in CC$;
(3) if $c \in C_t \cap F_{t_1,\ldots,t_p,t}$ and u are p metaterms of CC of respective types $t_1,\ldots,t_p$, then $c(u) \in CC$;
(4) if u and v are two metaterms of CC of respective types $s \to t$ and s then $@(u,v) \in CC$;
(5) if $u \in CC$ then $[x]u \in CC$;
(6) if $h \in F_{t_1,\ldots,t_p,t}$, $h <_F f$ and w are p metaterms of CC of respective types $t_1,\ldots,t_p$, then $h(w) \in CC$;
(7) if $g \in F_{t_1,\ldots,t_p,t}$, $g =_F f$ and u are $p \geq 1$ metaterms of CC of respective types $t_1,\ldots,t_p$ such that $u \mathrel{\widehat{\unlhd}^{>}_{stat_f}} l$, then $g(u) \in CC$.

Note that we do not consider in case (7) the notion of *critical interpretation* introduced in [5] for proving the termination of function definitions over strictly positive types (like Brouwer's ordinals or process algebra).

Definition 8 (General Schema). *A rewrite rule $f(l) \to r$ follows the* General Schema *GS if $r \in CC_f(l)$.*

A first example is given by the rule β itself: $@([x]Z(x), Z') \to Z(Z')$ (Z and Z' are both accessible).

$D([x]sin(F(x))) \to [x]@(D([y]F(y)), x) \times cos(F(x))$ also follows the General Schema since x and y belong to the computable closure of $[x]sin(F(x))$ by (2), hence $F(x)$ and $F(y)$ by (1) since F is accessible in $[x]sin(F(x))$, $[y]F(y)$ by (5), $D([y]F(y))$ by (7) since $[y]F(y)$ is a strict covered-subterm of $[x]sin(F(x))$, $@(D([y]F(y)), x)$ by (4), $cos(F(x))$ by (3), $@(D([y]F(y)), x) \times cos(F(x))$ by (6) and the whole right-hand side by (5).

5 Termination Proof

The termination proof follows Tait's technique of computability predicates [26, 9]. Computability predicates are sets of strongly normalizable terms satisfying appropriate conditions. For each type, we define an interpretation which is a computability predicate and we prove that every term is computable, *i.e.* it belongs to the interpretation of its type. For precise definitions, see [5].

The main things to know are:

– Computability implies strong normalizability.
– If u is a term of type $s \to t$, then it is computable iff, for every computable term v of type s, $@(u, v)$ is computable.
– Computability is preserved by reduction.
– A term is *neutral* if it is neither constructor-headed nor an abstraction. A neutral term u is computable if all its immediate reducts are computable.
– A constructor-headed term $c(u)$ is computable iff all the terms in u are computable.
– For basic types, computability is equivalent to strong normalizability.

Definition 9 (Computable Valuation). *A substitution is* computable *if all the terms of its codomain are computable. A substitute $\lambda(x).u$ is* computable *if, for any computable substitution θ such that $dom(\theta) \subseteq \{x\}$, $u\theta$ is computable. Finally, a valuation σ is* computable *if, for every metavariable Z, the substitute $\sigma(Z)$ is computable.*

Lemma 10 (Compatibility of Accessibility with Computability). *If $u \in Acc(v)$ and v is computable, then for any computable substitution θ such that $dom(\theta) \cap FV(v) = \emptyset$, $u\theta$ is computable.*

Proof. By induction on $Acc(v)$. Without loss of generality, we can assume that $dom(\theta) \subseteq FV(u)$ since $u\theta = u\theta|_{FV(u)}$.

(1) Immediate.
(2) θ is of the form $\theta' \uplus \{x \mapsto x\theta\}$ where $dom(\theta') \cap FV(v) = \emptyset$. By induction hypothesis, $([x]u)\theta'$ is computable. By taking x away from $FV(cod(\theta'))$, $([x]u)\theta' = [x]u\theta'$ and $u\theta = u\theta'\{x \mapsto x\theta\}$ is a reduct of $@([x]u\theta', x\theta)$, hence it is computable since $x\theta$ is computable.
(3) By induction hypothesis, $c(u)\theta = c(u\theta)$ is computable. Hence, by definition of the interpretation for inductive types, $u_i\theta$ is computable.
(4) By induction hypothesis, $f(u)\theta = f(u\theta)$ is computable. Hence $u_i\theta$ is strongly normalizable, and since, for terms of basic type, computability is equivalent to strong normalizability, $u_i\theta$ is computable.
(5) u must be of type $s \to t$. So, let w be a computable term of type s. Since $x \notin FV(u)$, $x \notin dom(\theta)$. Then, let $\theta' = \theta \uplus \{x \mapsto w\}$. θ' is computable and $dom(\theta') \cap FV(v) = \emptyset$ since $x \notin FV(v)$. Hence, by induction hypothesis, $@(u, x)\theta' = @(u\theta, w)$ is computable.

(6) Since $x \notin FV(u)$, $x \notin dom(\theta)$. Then, let $\theta' = \theta \uplus \{x \mapsto [\boldsymbol{y}]y_i\}$, $[\boldsymbol{y}]y_i$ being the i-th projection. θ' is computable and $dom(\theta') \cap FV(v) = \emptyset$ since $x \notin FV(v)$. Hence, by induction hypothesis, $@(x, \boldsymbol{u})\theta' = @([\boldsymbol{y}]y_i, \boldsymbol{u}\theta)$ is computable and its β-reduct $u_i\theta$ also.

Corollary 11. *Let l be a pattern, v a term and σ a valuation such that $l\sigma = v$. If Z is accessible in l and v is computable, then $\sigma(Z)$ is computable.*

For proving Lemma 14 below, we will reason by induction on $(f, \boldsymbol{u})$ with the ordering $\succeq = (\geq_{\mathcal{F}}, \rightarrow_{mul} \cup \widehat{\trianglerighteq}^{>}_{stat_f})_{lex}$, $\boldsymbol{u}$ being strongly normalizable arguments of f. Since $\widehat{\triangleright}$ commutes with $\rightarrow$, we can prove that $\widehat{\trianglerighteq}^{>}_{stat_f} \rightarrow_{mul}$ is included into $\rightarrow_{mul}^{0,1} \widehat{\trianglerighteq}^{>}_{stat_f}$ where $\rightarrow_{mul}^{0,1}$ means zero or one $\rightarrow_{mul}$-step. This implies that $\rightarrow_{mul} \cup \widehat{\trianglerighteq}^{>}_{stat_f}$ is well-founded since:

Lemma 12. *If a and b are two well-founded relations such that $ab \subseteq b^*a$ then $a \cup b$ is well-founded.*

Therefore the strict ordering $\succ$ associated to $\succeq$ is well-founded since $>_{\mathcal{F}}$ is assumed to be well-founded. Now, we can prove the correctness of the computable closure.

Lemma 13 (Computable Closure Correctness). *Let $f(l)$ be a pattern. Assume that σ is a computable valuation and that the terms in $l\sigma$ are computable. Assume also that, for every function symbol h and sequence of computable terms $\boldsymbol{w}$ such that $(f, l\sigma) \succ (h, \boldsymbol{w})$, $h(\boldsymbol{w})$ is computable. Then, for every $r \in \mathcal{CC}_f(l)$, $r\sigma$ is computable.*

Proof. The proof, by induction on $\mathcal{CC}_f(l)$, is quite similar to the one given in [5] except that, now, one has to deal with valuations instead of substitutions. The main difference is in case (1) for metavariables. We only give this case. A full proof can be found in Appendix C.

In fact, we prove that, for any computable valuation σ such that $FV(cod(\sigma)) \cap FV(r) = \emptyset$, for any computable substitution θ such that $dom(\theta) \subseteq FV(r)$ and for any $r \in \mathcal{CC}_f(l)$, $r\sigma\theta = r\theta\sigma$ is computable.

(1) $r = Z(\boldsymbol{v})$ where Z is a metavariable accessible in l and $\boldsymbol{v}$ are metaterms of $\mathcal{CC}$. We first prove it for a special case and then for the general case.

(a) $\boldsymbol{v}$ is a sequence of distinct bound variables, say $\boldsymbol{x}$. Without loss of generality, we can assume that $\sigma(Z) = \lambda(\boldsymbol{x}).w$. Then, $r\sigma\theta = w\theta$. Since σ is computable and $dom(\theta) \subseteq \{\boldsymbol{x}\} = FV(r)$, $w\theta$ is computable.

(b) $r\sigma\theta$ is a β-reduct of the term $@([\boldsymbol{x}]Z(\boldsymbol{x})\sigma\theta, \boldsymbol{v}\sigma\theta)$ where $\boldsymbol{x}$ are fresh distinct variables. By case (1a) and (5), $[\boldsymbol{x}]Z(\boldsymbol{x})\sigma\theta$ is computable and since, by induction hypothesis, the terms in $\boldsymbol{v}\sigma\theta$ are also computable, $r\sigma\theta$ is computable.

Lemma 14 (Computability of Function Symbols). *If all the rules satisfy the General Schema then, for every function symbol f, $f(\boldsymbol{u})$ is computable whenever the terms in $\boldsymbol{u}$ are computable.*

Proof. If f is a constructor then this is immediate since the terms in $\boldsymbol{u}$ are computable by assumption. Assume now that f is a function symbol. Since $f(\boldsymbol{u})$ is neutral, to prove that $f(\boldsymbol{u})$ is computable, it suffices to prove that all its immediate reducts are computable. We prove this by induction on $(f, \boldsymbol{u})$ with $\succ$ as well-founded ordering.

Let v be an immediate reduct of $f(\boldsymbol{u})$. v is either a head-reduct of $f(\boldsymbol{u})$ or of the form $f(u_1, \ldots, u_i', \ldots, u_n)$ with u_i' being an immediate reduct of u_i.

In the latter case, as computability predicates are stable by reduction, u_i' is computable. Hence, since $(f, u_1 \ldots u_i' \ldots u_n) \prec (f, \boldsymbol{u})$, by induction hypothesis, $f(u_1, \ldots, u_i', \ldots, u_n)$ is computable.

In the former case, there is a rule $f(\boldsymbol{l}) \to r$ and a valuation σ such that $\boldsymbol{u} = \boldsymbol{l}\sigma$ and $v = r\sigma$. By definition of the computable closure, and since $Var(r) \subseteq Var(l)$, every metavariable occuring in r is accessible in $\boldsymbol{l}$. Hence, since the terms in $\boldsymbol{l}\sigma$ are computable, by Corollary 11, $\sigma|_{Var(r)}$ is computable. Therefore, by Lemma 13, $r\sigma = r\sigma|_{Var(r)}$ is computable.

Theorem 15 (Strong Normalization). *Let $\mathcal{I} = (\mathcal{A}, \mathcal{R})$ be a β-IDTS satisfying the assumptions (A). If all the rules of $\mathcal{R}$ satisfy the General Schema, then $\to_{\mathcal{I}}$ is strongly normalizing.*

Proof. One can easily prove that, for every term u and computable substitution θ, $u\theta$ is computable. In case where $u = f(\boldsymbol{u})$, we conclude by Lemma 14. The theorem follows easily since the identity substitution is computable.

It is possible to improve this termination result as follows. After [12], if $\mathcal{R}$ follows the General Schema and $\mathcal{R}_1$ is a terminating set of non-duplicating[1] first-order rewrite rules, then $\mathcal{R} \cup \mathcal{R}_1$ is also terminating.

6   Application of the General Schema to HRSs

We just recall what is a HRS. The reader can find precise definitions in [18]. A HRS $\mathcal{H}$ is a pair $(\mathcal{A}, \mathcal{R})$ made of a HRS-alphabet $\mathcal{A}$ and a set $\mathcal{R}$ of HRS-rewrite rules over $\mathcal{A}$. A HRS-alphabet is a triple $(\mathcal{B}, \mathcal{X}, \mathcal{F})$ where $\mathcal{B}$ is a set of base types, $\mathcal{X}$ is a family $(X_s)_{s \in T(\mathcal{B})}$ of variables and $\mathcal{F}$ is a family $(F_s)_{s \in T(\mathcal{B})}$ of function symbols. The corresponding HRS-terms are the terms of the simply-typed λ-calculus built over $\mathcal{X}$ and $\mathcal{F}$ that are in η-long β-normal form.

So, a HRS $\mathcal{H}$ can be seen as an IDTS $\langle \mathcal{H} \rangle$ with the same symbols, the arity of which being determined by the maximum number of arguments they can take, plus the symbol @ for the application. Hence it is a β-IDTS. In [31], van Oostrom and van Raamsdonk studied this translation in detail and proved:

Lemma 16 (Van Oostrom and van Raamsdonk [31]). *Let $\mathcal{H}$ be a HRS. If $u \to_{\mathcal{H}} v$ then $\mathcal{I}(u) \to_{\mathcal{I}(\mathcal{H})} \to_{\beta}^* \mathcal{I}(v)$ where $\mathcal{I}(v)$ is in β-normal form.*

---

[1]   No metavariable occurs more often in the right-hand side than in the left-hand side.

As a consequence, $\mathcal{H}$ is strongly normalizing if $\langle \mathcal{H} \rangle$ so is. Thus, the General Schema can be used on $\langle \mathcal{H} \rangle$ for proving the termination of $\mathcal{H}$. In fact, it can be used directly on $\mathcal{H}$ if we adapt the notions of accessible subterm and computable closure to HRSs. See Appendix B for details.

Theorem 17 (Strong Normalization for HRSs). *Let $\mathcal{H} = (\mathcal{A}, \mathcal{R})$ be a HRS satisfying the assumptions (A). If all the rules of $\mathcal{R}$ satisfy the General Schema for HRSs, then $\to_{\mathcal{H}}$ is strongly normalizing.*

Proof. This results from the fact proved in Appendix B that, if $\mathcal{H}$ follows the General Schema for HRSs then $\langle \mathcal{H} \rangle$ follows the General Schema for IDTSs.

7   Confluence of IDTSs

First of all, since an IDTS is a sub-CRS, it is confluent whenever the underlying CRS is confluent. This is the case if it is weakly orthogonal, *i.e.* it is left-linear and all (higher-order) critical pairs are equal [29], or if it is left-linear and all critical pairs are development closed [30].

Now, one may wonder whether Nipkow's result for local confluence of HRSs [18] may be applied to IDTSs. To this end, we need to interpret an IDTS as a HRS. This can be done in the following natural way:

Definition 18 (Natural Translation of IDTSs into HRSs). *An IDTS-alphabet $\mathcal{A} = (\mathcal{B}, \mathcal{X}, \mathcal{F}, \mathcal{Z})$ can be naturally translated into the HRS-alphabet $\mathcal{H}(\mathcal{A}) = (\mathcal{B}, \mathcal{X}', \mathcal{F}')$ where:*
- $X'_{s_1 \to \ldots \to s_n \to \mathsf{s}} = X_{s_1 \to \ldots \to s_n \to \mathsf{s}} \cup \bigcup_{0 \le p \le n} Z_{s_1, \ldots, s_p, s_{p+1} \to \ldots \to s_n \to \mathsf{s}}$
- $F'_{s_1 \to \ldots \to s_n \to \mathsf{s}} = \bigcup_{0 \le p \le n} F_{s_1, \ldots, s_p, s_{p+1} \to \ldots \to s_n \to \mathsf{s}}$

An IDTS-metaterm u is naturally translated into a HRS-term $\mathcal{H}(u)$ as follows:
- $\mathcal{H}(x) = x \uparrow^{\eta}$ $\mathcal{H}(f(\boldsymbol{u})) = (f \; \mathcal{H}(\boldsymbol{u})) \uparrow^{\eta}$
- $\mathcal{H}([x]u) = \lambda x.\mathcal{H}(u)$ $\mathcal{H}(Z(\boldsymbol{u})) = (Z \; \mathcal{H}(\boldsymbol{u})) \uparrow^{\eta}$

Finally, an IDTS $\mathcal{I} = (\mathcal{A}, \mathcal{R})$ is translated into the HRS $\mathcal{H}(\mathcal{I}) = (\mathcal{H}(\mathcal{A}), \mathcal{H}(\mathcal{R}))$ where $\mathcal{H}(\mathcal{R}) = \{\mathcal{H}(l) \to \mathcal{H}(r) \mid l \to r \in \mathcal{R}\}$.

However, for Nipkow's result to hold, the rewrite rules must be of base type, which is not necessarily the case for IDTSs. This is why, in their study of the relations between CRSs and HRSs [31], van Oostrom and van Raamsdonk defined a translation from CRSs to HRSs, also denoted by $\langle \; \rangle$, which uses a new symbol Λ for forcing the translated terms to be of base type. Furthermore, they proved that (1) if $u \to_{\mathcal{I}} v$ then $\langle u \rangle \to_{\langle \mathcal{I} \rangle} \langle v \rangle$, and (2) if $\langle u \rangle \to_{\langle \mathcal{I} \rangle} v'$ then there is a term v such that $\langle v \rangle = v'$ and $u \to_{\mathcal{I}} v$. In fact, it is no more difficult to prove the same property for the translation $\mathcal{H}$. As a consequence, since $\langle \; \rangle$ (resp. $\mathcal{H}$) is injective, the (local) confluence of $\langle \mathcal{I} \rangle$ (resp. $\mathcal{H}(\mathcal{I})$) implies the (local) confluence of $\mathcal{I}$. Thus it is possible to deduce the local confluence of $\mathcal{I}$ from the analysis of the critical pairs of $\langle \mathcal{I} \rangle$ (resp. $\mathcal{H}(\mathcal{I})$), and indeed, it turns out that $\langle \mathcal{I} \rangle$ and $\mathcal{H}(\mathcal{I})$ have the "same" critical pairs (see the proof of Theorem 19 in Appendix C for details). Identifying $\mathcal{I}$ with its natural translation $\mathcal{H}(\mathcal{I})$, we claim that:

Theorem 19. *If every critical pair of* $\mathcal{I}$ *is confluent, then* $\mathcal{I}$ *is locally confluent.*

It could also have been possible to consider the translation $\mathcal{H}'$ which is identical to $\mathcal{H}$ but pulls down to base type the rewrite rules by taking $\mathcal{H}'(f(l) \rightarrow r) = (f\ \mathcal{H}(l)\ \boldsymbol{x}) \rightarrow v$ if $\mathcal{H}(r) = \lambda\boldsymbol{x}.v$ with v of base type. Note that the left-hand side is still a pattern. Then, it is possible to prove that $\mathcal{H}(\mathcal{I})$ and $\mathcal{H}'(\mathcal{I})$ have also the same critical pairs.

8   Conclusion

In Inductive Data Type Systems (IDTSs) [5], the use of first-order matching does not allow to define some functions as expected, resulting in non-confluent computations. By extending IDTS with the higher-order pattern-matching mechanism of Klop's Combinatory Reduction Systems (CRSs) [16], we solved this problem and made clear the relation between IDTSs and CRSs: IDTSs with higher-order pattern-matching are simply-typed CRSs.

We extended a decidable termination criterion defined for IDTSs with first-order matching and called the General Schema [5] to the case of higher-order pattern-matching, and we proved that a rewrite system following this schema is strongly-normalizing.

We also compared this unified approach to Nipkow's Higher-order Rewrite Systems (HRSs) [18]. First, we proved that the extended General Schema can be applied to HRSs. Second, we show how Nipkow's higher-order critical pair analysis technique for proving local confluence can be applied to IDTSs.

Now, several extensions should be considered.

We did not take into account the interpretation defined in [5] for dealing with definitions over strictly positive types (like Brouwer's ordinals or process algebra). However, we expect that it can also be adapted to higher-order pattern-matching.

It is also important to be able to relax the pattern condition which says that metavariables must be applied to distinct bound variables. But it is not clear how to prove the termination with Tait's computability predicates technique when this condition is not satisfied.

Another point is that some computations often need to be performed within some equational theories like commutativity or commutativity and associativity of some function symbols. It would be interesting to know if the General Schema technique can be adapted for dealing with such equational theories.

Finally, one may wonder whether all these results could be establish in the more general framework of van Oostrom and van Raamsdonk's Higher-Order Rewriting Systems (HORSs) [29, 32], under some suitable conditions over the substitution calculus.

Acknowledgments

I am very grateful to D. Kesner, A. Boudet and J.-P. Jouannaud for their suggestions and remarks on previous versions of this paper. I also thank the anonymous referees for their useful comments.

References

[1] P. Aczel. A general Church-Rosser theorem. Technical report, University of Manchester, United Kingdom, 1978.

[2] F. Barbanera, M. Fernández, and H. Geuvers. Modularity of strong normalization in the algebraic-λ-cube. *Journal of Functional Programming*, 7(6), 1997.

[3] H. Barendregt. Lambda calculi with types. In S. Abramski, D. M. Gabbai, and T. S. E. Maiboum, editors, *Handbook of logic in computer science*, volume 2. Oxford University Press, 1992.

[4] F. Blanqui, J.-P. Jouannaud, and M. Okada. The Calculus of Algebraic Constructions. In *Proc. of RTA'99, LNCS 1631*.

[5] F. Blanqui, J.-P. Jouannaud, and M. Okada. Inductive Data Type Systems, 1998. To appear in TCS. Available at http://www.lri.fr/~blanqui/.

[6] V. Breazu-Tannen. Combining algebra and higher-order types. In *Proc. of LICS'88, IEEE Computer Society*.

[7] V. Breazu-Tannen and J. Gallier. Polymorphic rewriting conserves algebraic strong normalization. In *Proc. of ICALP'89, LNCS 372*.

[8] V. Breazu-Tannen and J. Gallier. Polymorphic rewriting conserves algebraic strong normalization. *Theoretical Computer Science*, 83(1), 1991.

[9] J. R. Hindley and J. P. Seldin. *Introduction to combinators and λ-calculus*. London Mathematical Society, 1986.

[10] INRIA-Rocquencourt/CNRS/Université Paris-Sud/ENS Lyon, France. *The Coq Proof Assistant Reference Manual Version 6.3*, 1999. Available at http://pauillac.inria.fr/coq/.

[11] J.-P. Jouannaud and M. Okada. Executable higher-order algebraic specification languages. In *Proc. of LICS'91, IEEE Computer Society*.

[12] J.-P. Jouannaud and M. Okada. Abstract Data Type Systems. *Theoretical Computer Science*, 173(2), 1997.

[13] J.-P. Jouannaud and A. Rubio. The Higher-Order Recursive Path Ordering. In *Proc. of LICS'99, IEEE Computer Society*.

[14] Z. Khasidashvili. Expression Reduction Systems. In *Proc. of I. Vekua Institute of Applied Mathematics*, volume 36, 1990.

[15] J. W. Klop. *Combinatory Reduction Systems*. PhD thesis, University of Utrecht, Netherlands, 1980. Published as Mathematical Center Tract 129.

[16] J. W. Klop, V. van Oostrom, and F. van Raamsdonk. Combinatory reduction systems: introduction and survey. *Theoretical Computer Science*, 121(1-2), 1993.

[17] Z. Luo and R. Pollack. *LEGO Proof Development System: User's manual*. University of Edinburgh, Scotland, 1992.

[18] R. Mayr and T. Nipkow. Higher-order rewrite systems and their confluence. *Theoretical Computer Science*, 192, 1998.

[19] N. P. Mendler. *Inductive Definition in Type Theory*. PhD thesis, Cornell University, United States, 1987.

[20] D. Miller. A logic programming language with lambda-abstraction, function variables, and simple unification. In *Proc. of ELP'89, LNCS 475*.

[21] D. Miller and G. Nadathur. An overview of λProlog. In *Proc. of the 5th Int. Conf. on Logic Programming, 1988*.

[22] F. Müller. Confluence of the lambda calculus with left-linear algebraic rewriting. *Information Processing Letters*, 41, 1992.

[23] T. Nipkow. Higher-order critical pairs. In *Proc. of LICS'91, IEEE Computer Society*.

[24] M. Okada. Strong normalizability for the combined system of the typed lambda calculus and an arbitrary convergent term rewrite system. In *Proc. of ISSAC'89, ACM Press*.

[25] L. Paulson. Isabelle: a generic theorem prover. *LNCS 828*, 1994.

[26] W. W. Tait. Intensional interpretations of functionals of finite type I. *Journal of Symbolic Logic*, 32(2), 1967.

[27] M. Takahashi. λ-calculi with conditional rules. In *Proc. of TLCA'93, LNCS 664*.

[28] J. van de Pol and H. Schwichtenberg. Strict functionals for termination proofs. In *Proc. of TLCA'95, LNCS 902*.

[29] V. van Oostrom. *Confluence for Abstract and Higher-Order Rewriting*. PhD thesis, Vrije Universiteit, Netherlands, 1994.

[30] V. van Oostrom. Development closed critical pairs. In *Proc. of HOA'95, LNCS 1074*, 1995.

[31] V. van Oostrom and F. van Raamsdonk. Comparing Combinatory Reduction Systems and Higher-order Rewrite Systems. In *Proc. of HOA'93, LNCS 816*.

[32] F. van Raamsdonk. *Confluence and Normalization for Higher-Order Rewriting*. PhD thesis, Vrije Universiteit, Netherlands, 1996.

[33] D. Wolfram. *The clausal theory of types*. PhD thesis, University of Cambridge, United Kingdom, 1990.

A de Bruijn Notation for Higher-Order Rewriting

(Extended Abstract)

Eduardo Bonelli[1,2], Delia Kesner[2], and Alejandro Ríos[1]

[1] Departamento de Computación - Facultad de Ciencias Exactas y Naturales
Universidad de Buenos Aires, Pabellón I
Ciudad Universitaria (1428), Buenos Aires, Argentina
{ebonelli,rios}@dc.uba.ar
[2] LRI (UMR 8623) - Bât 490, Université de Paris-Sud
91405 Orsay Cedex, France
kesner@lri.fr

Abstract. We propose a formalism for higher-order rewriting in de Bruijn notation. This notation not only is used for terms (as usually done in the literature) but also for metaterms, which are the syntactical objects used to express general higher-order rewrite systems. We give formal translations from higher-order rewriting with names to higher-order rewriting with de Bruijn indices, and vice-versa. These translations can be viewed as an *interface* in programming languages based on higher-order rewrite systems, and they are also used to show some properties, namely, that both formalisms are operationally equivalent, and that confluence is preserved when translating one formalism into the other.

1 Introduction

Higher-order (term) rewriting concerns the transformation of terms in the presence of binding mechanisms for variables. Implementing higher-order rewriting requires, beforehand, taking care of a complex notion of substitution operation and of renaming of bound variables (α-conversion). As a paradigmatic example, the β-reduction axiom of λ-calculus [1], expressed $(\lambda x.M)N \longrightarrow_\beta M\{x \leftarrow N\}$, may be interpreted as: the result of executing function $\lambda x.M$ over argument N is obtained by substituting N for all (free) occurrences of x in M. Any implementation of higher-order rewriting must include instructions for computing this substitution. Although from the meta-level the execution of a substitution is atomic, the cost of computing it highly depends on the form of the terms, specially if unwanted variable capture conflicts must be avoided by renaming bound variables. De Bruijn indices take care of renaming because the representation of variables by indices completely eliminates unwanted capture of variables. However, de Bruijn formalisms have only been studied for particular systems (and only on the term level) and no general framework of higher-order rewriting with indices has been proposed. We address this problem here by focusing not only

L. Bachmair (Ed.): RTA 2000, LNCS 1833, pp. 62–79, 2000.

on de Bruijn terms (as usually done in the literature for λ-calculus [11]) but also on de Bruijn metaterms, which are the syntactical objects used to express any general higher-order rewrite system formulated in a de Bruijn context.

Many higher-order rewrite systems (HORS) exist and work in the area is currently very active: *CRS* [14], *ERS* [12], *CERS* [13], *HRS* [15], the systems in [22] and [20]. We choose in this work to use *ERS* because their syntax and semantics are simple and natural (they allow for example to write β-reduction in λ-calculus as usual while *CRS* do not) and the correspondence between *ERS* and *HRS* has already been established [21]. We shall begin with (a slightly simplified version of) the *ERS* formalism, that we shall call *SERS* (S for simplified) and introduce the de Bruijn index based higher-order rewrite system $SERS_{DB}$.

Our work is the first step in the construction of a formal interpretation of higher-order rewriting via a *first-order* theory. This kind of simulation would be possible with the aid of *explicit substitutions*. Indeed, this work follows, in some sense, the lines of [8] which interprets higher-order formalisms/problems into their respective first-order ones.

Our formalism is developed in order to be used as an *interface* of a programming language based on higher-order rewriting. Of course, the use of variable name based formalisms are necessary for humans to interact with computers in a user-friendly way. Clearly technical resources like de Bruijn indices and explicit substitutions should live behind the scene, in other words, should be implementation concerns. Moreover, it is required of whatever is behind the scene to be as faithful as possible as regards the formalism it is implementing. So a key issue shall be the detailed study of the relationship between SERS and $SERS_{DB}$. The translations we propose between them are extensions to higher-order of the translations studied in [11] and presented in [5].

As regards existing higher-order rewrite formalisms based on de Bruijn index notation and/or explicit substitutions to the best of the authors' knowledge there are but two: *Explicit CRS* [3] and *XRS* [16]. In [3] explicit substitutions a la λx [19,2] are added to the *CRS* formalism as a first step towards using higher-order rewriting with explicit substitutions for modeling the evaluation of functional programs in a faithful way. Since this is done in a variable name setting α-conversion must be dealt with as in *CRS*. Pagano's *XRS* constitutes the first HORS which fuses de Bruijn index notation and explicit substitutions. It is presented as a generalization of the $\lambda\sigma_{\Uparrow}$-calculus [6] but no connection has been established between *XRS* and well-known systems such as *CRS*, *ERS* and *HRS*. Indeed, it is not clear at all how some seemingly natural rules expressible, say, in the *ERS* formalism, may be written in an *XRS*. As an example, consider a rewrite system for logical expressions such that if $imply(e_1, e_2)$ reduces to the constant *true* then e_1 logically implies e_2 in classical first-order predicate logic. A possible rewrite rule could be:

$$(imp) \quad imply(\exists x \forall y M, \forall y \exists x M) \longrightarrow true$$

A naïve attempt might consider the rewrite rule

$$(imp_{db}) \quad imply(\exists \forall M, \forall \exists M) \longrightarrow true$$

as a possible representation of this rule in the *XRS* formalism, but it does not have the desired effect since $\exists\forall M$ and $\forall\exists M$ correspond to $\exists x\forall yM$ and $\forall x\exists yM$ but $\forall x\exists yM$ and $\forall y\exists xM$ are not equivalent. Note that regardless of the fact that *XRS* incorporate explicit substitutions, this problem arises already at the level of de Bruijn notation. Another example of interest is:

$$(\eta) \quad \lambda x.(Mx)\longrightarrow M \text{ if } x \text{ is not free in } M$$

which is usually expressed in a de Bruijn based system with explicit substitutions

$$(\eta_{db}) \quad \lambda(M1)\longrightarrow N \text{ if } M =_{\mathcal{C}} N[\uparrow]$$

where $M =_{\mathcal{C}} N$ means that M and N are equivalent modulo the theory of explicit substitutions $\mathcal{C}$. Neither the (imp) rule nor (η_{db}) is possible in the *XRS* formalism so that they do not have in principle the same expressive power as *ERS*. We shall propose de Bruijn based HORS that will allow such rules to be faithfully represented.

The main contribution of this paper is a general de Bruijn notation for higher-order syntax which bridges the gap between higher-order rewriting with names and with indices. This formalism suggests a first-order tool to implement HORS, which in contrast to [16] would represent *all* the HORS used in practice.

The rest of the paper is organized as follows. Section 2 introduces our work and study scenario, the *SERS* formalism. The de Bruijn based formalism $SERS_{DB}$ is defined in Section 3. Section 4 takes a close-up view of the relationship, via appropriate translations, between the formalisms *SERS* and $SERS_{DB}$. Also, preservation of confluence is considered. Finally, we conclude.

By lack of space we only present here an extended abstract, and therefore proofs, auxiliary lemmas and standard definitions are only hinted or just omitted, but the interested reader will find full details in [4].

2   Simplified Expression Reduction Systems

We introduce the variable name based higher-order rewrite formalism *SERS*.

2.1   Metaterms and Terms

Definition 1 (Signature). *Consider the denumerable and disjoint infinite sets:*

- $\mathcal{V} = \{x_1, x_2, x_3, \ldots\}$ *a set of* variables, *arbitrary variables are denoted* $x, y, \ldots$
- $\mathcal{B}_{\mathcal{MV}} = \{\alpha_1, \alpha_2, \alpha_3, \ldots\}$ *a set of* pre-bound o-metavariables *(o for object), denoted* $\alpha, \beta, \ldots$
- $\mathcal{F}_{\mathcal{MV}} = \{\widehat{\alpha_1}, \widehat{\alpha_2}, \widehat{\alpha_3}, \ldots\}$ *a set of* pre-free o-metavariables, *denoted* $\widehat{\alpha}, \widehat{\beta}, \ldots$
- $\mathcal{T}_{\mathcal{MV}} = \{X_1, X_2, X_3, \ldots\}$ *a set of* t-metavariables *(t for term), denoted* $X, Y, Z, \ldots$
- $\mathcal{F} = \{f_1, f_2, f_3, \ldots\}$ *a set of* function symbols *equipped with a fixed (possibly zero) arity, denoted* $f, g, h, \ldots$
- $\mathcal{B} = \{\lambda_1, \lambda_2, \lambda_3, \ldots\}$ *a set of* binder symbols *equipped with a fixed (non-zero) arity, denoted* $\lambda, \mu, \nu, \xi, \ldots$

The union of $\mathcal{B}_{\mathcal{MV}}$ and $\mathcal{F}_{\mathcal{MV}}$ is the set of o-metavariables of the signature. When speaking of metavariables without further qualifiers we refer to o and t-metavariables. Since all these alphabets are ordered, given any symbol s we shall denote $\mathcal{O}(s)$ its position in the corresponding alphabet.

Definition 2 (Labels). *A* label *is a finite sequence of symbols of an alphabet. We shall use $k, l, l_i, \ldots$ to denote arbitrary labels and ϵ for the empty label. If s is a symbol and l is a label then the notation $s \in l$ means that the symbol s appears in the label l, and also, we use sl to denote the new label whose head is s and whose tail is l. Other notations are $|l|$ for the length of l (number of symbols in l) and $\mathtt{at}(l, n)$ for the n-th element of l assuming $n \leq |l|$. Also, if s occurs (at least once) in l then $\mathtt{pos}(s, l)$ denotes the position of the first occurrence of s in l. If θ is a function defined on the alphabet of a label $l \equiv s_1 \ldots s_n$, then $\theta(l)$ denotes the label $\theta(s_1) \ldots \theta(s_n)$. In the sequel, we may use a label as a set (e.g. $S \cap l$ denotes the intersection of a set S with the set containing the elements of l) if no confusion arises. A* simple label *is a label without repeated symbols.*

Definition 3 (Pre-metaterms). *The set of SERS* pre-metaterms[1], *denoted $\mathcal{PMT}$, is defined by:*

$$M \quad ::= \quad \alpha \mid \widehat{\alpha} \mid X \mid f(M, \ldots, M) \mid \xi\alpha.(M, \ldots, M) \mid M[\alpha \leftarrow M]$$

Arities are supposed to be respected and we shall use $M, N, M_i, \ldots$ to denote pre-metaterms. The symbol $.[. \leftarrow .]$ in the pre-metaterm $M[\alpha \leftarrow M]$ is called metasubstitution operator. *The o-metavariable α in a pre-metaterm of the form $\xi\alpha.(M, \ldots, M)$ or $M[\alpha \leftarrow M]$ is referred to as the* formal parameter. *The set of binder symbols together with the metasubstitution operator are called* binder operators, *thus the metasubstitution operator is a binder operator (since it has binding power) but* not *a binder symbol since it is not an element of $\mathcal{B}$.*

A pre-metaterm M has an associated *tree*, denoted $tree(M)$, defined as expected. In the case of the metasubstitution operator we have: if T_1, T_2 are the trees of M_1, M_2, then the tree of $M_1[\alpha \leftarrow M_2]$ has root "sub", and sons "$[\,]\alpha$" (with son T_1) and T_2.

A position is a label over the alphabet $\mathbb{N}$. Given a pre-metaterm N appearing in M, the set of *occurrences* of N in M is the set of positions of $tree(M)$ where N occurs (positions in trees are defined as usual). The *parameter path* of an occurrence p in a tree T is the list containing all the (pre-bound) o-metavariables occuring in the path from p to the root of T.

---

[1]   The main difference between *SERS* and *ERS* is that in the latter binders and metasubstitutions are defined on *multiple* o-metavariables. Indeed, pre-metaterms like $\xi\alpha_1 \ldots \alpha_k.(M_1, \ldots, M_m)$ and $M[\alpha_1 \ldots \alpha_k \leftarrow M_1, \ldots, M_k]$ are possible in *ERS*, with the underlying hypothesis that $\alpha_1 \ldots \alpha_k$ are all distinct and with the underlying semantics that $M[\alpha_1 \ldots \alpha_k \leftarrow M_1, \ldots, M_k]$ denotes usual (parallel) substitution. It is well known that multiple substitution can be simulated by simple substitution. Furthermore, there is also a notion of scope indicator in *ERS*, used to express in which arguments the variables are bound. Scope indicators shall not be considered in *SERS* since they do not seem to contribute to the expressive power of *ERS*.

The following definition introduces the set of *metaterms*, which are pre-metaterms that are *well-formed* in the sense that all the formal parameters appearing in the same path of a pre-metaterm must be different and all the metavariables in $\mathcal{B}_{\mathcal{MV}}$ only occur bound.

Definition 4 (Metaterms). *A pre-metaterm M is a metaterm, denoted by $\mathcal{WF}(M)$, iff the predicate $\mathcal{WF}_\epsilon(M)$ holds, where $\mathcal{WF}_l(M)$ is defined as follows:*

- $\mathcal{WF}_l(\alpha)$ *iff $\alpha \in l$*
- $\mathcal{WF}_l(\widehat{\alpha})$ *and $\mathcal{WF}_l(X)$ are always true*
- $\mathcal{WF}_l(f(M_1, \ldots, M_n))$ *iff for all $1 \leq i \leq n$ we have $\mathcal{WF}_l(M_i)$*
- $\mathcal{WF}_l(\xi\alpha.(M_1, \ldots, M_n))$ *iff $\alpha \notin l$ and for all $1 \leq i \leq n$ we have $\mathcal{WF}_{\alpha l}(M_i)$*
- $\mathcal{WF}_l(M_1[\alpha \leftarrow M_2])$ *iff $\alpha \notin l$ and $\mathcal{WF}_l(M_2)$ and $\mathcal{WF}_{\alpha l}(M_1)$.*

For example, $f(\xi\alpha.(X), \lambda\alpha.(Y))$, $f(\widehat{\beta}, \lambda\alpha.(Y))$ and $g(\lambda\alpha.(\xi\beta.(h)))$ are meta-terms, while the pre-metaterms $f(\alpha, \xi\alpha.(X))$ and $f(\widehat{\beta}, \lambda\alpha.(\xi\alpha.(X)))$ are not.

In the sequel, pre-bound (free) o-metavariables occurring in metaterms shall simply be referred to as bound (free) o-metavariables. As we shall see, metaterms are used to specify rewrite rules.

Definition 5 (Free Metavariables of Pre-metaterms). *Let M be a pre-metaterm, then $FMVar(M)$ denotes the set of free metavariables of M, which is defined as follows:*

$$FMVar(X) \stackrel{\text{def}}{=} \{X\} \qquad FMVar(\alpha) \stackrel{\text{def}}{=} \{\alpha\} \qquad FMVar(\widehat{\alpha}) \stackrel{\text{def}}{=} \{\widehat{\alpha}\}$$

$$FMVar(f(M_1, \ldots, M_n)) \stackrel{\text{def}}{=} \bigcup_{i=1}^n FMVar(M_i)$$

$$FMVar(\xi\alpha.(M_1, \ldots, M_n)) \stackrel{\text{def}}{=} \left(\bigcup_{i=1}^n FMVar(M_i)\right) \setminus \{\alpha\}$$

$$FMVar(M_1[\alpha \leftarrow M_2]) \stackrel{\text{def}}{=} (FMVar(M_1) \setminus \{\alpha\}) \cup FMVar(M_2)$$

All metavariables which are not free are bound. We use $BMVar(M)$ to denote the bound metavariables of a metaterm M. Note that only o-metavariables may occur bound in a metaterm. We denote the set of metavariables of a metaterm or a pre-metaterm M by $MVar(M)$. Note that if M is a metaterm, then $FMVar(M)$ does not contain pre-bound o-metavariables.

Definition 6 (Terms and Contexts). *The set of SERS terms, denoted $\mathcal{T}$, and contexts are defined by:*

$$\text{Terms} \quad t ::= x \mid f(t, \ldots, t) \mid \xi x.(t, \ldots, t)$$
$$\text{Contexts} \quad C ::= \square \mid f(t, \ldots, C, \ldots, t) \mid \xi x.(t, \ldots, C, \ldots, t)$$

where $\square$ denotes a "hole". We shall use $s, t, t_i, \ldots$ for terms and C, D for contexts. We remark that in contrast to other formalisms dealing with higher-order rewriting, here the set of terms is not contained in the set of pre-metaterms since the set of variables and the set of o-metavariables are disjoint. The set of free (resp. bound) variables of a term t, denoted $FV(t)$ (resp $BV(t)$) are defined as usual.

With $C[t]$ we denote the term obtained by replacing the term t for the hole $\square$ in the context C. Note that this operation may introduce variable capture. We define the label *of a context as a sequence of variables as follows:*

$$\texttt{label}(\square) \quad \overset{\text{def}}{=} \epsilon$$

$$\texttt{label}(f(t_1, \ldots, C, \ldots, t_n)) \quad \overset{\text{def}}{=} \texttt{label}(C)$$

$$\texttt{label}(\xi x.(t_1, \ldots, C, \ldots, t_n)) \overset{\text{def}}{=} \texttt{label}(C)x$$

For example, the label of the context $C \equiv f(\lambda x.(z, \xi y.(h(y, \square))))$ is the sequence yx. The label of a context is a notion analogous to that of a parameter path of an occurrence, but defined for terms instead of pre-metaterms and where the only occurrence considered is that of the hole.

Definition 7 ((Restricted) Substitution of Terms). *The* (restricted) *substitution of a term t for a variable x in a term s, denoted $s\{x \leftarrow t\}$, is defined:*

$$x\{x \leftarrow t\} \qquad \overset{\text{def}}{=} t$$

$$y\{x \leftarrow t\} \qquad \overset{\text{def}}{=} y \qquad\qquad\qquad if\ x \not\equiv y$$

$$f(s_1, \ldots, s_n)\{x \leftarrow t\} \quad \overset{\text{def}}{=} f(s_1\{x \leftarrow t\}, \ldots, s_n\{x \leftarrow t\})$$

$$\xi x.(s_1, \ldots, s_n)\{x \leftarrow t\} \overset{\text{def}}{=} \xi x.(s_1, \ldots, s_n)$$

$$\xi y.(s_1, \ldots, s_n)\{x \leftarrow t\} \overset{\text{def}}{=} \xi y.(s_1\{x \leftarrow t\}, \ldots, s_n\{x \leftarrow t\})$$

$$if\ x \not\equiv y, \ and\ (y \notin FV(t)\ or\ x \notin FV(s))$$

Thus $.\{. \leftarrow .\}$ denotes the substitution operator on terms but it may *not* apply α-conversion (renaming of bound variables) in order to avoid unwanted variable captures. Therefore this notion of substitution is not defined for all terms (hence its name). When defining the notion of reduction relation on terms induced by rewrite rules we shall take α-conversion into consideration. We may define α-conversion on terms as the smallest reflexive, symmetric and transitive relation closed by contexts verifying the following equality:

$$(\alpha)\ \xi x.(s_1, \ldots, s_n) \equiv_\alpha \xi y.(s_1\{x \leftarrow y\}, \ldots, s_n\{x \leftarrow y\}) \quad y\ \text{not in } s_1, \ldots, s_n$$

Note that since y does not occur in $s_1, \ldots, s_n$ substitution is defined. We shall use $t \equiv_\alpha s$ to denote that the terms t and s are α-convertible. This conversion is sound in the sense that $t \equiv_\alpha s$ implies $FV(t) = FV(s)$.

The notion of α-conversion for terms has a symmetrical one for pre-metaterms which we call v-*equivalence* (v for variant). The intuitive meaning of two v-equivalent pre-metaterms is that they are able to receive the *same* set of potential "valuations" (c.f. Definition 10). Thus for example, as one would expect, $\lambda \alpha.(X) \neq_v \lambda \beta.(X)$ because when α and X are replaced by x and β is replaced by y, one obtains $\lambda x.(x)$ and $\lambda y.(x)$, which are not α-convertible. However, since pre-metaterms contain t-metavariables, the notion of v-equivalence is not straightforward as the notion of α-conversion in the case of terms.

Definition 8 (v-Equivalence for Pre-metaterms). *Given pre-metaterms M and N, we say that M is v-equivalent to N, iff $M =_v N$ where $=_v$ is the smallest reflexive, symmetric and transitive relation closed by metacontexts[2] verifying:*

$$(v1) \quad \xi \alpha.(P_1, \ldots, P_n) =_v \xi \beta.(P_1 \ll\!\alpha\!\leftarrow\!\beta\!\gg \ldots P_n \ll\!\alpha\!\leftarrow\!\beta\!\gg)$$

$$(v2) \quad P_1[\alpha \leftarrow P_0] \quad =_v P_1 \ll\!\alpha\!\leftarrow\!\beta\!\gg [\beta \leftarrow P_0]$$

---

[2] Metacontexts are defined analogously to contexts. The notion of "label of a context" is extended to metacontexts as expected.

where β is a pre-bound o-metavariable which does not occur in $P_1, \ldots, P_n$ in (v1) *and does not occur in P_1 in* (v2), *P_i does not contain t-metavariables for* $1 \le i \le n$, *and $P \ll \alpha \leftarrow Q \gg$ is the restricted substitution for pre-metaterms:*

$$\alpha \ll \alpha \leftarrow Q \gg \quad \overset{\text{def}}{=} Q$$
$$\alpha' \ll \alpha \leftarrow Q \gg \quad \overset{\text{def}}{=} \alpha' \quad \alpha \not\equiv \alpha'$$
$$\widehat{\alpha'} \ll \alpha \leftarrow Q \gg \quad \overset{\text{def}}{=} \widehat{\alpha'}$$
$$X \ll \alpha \leftarrow Q \gg \quad \overset{\text{def}}{=} X$$
$$f(M_1, \ldots, M_n) \ll \alpha \leftarrow Q \gg \quad \overset{\text{def}}{=} f(M_1 \ll \alpha \leftarrow Q \gg, \ldots, M_n \ll \alpha \leftarrow Q \gg)$$
$$(\xi \alpha.(M_1, \ldots, M_n)) \ll \alpha \leftarrow Q \gg \quad \overset{\text{def}}{=} \xi \alpha.(M_1, \ldots, M_n)$$
$$(\xi \alpha'.(M_1, \ldots, M_n)) \ll \alpha \leftarrow Q \gg \quad \overset{\text{def}}{=} \xi \alpha'.(M_1 \ll \alpha \leftarrow Q \gg, \ldots, M_n \ll \alpha \leftarrow Q \gg)$$
$$\alpha \not\equiv \alpha', \; (\alpha' \notin FMVar(Q) \text{ or } \alpha \notin FMVar(P))$$
$$(M_1[\alpha \leftarrow M_2]) \ll \alpha \leftarrow Q \gg \quad \overset{\text{def}}{=} M_1[\alpha \leftarrow M_2 \ll \alpha \leftarrow Q \gg]$$
$$(M_1[\alpha' \leftarrow M_2]) \ll \alpha \leftarrow Q \gg \quad \overset{\text{def}}{=} (M_1 \ll \alpha \leftarrow Q \gg)[\alpha' \leftarrow M_2 \ll \alpha \leftarrow Q \gg]$$
$$\alpha \not\equiv \alpha', \; (\alpha' \notin FMVar(Q) \text{ or } \alpha \notin FMVar(M_1))$$

Example 1. $\lambda \alpha.(\alpha) =_v \lambda \beta.(\beta)$, $\lambda \alpha.(f) =_v \lambda \beta.(f)$, but $\lambda \alpha.(X) \neq_v \lambda \beta.(X)$, $\lambda \beta.(\lambda \alpha.(X)) \neq_v \lambda \alpha.(\lambda \beta.(X))$.

2.2   Reduction

Whereas the rewrite rules are specified by using metaterms, the reduction relation is defined on terms.

Definition 9 (*SERS* **Rewrite Rule**). *An SERS rewrite rule is a pair of metaterms* (G, D) *(also written $G \longrightarrow D$) such that*

- *the first symbol in G is a function symbol or a binder symbol*
- *$FMVar(D) \subseteq FMVar(G)$*
- *G contains no occurrence of the metasubstitution operator*

Example 2. The λx-calculus [3,19] is defined by the following *SERS* rewrite rules:

$$
\begin{array}{lll}
@(\lambda \alpha.(X), Z) & \longrightarrow_{Beta} & \Sigma(\sigma \alpha.(X), Z) \\
\Sigma(\sigma \alpha.(@(X, Y)), Z) & \longrightarrow_{App} & @(\Sigma(\sigma \alpha.(X), Z), \Sigma(\sigma \alpha.(Y), Z)) \\
\Sigma(\sigma \alpha.(\lambda \beta.(X)), Z) & \longrightarrow_{Lambda} & \lambda \beta.(\Sigma(\sigma \alpha.(X), Z)) \\
\Sigma(\sigma \alpha.(\alpha), Z) & \longrightarrow_{Var1} & Z \\
\Sigma(\sigma \alpha.(\widehat{\beta}), Z) & \longrightarrow_{Var2} & \widehat{\beta}
\end{array}
$$

Note that our formalism allows us to specify the Var2 rule as originally done in [19], while formalisms such as *CRS* force one to change this rule to a stronger one, called *gc*, written as $\Sigma(\sigma \alpha.(X), Z) \longrightarrow_{gc} X$, where the admissibility condition on valuations guarantees that if X/t is part of the valuation θ, then $\theta(\alpha)$ cannot be in $FV(t)$.

Example 3. The $\lambda\Delta$-calculus [18] is defined by the following *SERS* rewrite rules:

$$
\begin{array}{ll}
@(\lambda\alpha.(X), Z) & \longrightarrow_{Beta} \; X[\alpha \leftarrow Z] \\
@(\Delta\alpha.(X), Z) & \longrightarrow_{\Delta1} \; \Delta\beta.(X[\alpha \leftarrow \lambda\gamma.(@(\beta, @(\gamma, Z)))]) \\
\Delta\alpha.(@(\alpha, X)) & \longrightarrow_{\Delta2} \; X \\
\Delta\alpha.(@(\alpha, (\Delta\beta.(@(\alpha, X))))) & \longrightarrow_{\Delta3} \; X
\end{array}
$$

Definition 10 (Valuation). *A* variable assignment *is a (partial) function θ_v from o-metavariables to variables with finite domain, such that for every pair of o-metavariables $\alpha, \widehat{\beta}$ we have $\theta_v\alpha \not\equiv \theta_v\widehat{\beta}$ (pre-bound and pre-free o-metavariables are assigned different values).*

A valuation θ *is a pair of (partial) functions (θ_v, θ_t) where θ_v is a variable assignment and θ_t maps t-metavariables to terms. We write $Dom(\theta)$ for $Dom(\theta_v) \cup Dom(\theta_t)$.[3] A valuation θ may be extended in a unique way to the set of pre-metaterms M such that $MVar(M) \subseteq Dom(\theta)$ as follows:*

$$
\begin{array}{ll}
\overline{\theta}\alpha \stackrel{def}{=} \theta_v\alpha & \overline{\theta}f(M_1, \ldots, M_n) \stackrel{def}{=} f(\overline{\theta}M_1, \ldots, \overline{\theta}M_n) \\
\overline{\theta}\widehat{\alpha} \stackrel{def}{=} \theta_v\widehat{\alpha} & \overline{\theta}(\xi\alpha.(M_1, \ldots, M_n)) \stackrel{def}{=} \xi\theta_v\alpha.(\overline{\theta}M_1, \ldots, \overline{\theta}M_n) \\
\overline{\theta}X \stackrel{def}{=} \theta_t X & \overline{\theta}(M_1[\alpha \leftarrow M_2]) \stackrel{def}{=} \overline{\theta}(M_1)\{\theta_v\alpha \leftarrow \overline{\theta}M_2\}
\end{array}
$$

We shall not distinguish between θ and $\overline{\theta}$ if no ambiguities arise. Also, we sometimes write $\theta(M)$ thereby implicitly assuming that $MVar(M) \subseteq Dom(\theta)$.

Returning to the intuition behind *v-equivalence* the idea is that it can be translated into α-conversion in the sense that $M =_v N$ implies $\theta M \equiv_\alpha \theta N$ for any valuation θ such that θM and θN are defined. Indeed, coming back to Example 1 and taking $\theta = \{\alpha/x, \beta/y, X/x\}$, we have $\theta\lambda\alpha.(\alpha) \equiv \lambda x.(x) \equiv_\alpha \lambda y.(y) \equiv \theta\lambda\beta.(\beta), \theta\lambda\alpha.(f) \equiv \lambda x.(f) \equiv_\alpha \lambda y.(f) \equiv \theta\lambda\beta.(\beta), \theta\lambda\alpha.(X) \equiv \lambda x.(x) \not\equiv_\alpha \lambda y.(x) \equiv \theta\lambda\beta.(X), \theta\lambda\beta.(\lambda\alpha.(X)) \equiv \lambda y.(\lambda x.(x)) \not\equiv_\alpha \lambda x.(\lambda y.(x)) \equiv \theta\lambda\alpha.(\lambda\beta.(X))$.

Definition 11 (Safe Valuations). *Let $M \in \mathcal{PMT}$ and θ a valuation with $MVar(M) \subseteq Dom(\theta)$. We say that θ is safe for M if $\overline{\theta}M$ is defined. Likewise, if (G, D) is a rewrite rule, we say that θ is safe for (G, D) if $\overline{\theta}D$ is defined.*

Note that if the notion of substitution we are dealing with were not restricted then α-conversion could be required in order to apply a valuation to a premetaterm. Also, for any valuation θ and pre-metaterm M with $MVar(M) \subseteq Dom(\theta)$ that contains no occurrences of the metasubstitution operator θ is safe for M. Thus, we only ask θ to be safe for D (not G) in the previous definition.

The following condition is the classical notion of *admissibility* used in higher-order rewriting [21] to avoid inconsistencies in rewrite steps.

Definition 12 (Path Condition for T-Metavariables). *Let X be a t-metavariable. Consider all the occurrences $p_1, \ldots, p_n$ of X in (G, D), and their respective parameter paths $l_1, \ldots, l_n$ in the trees corresponding to G and D. A valuation θ verifies the path condition for X in (G, D) if for every $x \in FV(\theta X)$, either $(\forall 1 \leq i \leq n$ we have $x \in \theta l_i)$ or $(\forall 1 \leq i \leq n$ we have $x \notin \theta l_i)$.*

---

[3] As usual, $Dom(\psi)$ denotes the domain of the partial function ψ.

This definition may be read as: one occurrence of $x \in FV(\theta X)$ with X in (G, D) is in the scope of some binding occurrence of x iff every occurrence of X in (G, D) is in the scope of a bound o-metavariable α with $\theta \alpha \equiv x$. For example, consider the *SERS* rule $\lambda \alpha.(\xi \beta.(X)) \longrightarrow \xi \beta.(X)$ and the valuations $\theta_1 = \{\alpha/x, \beta/y, X/z\}$ and $\theta_2 = \{\alpha/x, \beta/y, X/x\}$. Then θ_1 verifies the path condition for X, but θ_2 does not since when instantiating the rewrite rule with θ_2 the variable x shall occur both bound (on the LHS) and free (on the RHS).

Definition 13 (Admissible Valuations). *A valuation θ is said to be* admissible *for a rewrite rule (G, D) iff*

- θ *is safe for (G, D)*
- *if α and β occur in (G, D) with $\alpha \not\equiv \beta$ then $\theta_v \alpha \not\equiv \theta_v \beta$*
- θ *verifies the path condition for every t-metavariable in (G, D)*

Note that an admissible valuation is safe by definition, but a safe valuation may not be admissible: consider the rule $\lambda \alpha.app(X, \alpha) \longrightarrow X$, the valuation $\theta = \{\alpha/x, X/x\}$ is trivially safe but is not admissible since the path condition is not verified: $x \in \theta(\alpha)$ but $x \notin \theta(\epsilon)$ (x occurs bound on the LHS and free on the RHS).

Now, there are two possible and equivalent ways to define reduction in a higher-order framework. One can either define reduction via a notion of substitution which makes explicit use of α-conversion, as it is usually done [10], or, as it is done here, reduction is explicitly defined as reduction modulo α-conversion and using a notion of restricted substitution which does not make use of α-conversion. We choose this second (and more involved) approach since we prefer to have a notion of reduction on terms in both formalisms (with names and de Bruijn indices), which is similar enough to make technical proofs work easily.

Definition 14 (Reduction on Terms). *Let $\mathcal{R}$ be a set of SERS rewrite rules and s, t terms. We say that s $\mathcal{R}$-reduces to t, written $s \longrightarrow_{\mathcal{R}} t$, iff there exists a rewrite rule $(G, D) \in \mathcal{R}$, an admissible valuation θ for (G, D) and a context C such that $s \equiv_\alpha C[\theta G]$ and $t \equiv_\alpha C[\theta D]$.*

3   Simplified Expression Reduction Systems with Indices

We introduce de Bruijn indices based higher-order rewrite formalism $SERS_{DB}$.

3.1   De Bruijn Metaterms and Terms

A classical way to avoid α-conversion is to use de Bruijn index notation [7], where names of variables are replaced by natural numbers. When talking about a set N of de Bruijn indices we may refer to $\texttt{Names}(N)$ as the set of names of N given by the order on the set of variables $\mathcal{V}$ introduced in Section 2. Indeed, if $N = \{n_1, \ldots, n_m\}$, then $\texttt{Names}(N) = \{x_{n_1}, \ldots, x_{n_m}\}$.

In the sequel, in order to distinguish a concept defined for the *SERS* formalism from its corresponding version (if it exists) in the $SERS_{DB}$ formalism we may prefix it using the qualifying term "de Bruijn", eg. "de Bruijn metaterms".

Definition 15 (de Bruijn Signature). *Consider the denumerable and disjoint infinite sets:*

- $\{\alpha_1, \alpha_2, \alpha_3, \ldots\}$ *a set of symbols called* binder indicators, *denoted* $\alpha, \beta, \ldots$,
- $\mathcal{I}_{\mathcal{MV}} = \{\widehat{\alpha_1}, \widehat{\alpha_2}, \ldots\}$ *a set of* i-metavariables *(i for index), denoted* $\widehat{\alpha}, \widehat{\beta}, \ldots$,
- $\mathcal{T}_{\mathcal{MV}} = \{X_l^1, X_l^2, X_l^3, \ldots\}$ *a set of* t-metavariables *(t for term), where l ranges over the set of labels built over binder indicators, denoted* $X_l, Y_l, Z_l, \ldots$,
- $\mathcal{F} = \{f_1, f_2, f_3, \ldots\}$ *a set of* function symbols *equipped with a fixed (possibly zero) arity, denoted* $f, g, h, \ldots$,
- $\mathcal{B} = \{\lambda_1, \lambda_2, \lambda_3, \ldots\}$ *a set of binder symbols equipped with a fixed (non-zero) arity, denoted* $\lambda, \mu, \nu, \xi, \ldots$.

We remark that the set of binder indicators is exactly the set of pre-bound o-metavariables introduced in Definition 1. The reason for using the same alphabet in both formalisms shall become clear in Section 4, but intuitively, we need a mechanism to annotate binding paths in the de Bruijn setting to distinguish metaterms like $\xi\beta.(\xi\alpha.(X))$ and $\xi\alpha.(\xi\beta.(X))$ appearing in the same rule when translated into an $SERS_{DB}$ system.

Definition 16 (de Bruijn Pre-metaterms). *The set of* de Bruijn pre-meta-terms, *denoted* $\mathcal{PMT}_{db}$, *is defined by the following two-sorted grammar:*

$$metaindices \quad I ::= 1 \mid \mathtt{S}(I) \mid \widehat{\alpha}$$
$$pre\text{-}metaterms \quad A ::= I \mid X_l \mid f(A, \ldots, A) \mid \xi(A, \ldots, A) \mid A[\![A]\!]$$

The symbol $.[\![.]\!]$ *in a pre-metaterm* $A[\![A]\!]$ *is called* de Bruijn metasubstitution operator. *The binder symbols together with the de Bruijn metasubstitution operator are called* binder operators, *and the same remark of Definition 3 applies.*

We shall use $A, B, A_i, \ldots$ to denote de Bruijn pre-metaterms and the convention that $\mathtt{S}^0(1) = 1$, $\mathtt{S}^0(\widehat{\alpha}) = \widehat{\alpha}$ and $\mathtt{S}^{j+1}(n) = \mathtt{S}(\mathtt{S}^j(n))$. As usually done for indices, we shall abbreviate $\mathtt{S}^{j-1}(1)$ as j.

Even if the formal mechanism used to translate pre-metaterms with names into pre-metaterms with de Bruijn indices will be given in Section 4, let us introduce intuitively some ideas in order to justify the syntax used for i-metavariables. In the formalism $SERS$ there is a clear distinction between free and bound o-metavariables. This fact must also be reflected in the formalism $SERS_{DB}$, where bound o-metavariables are represented with indices and free o-metavariables are represented with i-metavariables (this distinction between free and bound variables is also used in some formalizations of λ-calculus [17]). However, free variables in $SERS_{DB}$ appear always in a binding context, so that a de Bruijn valuation of such kind of variables has to reflect the adjustment needed to represent the same variables but in a different context. This can be done by surrounding the i-metavariable by as many operators $\mathtt{S}$ as necessary. As an example consider the pre-metaterm $\xi\alpha.(\widehat{\beta})$. If we translate it to $\xi(\widehat{\beta})$, then a de Bruijn valuation like $\kappa = \{\widehat{\beta}/1\}$ binds the variable whereas this is completely impossible in the name formalism thanks to the conditions imposed on a name valuation (c.f. condition on variable assignments in Definition 10). Our solution is then to translate the pre-metaterm $\xi\alpha.(\widehat{\beta})$ by $\xi(\mathtt{S}(\widehat{\beta}))$ in such a way that there is no capture of variables since $\kappa(\xi(\mathtt{S}(\widehat{\beta})))$ is exactly $\xi(2)$. The solution adopted here is in some sense what is called *pre-cooking* in [9].

We use $MVar(A)$ (resp. $MVar_i(A)$ and $MVar_t(A)$) to denote the set of all metavariables (resp. i- and t-metavariables) of the de Bruijn pre-metaterm A.

As in the *SERS* formalism, we also need here a notion of well-formed pre-metaterm. The first motivation is to guarantee that labels of t-metavariables are correct w.r.t the context in which they appear, the second one is to ensure that indices like $S^i(1)$ (resp. $S^i(\widehat{\alpha})$) correspond to bound (resp. free) variables. Indeed, the pre-metaterms $\xi(X_{\alpha\beta})$, $\xi(\xi(4))$ and $\xi(\widehat{\alpha})$ shall not make sense for us, and hence shall not be considered well-formed.

Definition 17 (de Bruijn Metaterms). *A pre-metaterm $A \in \mathcal{PMT}_{db}$ is said to be a metaterm iff the predicate $\mathcal{WF}(A)$ holds, where $\mathcal{WF}(A)$ iff $\mathcal{WF}_\epsilon(A)$, and $\mathcal{WF}_l(A)$ is defined as follows:*

- $\mathcal{WF}_l(S^j(1))$ *iff* $j + 1 \leq |l|$
- $\mathcal{WF}_l(S^j(\widehat{\alpha}))$ *iff* $j = |l|$
- $\mathcal{WF}_l(X_k)$ *iff* $l = k$ *and l is a simple label*
- $\mathcal{WF}_l(f(A_1, \ldots, A_n))$ *iff for all $1 \leq i \leq n$ we have $\mathcal{WF}_l(A_i)$*
- $\mathcal{WF}_l(\xi(A_1, \ldots, A_n))$ *iff there exists $\alpha \notin l$ such that for all $1 \leq i \leq n$ we have $\mathcal{WF}_{\alpha l}(A_i)$*
- $\mathcal{WF}_l(A_1[\![A_2]\!])$ *iff $\mathcal{WF}_l(A_2)$ and there exists $\alpha \notin l$ such that $\mathcal{WF}_{\alpha l}(A_1)$*

Therefore indices of the form $S^j(1)$ may only occur in metaterms if they represent bound variables and well-formed metaindices of the form $S^j(\widehat{\alpha})$ always represent a free variable. Note that when considering $\mathcal{WF}_l(M)$ and $\mathcal{WF}_l(A)$ it is Definitions 4 and 17 which are referenced, respectively.

Example 4. Pre-metaterms $\xi(X_\alpha, \lambda(Y_{\beta\alpha}, S(1)))$, $f(\widehat{\beta}, \lambda(Y_\alpha, S(\widehat{\alpha})))$, $g(\lambda(\xi(h)))$ are metaterms, while $f(S(\widehat{\alpha}), \xi(X_\beta))$, $\lambda(\xi(X_{\alpha\alpha}))$, $f(\widehat{\beta}, \lambda(\xi(S(\widehat{\beta}))))$ are not.

Definition 18 (de Bruijn Terms and de Bruijn Contexts). *The set of* de Bruijn terms, *denoted $\mathcal{T}_{db}$, and the set of de Bruijn contexts are defined by:*

de Bruijn indices $n ::= 1 \mid S(n)$
de Bruijn terms $a ::= n \mid f(a, \ldots, a) \mid \xi(a, \ldots, a)$
de Bruijn contexts $E ::= \Box \mid f(a, \ldots, E, \ldots, a) \mid \xi(a, \ldots, E, \ldots, a)$

We use $a, b, a_i, b_i, \ldots$ for de Bruijn terms and $E, F, \ldots$ for de Bruijn contexts. We may refer to the *binder path number* of a context, which is the number of binders between the $\Box$ and the root.

We use $FV(a)$ to denote the set of free variables (indices) in a; the result of substituting a term b for the index $n \geq 1$ in a term a is denoted $a\{\!\{n \leftarrow b\}\!\}$; the updating functions are denoted $\mathcal{U}_i^n(.)$ for $i \geq 0$ and $n \geq 1$. All these concepts are defined as usual.

Definition 19 (Free de Bruijn Metavariables). *Let A be a de Bruijn pre-metaterm. The set of free metavariables of A, $FMVar(A)$, is defined as:*

$$FMVar(1) \overset{\mathrm{def}}{=} \emptyset$$
$$FMVar(S(I)) \overset{\mathrm{def}}{=} FMVar(I)$$
$$FMVar(\widehat{\alpha}) \overset{\mathrm{def}}{=} \{\widehat{\alpha}\}$$
$$FMVar(X_l) \overset{\mathrm{def}}{=} \{X_l\}$$
$$FMVar(f(A_1, \ldots, A_n)) \overset{\mathrm{def}}{=} \bigcup_{i=1}^n FMVar(A_i)$$
$$FMVar(\xi(A_1, \ldots, A_n)) \overset{\mathrm{def}}{=} \bigcup_{i=1}^n FMVar(A_i)$$
$$FMVar(A_1[\![A_2]\!]) \overset{\mathrm{def}}{=} FMVar(A_1) \cup FMVar(A_2)$$

Note that this definition also applies to de Bruijn metaterms. The set of names of free metavariables of A is the set of free metavariables of A where each X_l is replaced simply by X. This notion will be used in Definition 20.

3.2   Reduction

We define rewrite rules, valuations, their validity, and reduction in $SERS_{DB}$.

Definition 20 (de Bruijn Rewrite Rule). *A de Bruijn rewrite rule is a pair of de Bruijn metaterms (L, R) (also written $L \longrightarrow R$) such that*

- *the first symbol in L is a function symbol or a binder symbol*
- *the set of names of $FMVar(R)$ is included in the set of names of $FMVar(L)$*
- *the metasubstitution operator does not occur in L*

Definition 21 (de Bruijn Valuation). *A de Bruijn valuation κ is a pair of (partial) functions (κ_i, κ_t) where κ_i is a function from i-metavariables to integers, and κ_t is a function from t-metavariables to de Bruijn terms. We denote by $Dom(\kappa)$ the set $Dom(\kappa_i) \cup Dom(\kappa_t)$. A valuation κ determines in a unique way a function $\overline{\kappa}$ from the set of pre-metaterms A with $FMVar(A) \subseteq Dom(\kappa)$ to the set of terms as follows:*

$$\overline{\kappa}1 \stackrel{def}{=} 1 \qquad \overline{\kappa}f(A_1, \ldots, A_n) \stackrel{def}{=} f(\overline{\kappa}A_1, \ldots, \overline{\kappa}A_n)$$

$$\overline{\kappa}\mathsf{S}(I) \stackrel{def}{=} \mathsf{S}(\overline{\kappa}I) \qquad \overline{\kappa}\xi(A_1, \ldots, A_n) \stackrel{def}{=} \xi(\overline{\kappa}A_1, \ldots, \overline{\kappa}A_n)$$

$$\overline{\kappa}\widehat{\alpha} \stackrel{def}{=} \kappa_i\widehat{\alpha}$$

$$\overline{\kappa}X_l \stackrel{def}{=} \kappa_t X_l \qquad \overline{\kappa}(A_1[\![A_2]\!]) \stackrel{def}{=} \overline{\kappa}(A_1)\{\!\{1 \leftarrow \overline{\kappa}A_2\}\!\}$$

Note that in the above definition the substitution operator $.\{\!\{. \leftarrow .\}\!\}$ refers to the usual substitution defined on terms with de Bruijn indices.

We now introduce the notion of *value function* which is used to give semantics to metavariables with labels in the $SERS_{DB}$ formalism. The goal pursued by the labels of metavariables is that of incorporating "context" information as a defining part of a metavariable. As a consequence, we must verify that the terms substituted for every occurrence of a fixed metavariable coincide "modulo" their corresponding context. Dealing with such notion of "coherence" of substitutions in a de Bruijn formalism is also present in other formalisms but in a more restricted form. Thus for example, as mentioned before, a pre-cooking function is used in [9] in order to avoid variable capture in the higher-order unification procedure. In XRS [16] the notions of binding arity and pseudo-binding arity are introduced in order to take into account the parameter path of the different occurrences of t-metavariables appearing in a rewrite rule. Our notion of "coherence" is implemented with *valid valuations* (cf. Definition 23) and it turns out to be more general than the solutions proposed in [9] and [16].

Definition 22 (Value Function). *Let $a \in \mathcal{T}_{db}$ and l be a label of binder indicators. Then we define the value function $Value(l, a)$ as $Value^0(l, a)$ where*

$$Value^i(l,n) \quad \stackrel{\text{def}}{=} \quad \begin{cases} n & if\ n \le i \\ \mathtt{at}(l, n-i) & if\ 0 < n-i \le |l| \\ x_{n-i-|l|} & if\ n-i > |l| \end{cases}$$

$$Value^i(l, f(a_1, \dots, a_n)) \stackrel{\text{def}}{=} f(Value^i(l, a_1), \dots Value^i(l, a_n))$$

$$Value^i(l, \xi(a_1, \dots, a_n)) \stackrel{\text{def}}{=} \xi(Value^{i+1}(l, a_1), \dots, Value^{i+1}(l, a_n))$$

It is worth noting that $Value^i(l,n)$ may give three different kinds of results. This is just a technical trick to make easier later proofs. Indeed, we have for example $Value(\alpha\beta, \xi(f(3,1))) \equiv \xi(f(\beta,1)) \equiv Value(\beta\alpha, \xi(f(2,1)))$ and $Value(\epsilon, f(\xi(1), \lambda(2))) \equiv f(\xi(1), \lambda(x_1)) \not\equiv f(\xi(1), \lambda(\alpha)) \equiv Value(\alpha, f(\xi(1), \lambda(2)))$. Thus the function $Value(l,a)$ interprets the de Bruijn term a in an l-context: bound indices are left untouched, free indices referring to the l-context are replaced by the corresponding binder indicator and the remaining free indices are replaced by their corresponding variable names.

In order to introduce the notion of valid de Bruijn valuations let us consider the following rule:

$$\xi\alpha.(\xi\beta.(X)) \longrightarrow_r \xi\beta.(\xi\alpha.(X))$$

Even if translation of rewrite rules into de Bruijn rewrite rules has not been defined yet (Section 4), one may guess that a reasonable translation would be the following rule (called r_{DB}):

$$\xi(\xi(X_{\beta\alpha})) \longrightarrow_{r_{DB}} \xi(\xi(X_{\alpha\beta}))$$

which indicates that β (resp. α) is the first bound occurrence in the LHS (resp. RHS) while α (resp. β) is the second bound occurrence in the LHS (resp. RHS). Now, if X is instantiated by x, α by x and β by y in the $SERS$ system, then we have a r-reduction step $\xi x.(\xi y.(x)) \longrightarrow \xi y.(\xi x.(x))$. However, to reflect this fact in the corresponding $SERS_{DB}$ system we need to instantiate $X_{\beta\alpha}$ by 2 and $X_{\alpha\beta}$ by 1, thus obtaining a r_{DB}-reduction step $\xi(\xi(2)) \longrightarrow \xi(\xi(1))$. This clearly shows that de Bruijn t-metavariables having the same name but different label cannot be instantiated arbitrarily as they have to reflect the renaming of variables which is indicated by their labels. This is exactly the role of the property of validity:

Definition 23 (Valid de Bruijn Valuation). *A de Bruijn valuation κ is said to be valid if for every pair of t-metavariables X_l and $X_{l'}$ in $Dom(\kappa)$ we have $Value(l, \kappa X_l) \equiv Value(l', \kappa X_{l'})$. Likewise, we say that a de Bruijn valuation κ is valid for a rewrite rule (L, R) if for every pair of t-metavariables X_l and $X_{l'}$ in (L, R) we have $Value(l, \kappa X_l) \equiv Value(l', \kappa X_{l'})$.*

It is interesting to note that there is no concept analogous to safeness (cf. Definition 11) as used for named $SERS$ due to the use of de Bruijn indices. Also, the last condition in the definition of an admissible valuation (cf. Definition 13) is subsumed by the above Definition 23 in the setting of $SERS_{DB}$.

Example 5. Returning to the example just after Definition 22 we have that $\kappa = \{X_{\beta\alpha}/2, X_{\alpha\beta}/1\}$ is valid since $Value(\beta\alpha, 2) \equiv \alpha \equiv Value(\alpha\beta, 1)$.

Another interesting example is the well-known η-contraction rule $\lambda x.@(X, x)$- $\longrightarrow X$ if $x \notin FV(X)$. It can be expressed in the *SERS* formalism as the rule (η_n) $\lambda\alpha.@(X, \alpha)\longrightarrow X$, and in the $SERS_{DB}$ formalism as the rule (η_{db}) - $\lambda(@(X_\alpha, 1))\longrightarrow X_\epsilon$.

Remark that this kind of rule cannot be expressed in the *XRS* formalism [16] since it does not verify the binding arity condition. Our formalism allows us to write rules like η_{db} because valid valuations will test for coherence of values. Indeed, an admissible valuation for η_n is a valuation θ such that θX does not contain a free occurrence of $\theta(\alpha)$. This is exactly the condition used in any usual formalization of the η-rule.

Definition 24 (Reduction on de Bruijn Terms). *Let $\mathcal{R}$ be a set of de Bruijn rules and a, b de Bruijn terms. We say that a $\mathcal{R}$-reduces to b, written $a\longrightarrow_\mathcal{R} b$, iff there is a de Bruijn rule $(L, R) \in \mathcal{R}$ and a de Bruijn valuation κ valid for (L, R) such that $a \equiv E[\kappa L]$ and $b \equiv E[\kappa R]$, where E is a de Bruijn context.*

Thus, the term $\lambda(app(\lambda(app(1, 3)), 1))$ rewrites by the η_{db} rule to $\lambda(app(1, 2))$, using the (valid) valuation $\kappa = \{X_\alpha/\lambda(app(1, 3), X_\epsilon/\lambda(app(1, 2))\}$.

4   Relating *SERS* and $SERS_{DB}$

In this section we show how reduction in the *SERS* formalism may be simulated in the $SERS_{DB}$ formalism and vice-versa.

Definition 25 (From Terms (and Contexts) to de Bruijn Terms (and Contexts)). *The translation of a term t, denoted $T(t)$, is defined as $T_\epsilon(t)$ where*

$$T_k(x) \stackrel{\text{def}}{=} \begin{cases} \mathsf{pos}(x, k) & \text{if } x \in k \\ \mathcal{O}(x) + |k| & \text{if } x \notin k \end{cases}$$

$$T_k(f(t_1, \ldots, t_n)) \stackrel{\text{def}}{=} f(T_k(t_1), \ldots, T_k(t_n))$$

$$T_k(\xi x.(t_1, \ldots, t_n)) \stackrel{\text{def}}{=} \xi(T_{xk}(t_1), \ldots, T_{xk}(t_n))$$

The translation of a context, denoted $T(C)$, adds the clause $T_k(\square) \stackrel{\text{def}}{=} \square$.

Definition 26 (From Pre-metaterms to de Bruijn Pre-metaterms). *The translation of a pre-metaterm M, denoted $T(M)$, is defined as $T_\epsilon(M)$ where:*

$T_k(\alpha) \stackrel{\text{def}}{=} \mathsf{pos}(\alpha, k)$, *if* $\alpha \in k$ $T_k(f(M_1, \ldots, M_n)) \stackrel{\text{def}}{=} f(T_k(M_1), \ldots, T_k(M_n))$

$T_k(\widehat{\alpha}) \stackrel{\text{def}}{=} \mathsf{s}^{|k|}(\widehat{\alpha})$ $T_k(\xi\alpha.(M_1, \ldots, M_n)) \stackrel{\text{def}}{=} \xi(T_{\alpha k}(M_1), \ldots, T_{\alpha k}(M_n))$

$T_k(X) \stackrel{\text{def}}{=} X_k$ $T_k(M_1[\alpha \leftarrow M_2]) \stackrel{\text{def}}{=} T_{\alpha k}(M_1)[\![T_k(M_2)]\!]$

Note that if M is a metaterm, then $T(M)$ will be a de Bruijn metaterm and only have t-metavariables with simple labels. Note also that, for some pre-metaterms, such as $\xi\alpha.(\beta)$, the translation $T(.)$ is not defined.

Lemma 1 (T Preserves Well-Formedness). *If M is a metaterm, then $T(M)$ is a de Bruijn metaterm.*

Definition 27 (From _SERS_ Rewrite Rules to _SERS_$_{DB}$ Rewrite Rules).
Let (G, D) be a rewrite rule in the SERS formalism. Then $T(G, D)$ denotes the translation of the rewrite rule, defined as $(T(G), T(D))$.

As an immediate consequence of Lemma 1 and Definition 27, if (G, D) is an _SERS_ rewrite rule, then $T(G, D)$ is an _SERS_$_{DB}$ rewrite rule.

Example 6. Following Example 2, the specification of $\lambda\mathbf{x}$ in the _SERS_$_{DB}$ formalism is given below.

$$
\begin{aligned}
@(\lambda(X_\alpha), Z_\epsilon) &\longrightarrow \Sigma(\sigma(X_\alpha), Z_\epsilon) \\
\Sigma(\sigma(@(X_\alpha, Y_\alpha)), Z_\epsilon) &\longrightarrow @(\Sigma(\sigma(X_\alpha), Z_\epsilon), \Sigma(\sigma(Y_\alpha), Z_\epsilon)) \\
\Sigma(\sigma(\lambda(X_{\beta\alpha})), Z_\epsilon) &\longrightarrow \lambda(\Sigma(\sigma(X_{\alpha\beta}), Z_\beta)) \\
\Sigma(\sigma(1), Z_\epsilon) &\longrightarrow Z_\epsilon \\
\Sigma(\sigma(\mathbf{S}(\widehat{\beta})), Z_\epsilon) &\longrightarrow \widehat{\beta}
\end{aligned}
$$

The rule $\Sigma(\sigma(\lambda(X_{\beta\alpha})), Z_\epsilon) \longrightarrow \lambda(\Sigma(\sigma(X_{\alpha\beta}), Z_\beta))$ is interesting since it illustrates the use of binder commutation from $X_{\beta\alpha}$ to $X_{\alpha\beta}$ and shows how some index adjustment shall be necessary when going from Z_ϵ to Z_β.

Example 7. The translation of the $\lambda\Delta$-calculus (Example 3) yields the following rewrite rules in the _SERS_$_{DB}$ formalism

$$
\begin{aligned}
@(\lambda(X_\alpha), Z_\epsilon) &\longrightarrow X_\alpha[\![Z_\epsilon]\!] \\
@(\Delta(X_\alpha), Z_\epsilon) &\longrightarrow \Delta(X_{\alpha\beta}[\![\lambda(@(\mathbf{S}(1), @(1, Z_{\gamma\beta})))]\!]) \\
\Delta(@(1, X_\alpha)) &\longrightarrow X_\epsilon \\
\Delta(@(1, (\Delta(@(\mathbf{S}(1), X_{\beta\alpha}))))) &\longrightarrow X_\epsilon
\end{aligned}
$$

We remark that the translation of Δ_1, Δ_2 and Δ_3 would not be possible in XRS [16].

Proposition 1 (Simulating _SERS_ Reduction via _SERS_$_{DB}$ Reduction).
Suppose $s \longrightarrow t$ in the SERS formalism using the rewrite rule (G, D). Then we have $T(s) \longrightarrow T(t)$ in the SERS$_{DB}$ formalism using the rule $T(G, D)$.

We now consider how reduction in _SERS_$_{DB}$ may be simulated in _SERS_.

Definition 28 (From de Bruijn Terms (Contexts) to Terms (Contexts)).
We define the translation of $a \in \mathcal{T}_{db}$, denoted $U(a)$, as $U_\epsilon^{\mathtt{Names}(FV(a))}(a)$ where, for every finite set of variables S, and label of variables k, $U_k^S(a)$ is defined by:

$$
\begin{aligned}
U_k^S(n) &\stackrel{\text{def}}{=} \begin{cases} \mathtt{at}(k, n) & \text{if } n \leq |k| \\ x_{n-|k|} & \text{if } n > |k| \text{ and } x_{n-|k|} \in S \end{cases} \\
U_k^S(f(a_1, \ldots, a_n)) &\stackrel{\text{def}}{=} f(U_k^S(a_1), \ldots, U_k^S(a_n)) \\
U_k^S(\xi(a_1, \ldots, a_n)) &\stackrel{\text{def}}{=} \xi x.(U_{xk}^S(a_1), \ldots, U_{xk}^S(a_n)) \quad \text{for any } x \notin k \cup S
\end{aligned}
$$

The translation of a de Bruijn context E, denoted $U(E)$, is defined as above but adding the clause $U_k^S(\Box) \stackrel{\text{def}}{=} \Box$. We remark that we can always choose $x \notin k \cup S$ since both k and S are finite.

Note that $U(.)$ is not a function in the sense that the choice of bound variables is non-deterministic. However, if t and t' belong both to $U(a)$, then $t \equiv_\alpha t'$. Thus, $U(.)$ can be seen as a function from de Bruijn terms to α-equivalence classes.

Definition 29 (From de Bruijn Pre-metaterms to Pre-metaterms). *The translation of a de Bruijn pre-metaterm A, denoted $U(A)$, is defined as $U_\epsilon(A)$, where $U_l(A)$ is defined as follows:*

$$U_l(\mathsf{S}^i(1)) \stackrel{\text{def}}{=} \texttt{at}(l, i+1) \qquad\qquad\qquad if\ i+1 \leq |l|$$

$$U_l(\mathsf{S}^{|l|}(\widehat{\alpha})) \stackrel{\text{def}}{=} \widehat{\alpha}$$

$$U_l(X_l) \stackrel{\text{def}}{=} X$$

$$U_l(f(A_1, \ldots, A_n)) \stackrel{\text{def}}{=} f(U_l(A_1), \ldots, U_l(A_n))$$

$$U_l(\xi(A_1, \ldots, A_n)) \stackrel{\text{def}}{=} \xi\alpha.(U_{\alpha l}(A_1), \ldots, U_{\alpha l}(A_n))$$
$$\qquad\qquad if\ 1 \leq i \leq n\ \ \mathcal{WF}_{\alpha l}(A_i)\ for\ some\ \alpha \notin l$$

$$U_l(A_1[\![A_2]\!]) \stackrel{\text{def}}{=} U_{\alpha l}(A_1)[\alpha \leftarrow U_l(A_2)]$$
$$\qquad\qquad if\ \mathcal{WF}_{\alpha l}(A_1)\ for\ some\ \alpha \notin l$$

As in Definition 28 we remark that the translation of a de Bruijn pre-metaterm is not a function since it depends on the choice of the names for o-metavariables. Indeed, two different pre-metaterms obtained by this translation will be v-equivalent. Also, for some de Bruijn pre-metaterms such as $\xi(2)$, the translation may not be defined. However, it is defined on de Bruijn metaterms.

Definition 30 (From $SERS_{DB}$ Rewrite Rules to $SERS$ Rewrite Rules). *Let (L, R) be a de Bruijn rewrite rule then its translation, denoted $U(L, R)$, is the pair of metaterms $(U_\epsilon(L), U_\epsilon(R))$.*

Note that if $\mathcal{WF}_l(A)$ holds then $U_l(A)$ is also a named metaterm, that is, $\mathcal{WF}_l(U_l(A))$ also holds. Therefore, by Definition 9 the translation of a de Bruijn rule is a rule in $SERS$. As mentioned above, if a de Bruijn pre-metaterm A is not a de Bruijn metaterm then $U_l(A)$ may not be defined.

Example 8. Consider the rule $@(\Delta(X_\alpha), Z_\epsilon) \longrightarrow \Delta(X_{\alpha\beta}[\![\lambda(@(\mathsf{S}(1), @(1, Z_{\gamma\beta})))]\!])$ from Example 7. The translation in Definition 29 yields the rule $@(\Delta\alpha.(X), Z) \longrightarrow \Delta\beta.(X[\alpha \leftarrow \lambda\gamma.(@(\beta, @(\gamma, Z)))])$ and the translation in Definition 29 on the rule $\Sigma(\sigma(\mathsf{S}(\widehat{\beta})), Z_\epsilon) \longrightarrow \widehat{\beta}$ yields $\Sigma(\sigma\gamma.(\widehat{\beta}), Z) \longrightarrow \widehat{\beta}$ for some bound metavariable γ.

Proposition 2 (Simulating $SERS_{DB}$ Reduction via $SERS$ Reduction). *Suppose $a \longrightarrow b$ in the $SERS_{DB}$ formalism using rewrite rule (L, R). Then we have $U(a) \longrightarrow U(b)$ in the $SERS$ formalism using rule $U(L, R)$.*

As regards the relationship between the translations over pre-metaterms and terms introduced above we may obtain two results stating, respectively, that given a metaterm M then $U(T(M))$ is v-equivalent to M and that given a de Bruijn metaterm A then $T(U(A))$ is identical to A. These results are used to show that confluence is preserved when translating in both directions.

Theorem 1 (Preservation of Confluence).
1. *If $\mathcal{R}$ is a confluent SERS then $T(\mathcal{R})$ is a confluent $SERS_{DB}$.*
2. *If $\mathcal{R}$ is a confluent $SERS_{DB}$ then $U(\mathcal{R})$ is a confluent SERS.*

5   Conclusions

We have proposed a formalism for higher-order rewriting with de Bruijn notation and we have shown that rewriting with names and rewriting with indices are
semantically equivalent. We have given formal translations from one formalism
into the other which can be viewed as an *interface* in programming languages
based on higher-order rewrite systems. This work fills the gap between classical
presentations of higher-order rewriting with names existing in the literature and
first-order presentations of higher-order rewriting such as [16]. Moreover, it explicitly suggests that *XRS* are not sufficient to express *an arbitrary* higher-order
rewrite system.

Further ongoing work uses the formalism presented here to propose a tool for
implementing higher-order rewrite systems via first-order ones. This tool would
incorporate not only de Bruijn notation but also explicit substitutions in a very
general form.

References

1. H.P. Barendregt. *The Lambda Calculus: Its Syntax and Semantics*. Studies in Logic
   and the Foundations of Mathematics. North-Holland, Amsterdam, 1984. Revised
   edition.
2. R. Bloo. *Preservation of Termination for Explicit Substitution*. PhD thesis, Eindhoven University, 1997.
3. R. Bloo and K. Rose. Combinatory reduction systems with explicit substitution
   that preserve strong normalisation. In RTA, LNCS 1103, pages 169-183. 1996.
4. E. Bonelli, D. Kesner, and A. Ríos. A de Bruijn notation for higher-order rewriting,
   2000. Available as `ftp://ftp.lri.fr/LRI/articles/kesner/dBhor.ps.gz`.
5. P.-L. Curien. *Categorical combinators, sequential algorithms and functional programming*. Progress in Theoretical Computer Science. Birkhaüser, 1993. Second
   edition.
6. P.-L. Curien, T. Hardin, and J.-J. Lévy. Confluence properties of weak and strong
   calculi of explicit substitutions. *Journal of the ACM*, 43(2):362–397, march 1996.
7. N. de Bruijn. Lambda-calculus notation with nameless dummies, a tool for automatic formula manipulation, with application to the church-rosser theorem. *Indag.
   Mat.*, 5(35):381–392, 1972.
8. G. Dowek. *La part du calcul*. Univesité de Paris VII, 1999. Thèse d'Habilitation à
   diriger des recherches.
9. G. Dowek, T. Hardin, and C. Kirchner. Higher-order unification via explicit substitutions. In LICS, 1995.
10. R. Hindley and J.P. Seldin. *Introduction to Combinators and λ-calculus*. London
    Mathematical Society. 1980.
11. F. Kamareddine and A. Ríos. Bridging de bruijn indices and variable names in
    explicit substitutions calculi. *Logic Journal of the Interest Group of Pure and
    Applied Logic (IGPL)*, 6(6):843–874, 1998.
12. Z. Khasidashvili. Expression reduction systems. In *Proceedings of I. Vekua Institute
    of Applied Mathematics*, volume 36, Tbilisi, 1990.

13. Z. Khasidashvili and V. van Oostrom. Context-sensitive Conditional Expression Reduction Systems. In ENTCS, Vol.2, Proceedings of the Joint COMPU-GRAPH/SEMAGRAPH Workshop on Graph Rewriting and Computation (SEG-RAGRA'95), Volterra, 1995.

14. J. W. Klop. *Combinatory Reduction Systems*, vol. 127 of *Mathematical Centre Tracts*. CWI, Amsterdam, 1980. PhD Thesis.

15. T. Nipkow. Higher order critical pairs. In LICS, pages 342–349, 1991.

16. B. Pagano. *Des Calculs de Susbtitution Explicite et leur application à la compilation des langages fonctionnels*. PhD thesis, Université Paris VI, 1998.

17. R. Pollack. Closure under alpha-conversion. In TYPES, LNCS 806. 1993.

18. J. Rehof and M. H. Sørensen. The λ_Δ calculus. In TACS, LNCS 789, pages 516–542. 1994.

19. K. Rose. Explicit cyclic substitutions. In CTRS, LNCS 656, pages 36–50, 1992.

20. V. van Oostrom and F. van Raamsdonk. Weak orthogonality implies confluence: the higher-order case. In LFCS, LNCS 813, pages 379–392, 1994.

21. F. van Raamsdonk. *Confluence and Normalization for higher-Order Rewriting*. PhD thesis, Amsterdam University, The Netherlands, 1996.

22. D. Wolfram. *The Clausal Theory of Types*, volume 21 of *Cambridge Tracts in Theoretical Computer Science*. Cambridge University Press, 1993.

Rewriting Techniques in Theoretical Physics*

Evelyne Contejean[1], Antoine Coste[2], and Benjamin Monate[1]

[1] LRI, CNRS UMR 8623, Bât. 490
{contejea,monate}@lri.fr
[2] Laboratoire de Physique Théorique, CNRS UMR 8627, Bât. 211
Université Paris-Sud, Centre d'Orsay
91405 Orsay Cedex, France
coste@th.u-psud.fr

Abstract. This paper presents a general method for studying some quotients of the special linear group SL_2 over the integers, which are of fundamental interest in the field of statistical physics. Our method automatically helps in validating some conjectures due to physicists, such as conjectures stating that a set of equations completely describes a finite given quotient of SL_2. In a first step, we show that in the cases we are interested in, the usual presentation of finitely generated groups with some constant generators and a binary concatenation can be turned into an equivalent one with unary generators. In a second step, when the completion of the transformed set of equations terminates, we show how to compute *directly* the associated normal forms automaton. According to the presence of loops, we are able to decide the finiteness of the quotient, and to compute its cardinality. When the quotient is infinite, the automaton gives some hints on what kind of equations are needed in order to insure the finiteness of the quotient.

Introduction

In the field of statistical physics in dimension 2, conformal invariance plays a crucial role [9, 5, 4]. It permits to express a physical theory which is invariant by geometrical transformations preserving the angles. Among the possible applications, one has to mention the study of demagnetization phenomena (Ising model), of polymer films and of string theory.

In this paper, we provide some tools for studying the so called rational conformal theories for which appear a finite dimensional representation of $SL_2(\mathbb{Z})$.

The mathematical formulation of the problem is to investigate the structure of the infinite group $SL_2(\mathbb{Z})$ generated by two $2{\times}2$ matrices usually denoted by S and T. In particular, we want to study some quotients of $SL_2(\mathbb{Z})$, and determine whether they are finite or not, and if they are finite, whether they are quotients of $SL_2(\mathbb{Z}/n\mathbb{Z})$ for some n. Rewriting techniques (see [6] for some backgound) are used here to study a class of finitely generated groups, quotiented by some additional ground equations, which are actually isomorphic to some quotients of $SL_2(\mathbb{Z})$. The study of finitely presented groups goes back to the end of the

---

* An extended version with complete proofs of this paper available at
http://www.lri.fr/~monate/rtatp-ext.ps.gz

19th century; the problem of deciding whether 2 such groups are isomorphic was first stated by Tietze, and the word problem by Dehn. Unfortunately, as many interesting problems, they have been proven to be undecidable 50 years later (see [13] for the historical background). However, there are some positive results in some sub-cases, mainly for automatic groups [8] or when Knuth-Bendix completion [10] terminates [7].

In a first step, we show that in this particular framework, the quotient algebras we are dealing with, are isomorphic to quotient algebras where the signature contains only one constant function symbol, and some unary function symbols, i.e. a monoid. This is done by a translation which eliminates the inverse function symbol. Compared with the method of Bündgen [2], ours does not use any term ordering nor completion. Then, we use Knuth-Bendix completion over the monoid in order to get a convergent rewrite system, and eventually, in the case when the completion succeeds, we build an automaton which accepts the terms in normal form.

We present here a specialized version of a general algorithm [12] which computes the bottom-up tree automaton directly, and not as the complement of the union of the automata for the instances of the left hand sides of the rewriting rules. In this paper, the automaton is not a fully general tree automaton, since we only have function symbols of arity ≤ 1. Gilman [7] proposed a similar algorithm building a directed graph for the case of finitely presented groups which can be interpreted as a top-down (finite states) automaton.

Eventually, our automaton is used to determine the cardinality of the quotient algebra: if it contains a loop, then there are infinitely many distinct elements in the algebra, otherwise, there are finitely many, and we can count them as the number of paths from the initial state to final states in the automaton.

The paper is organized as follows: in the first section, we exhibit a finite group, $\mathbb{P}SL_2(\mathbb{Z}/5\mathbb{Z})$ and we show how our method can be used in order to prove that it is isomorphic to a quotient term algebra. In the following sections, we give the theoretical results which justify the correctness of our method; in section 2, we show the isomorphism between two presentations of finitely generated groups, the first one where the generators are constant function symbols (with an additional binary concatenation symbol), the second one where the generators are unary function symbols. In section 3, we present a direct construction of a bottom-up tree automaton recognizing the terms which are in normal form w.r.t. a convergent rewrite system, where the function symbols are of arity ≤ 1.

1   A Motivating Example: Klein's Dodecahedron

In this section, we consider the finitely generated group $T(\{\cdot, 1, \text{inv}, S, T\})/\mathcal{G} \cup \mathcal{N} \cup \mathcal{E}$, where $\cdot$ is the binary associative operation of the group, 1 is the unit, inv is the unary inverse, S and T are the generators, $\mathcal{G}$ is an equational presentation of groups, $\mathcal{N} = \{S \cdot S = 1, \ T \cdot T \cdot T \cdot T \cdot T = 1\}$ and $\mathcal{E} = \{S \cdot T \cdot S \cdot T \cdot S \cdot T = 1\}$.

We show how our general method can be used in order to prove that the above group is isomorphic to $\mathbb{P}SL_2(\mathbb{Z}/5\mathbb{Z})$, that is the group of 2×2 matrices over $\mathbb{Z}/5\mathbb{Z}$, the determinant of which is equal to 1, quotiented by the relation $\begin{pmatrix} 1 & 0 \\ 0 & 1 \end{pmatrix} = \begin{pmatrix} -1 & 0 \\ 0 & -1 \end{pmatrix}$. $\mathbb{P}SL_2(\mathbb{Z}/5\mathbb{Z})$ admits two generators $S = \begin{pmatrix} 0 & -1 \\ 1 & 0 \end{pmatrix}$, $T = \begin{pmatrix} 1 & 1 \\ 0 & 1 \end{pmatrix}$, and can be seen as a dodecahedron where each edge is duplicated into two "oriented" edges. Right multiplying by T an element, that is an oriented edge, amounts to follow this oriented edge to the next one, keeping the same orientation, and right multiplying by S amounts to take the same edge with the converse orientation. This representation allows to "prove" by hand that this group is finite [5].

First, according to the main theorem of section 2, we have that $\mathcal{T}(\{\cdot, 1, \mathrm{inv}, S, T\})/\mathcal{G} \cup \mathcal{N} \cup \mathcal{E}$ is isomorphic to $\mathcal{T}(\{\overline{1}, \overline{S}, \overline{T}\})/\tau(\mathcal{N} \cup \mathcal{E})$, where $\overline{1}$ is a constant symbol, $\overline{S}$ and $\overline{T}$ are unary function symbols:

$$\tau(\mathcal{N}) = \{\overline{S}^2(x) = x,\ \overline{T}^5(x) = x\},\ \text{and}$$
$$\tau(\mathcal{E}) = \{\overline{S}(\overline{T}(\overline{S}(\overline{T}(\overline{S}(\overline{T}(x)))))) = x\}.$$

Then, we complete $\tau(\mathcal{N} \cup \mathcal{E})$ into the convergent rewrite system R (with respect to the RPO defined by the precedence $\overline{S} > \overline{T} > \overline{1}$)

$$
\begin{aligned}
\overline{S}^2(x) &\longrightarrow x \\
\overline{T}^5(x) &\longrightarrow x \\
\overline{S}(\overline{T}^4(\overline{S}(x))) &\longrightarrow \overline{T}(\overline{S}(\overline{T}(x))) \\
\overline{S}(\overline{T}(\overline{S}(x))) &\longrightarrow \overline{T}^4(\overline{S}(\overline{T}^4(x))) \\
\overline{S}(\overline{T}^3(\overline{S}(\overline{T}^4(x)))) &\longrightarrow \overline{T}(\overline{S}(\overline{T}^2(\overline{S}(x)))) \\
\overline{S}(\overline{T}^2(\overline{S}(\overline{T}(x)))) &\longrightarrow \overline{T}^4(\overline{S}(\overline{T}^3(\overline{S}(x)))) \\
\overline{S}(\overline{T}^3(\overline{S}(\overline{T}^3(\overline{S}(x))))) &\longrightarrow \overline{T}(\overline{S}(\overline{T}^3(\overline{S}(\overline{T}(x)))))
\end{aligned}
$$

The above system has 7 rules. Note that if we complete directly $\mathcal{G} \cup \mathcal{N} \cup \mathcal{E}$, we get a convergent system with 26 rules; 10 rules for $\mathcal{G}$, 2 rules for rewriting $\mathrm{inv}(S)$ and $\mathrm{inv}(T)$, and $14 = 2 \times 7$ rules, each rule $C_1[x] \longrightarrow C_2[x]$ of the above system being duplicated into $\mathscr{C}(C_1[\bullet]) \longrightarrow \mathscr{C}(C_2[\bullet])$, and $\mathscr{C}(C_1[\bullet]) \cdot x \longrightarrow \mathscr{C}(C_2[\bullet]) \cdot x$, where x is a variable ($\mathscr{C}(C_i[\bullet])$ is the canonical term associated with the context $C_i[\bullet]$, see definition 3 below).

Then we use the 7 rules system for building an automaton recognizing the terms of $\mathcal{T}(\{\overline{1}, \overline{S}, \overline{T}\})$ which are in normal form. Since the automaton is a restriction of a tree automaton, a term as a tree, is read bottom-up, hence as a string, from right to left. We get the automaton given in figure 1. Eventually, since there are no loops in the automaton, we know that the quotient algebra is finite, and counting the number of distinct paths in this automaton, we get that the quotient algebra has 60 distinct elements. Note that the starting equalities $\mathcal{N} \cup \mathcal{E}$ are valid in $\mathbb{P}SL_2(\mathbb{Z}/5\mathbb{Z})$, hence $\mathbb{P}SL_2(\mathbb{Z}/5\mathbb{Z})$ is isomorphic to a subgroup of $\mathcal{T}(\{\overline{1}, \overline{S}, \overline{T}\})/\tau(\mathcal{N} \cup \mathcal{E})$ and since we know that $\mathbb{P}SL_2(\mathbb{Z}/5\mathbb{Z})$ has 60 elements, actually $\mathbb{P}SL_2(\mathbb{Z}/5\mathbb{Z})$ is isomorphic to $\mathcal{T}(\{\overline{1}, \overline{S}, \overline{T}\})/\tau(\mathcal{N} \cup \mathcal{E})$.

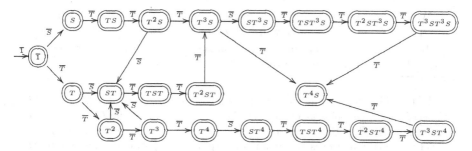

Fig. 1. A normal forms automaton for the Klein's dodecahedron.

2   Eliminating the Inverse from a Presentation of a Group

It is well-known folklore that the ground terms of a finitely generated monoid may be seen as built over a signature containing the generators considered either as constants or as unary function symbols. In the first case, the signature has to contain a binary associative symbol for concatenation, in the second case it has to contain a constant which is the common leaf of all ground terms. In this section, we generalize this view to a class of finitely generated groups, which are defined as quotient algebras $T(\mathcal{F})/\mathcal{G} \cup \mathcal{N} \cup \mathcal{E}$ defined over a signature $\mathcal{F} = \{\cdot, 1, inv, L_1, \ldots, L_n\}$ where $\cdot$ is the associative binary concatenation, 1 is a constant intended to represent the unit of the group, inv is a unary function symbol the meaning of which is the inverse of an element in the group and $L_1, \ldots, L_n$ are the constant generators. The algebra is quotiented by a set of equations divided into three subsets; $\mathcal{G}$ is a presentation of (non-commutative) groups,

$$\mathcal{N} = \{L_1^{j_{L_1}} = 1, \ldots, L_n^{j_{L_n}} = 1\}$$

with $j_{L_i} > 0$ for all i, where L_i^j denotes $\underbrace{L_i \cdot \ldots \cdot L_i}_{j \text{ times}}$ and $\mathcal{E}$ contains only equations between ground terms. In the following, $\mathcal{G}$ will be either $\mathcal{G}_{min}$ a minimal presentation of groups, or $\mathcal{G}_c$, an equivalent convergent presentation obtained from $\mathcal{G}_{min}$ by completion:

$$\mathcal{G}_{min} = \begin{cases} (x \cdot y) \cdot z = x \cdot (y \cdot z) \\ x \cdot 1 \quad\quad = x \\ x \cdot inv(x) = 1 \end{cases} \qquad \mathcal{G}_c = \begin{cases} (x \cdot y) \cdot z \quad\quad = x \cdot (y \cdot z) \\ x \cdot 1 \quad\quad\quad\quad = x \\ 1 \cdot x \quad\quad\quad\quad = x \\ x \cdot inv(x) \quad\quad = 1 \\ inv(x) \cdot x \quad\quad = 1 \\ x \cdot (inv(x) \cdot y) = y \\ inv(x) \cdot (x \cdot y) = y \\ inv(x \cdot y) \quad\quad = inv(y) \cdot inv(x) \\ inv(1) \quad\quad\quad\quad = 1 \\ inv(inv(x)) \quad\quad = x \end{cases}$$

We will also consider $T(\overline{\mathcal{F}})$, the algebra built over the "unary" signature $\overline{\mathcal{F}} = \{\overline{1}, \overline{L_1}, \ldots, \overline{L_n}\}$ associated with $\{\cdot, 1, inv, L_1, \ldots, L_n\}$, where $\overline{1}$ is a constant symbol, and $\overline{L_1}, \ldots, \overline{L_n}$ are unary function symbols. As usual, $\overline{L_i}^j(w)$ denotes $\underbrace{\overline{L_i}(\ldots(\overline{L_i}(w))\ldots)}_{j \text{ times}}$, when w is a term in $T(\overline{\mathcal{F}})$.

In the following, we define a translation from the terms of $T(\mathcal{F})$ into the terms of $T(\overline{\mathcal{F}})$. Our first attempt leads to translate for example $L_1 \cdot L_1 \cdot L_2$ into $\overline{L_1}(\overline{L_1}(\overline{L_2}(\overline{1})))$. But, this translation has to be compatible with the term algebra structure since our aim is to get an isomorphism. In particular, if we add the context $L_3 \cdot \bullet \cdot L_4$ to $L_1 \cdot L_1 \cdot L_2$, this has to be reflected by the translation. A part of the context, namely $L_3 \cdot \bullet$, can be translated into the context $\overline{L_3}(\bullet)$, but the other part, $\overline{L_4}$, cannot be considered as a context in the term $\overline{L_3}(\overline{L_1}(\overline{L_1}(\overline{L_2}(\overline{L_4}(\overline{1})))))$. The point is that a ground context in $T(\mathcal{F})$ has to be divided into a context part and a substitution part in $T(\overline{\mathcal{F}})$: the translation of $L_1 \cdot L_1 \cdot L_2$ should be $\overline{L_1}(\overline{L_1}(\overline{L_2}(x)))$ where x is a variable, and adding the context $L_3 \cdot \bullet \cdot L_4$ amounts to adding the context $\overline{L_3}(\bullet)$, and to applying the substitution $x \mapsto \overline{L_4}(x)$.

Eventually we come to the idea of defining a translation τ from the pairs of $T(\mathcal{F}) \times T(\overline{\mathcal{F}}, \mathcal{X})$ into $T(\overline{\mathcal{F}}, \mathcal{X})$. The intuitive meaning of τ is to concatenate the translation of its first argument to its second argument. We also take into account the equations of $\mathcal{G}$ and $\mathcal{N}$ during the translation, since there is no direct way for expressing the inverse of an element in $T(\overline{\mathcal{F}}, \mathcal{X})$. Hence τ is defined as

$$\tau(1, w) = w$$
$$\tau(inv(1), w) = w$$
$$\forall i \ \tau(L_i, w) = \overline{L_i}(w)$$
$$\forall i \ \tau(inv(L_i), w) = \overline{L_i}^{j_{L_i}-1}(w)$$
$$\tau(t_1 \cdot t_2, w) = \tau(t_1, \tau(t_2, w))$$
$$\tau(inv(t_1 \cdot t_2), w) = \tau(inv(t_2), \tau(inv(t_1), w))$$
$$\tau(inv(inv(t)), w) = \tau(t, w)$$

where t, t_1 and t_2 are any ground terms of $T(\mathcal{F})$, and w any term in $T(\overline{\mathcal{F}}, \mathcal{X})$. It is quite obvious that τ is completely defined.

Definition 1. *The* translation *of an equation $l = r$ where l and r are ground terms in $T(\mathcal{F})$ is the equation $\tau(l, x) = \tau(r, x)$, where x is a given variable of $\mathcal{X}$. It is denoted by $\tau(l = r)$. The translation is extended to a set of ground equations in the obvious way.*

Some of the proofs in the following are by induction over the size of terms and contexts. This leads to introduce the following definition:

Definition 2. *The size of a term t of $T(\mathcal{F})$, $\|t\|$, is defined as follows:*
$$\|1\| = 1 \qquad \|t_1 \cdot t_2\| = \|t_1\| + \|t_2\|$$
$$\forall i \ \|L_i\| = 1 \qquad \|inv(t_1)\| = \max_i\{j_{L_i}\} \times \|t_1\|$$
and the size of a context $C[\bullet]$ in $T(\mathcal{F})$, as follows:
$$\|\bullet\| = 0 \qquad \forall i \ \|\overline{L_i}(C[\bullet])\| = 1 + \|C[\bullet]\|$$

Lemma 1. *Let t be any ground term in $T(\mathcal{F})$. Then there exists a unique context $C_t[\bullet]$ in $T(\overline{\mathcal{F}})$ such that for all term w in $T(\overline{\mathcal{F}}, \mathcal{X})$, $\tau(t, w) = C_t[w]$ and moreover $\|C_t[\bullet]\| \leq \|t\|$.*

Proof. By induction on the term structure of t.

Proposition 1. *If t and t' are two ground terms in $T(\mathcal{F})$, such that $t =_{\mathcal{G}_c \cup \mathcal{N}} t'$ (or equivalently $t =_{\mathcal{G}_{min} \cup \mathcal{N}} t'$), then for all term w in $T(\overline{\mathcal{F}}, \mathcal{X})$, the following equality holds:*

$$\tau(t, w) =_{\tau(\mathcal{N})} \tau(t', w)$$

Proof. Since $\mathcal{G}_c$ and $\mathcal{G}_{min}$ are two equivalent presentations of groups, if $t =_{\mathcal{G}_c \cup \mathcal{N}} t'$, then $t =_{\mathcal{G}_{min} \cup \mathcal{N}} t'$, hence there exists an equational proof

$$t \equiv t_0 \longleftrightarrow_{\mathcal{G}_{min} \cup \mathcal{N}} t_1 \ldots t_{m-1} \longleftrightarrow_{\mathcal{G}_{min} \cup \mathcal{N}} t_m \equiv t'$$

The proof of the lemma is by induction on the multiset $\{\|t_0\|, \|t_1\|, \ldots, \|t_m\|\}$.

This is obvious when the proof has length zero. Let us assume that $m \geq 1$. We shall show that $\tau(t_0, w) =_{\tau(\mathcal{N})} \tau(t_1, w)$, then we can conclude by induction that $\tau(t_1, w) =_{\tau(\mathcal{N})} \tau(t_m, w)$, hence by transitivity of $=_{\tau(\mathcal{N})}$ that $\tau(t_0, w) =_{\tau(\mathcal{N})} \tau(t_m, w)$.

- Let us assume that the first equational step occurs at the top of the terms t_0 and t_1. We reason by cases over the equation that has been applied. If the equation is $(x \cdot y) \cdot z = x \cdot (y \cdot z)$ or $x \cdot 1 = x$, we get the desired result by applying the definition of τ.
  - Let us assume that the equation is $x \cdot \mathrm{inv}(x) = 1$. Since we only consider ground terms, we reason by cases over x: x may be of the form $1, L_1, \ldots, L_n, u_1 \cdot u_2$ or $\mathrm{inv}(u)$, where u_1, u_2 and u are closed terms. We treat into details the cases L_i and $u_1 \cdot u_2$. The other cases are similar.

$$
\begin{aligned}
\tau(L_i \cdot \mathrm{inv}(L_i), w) &=_{def} \tau(L_i, \tau(\mathrm{inv}(L_i), w)) \\
&=_{def} \overline{L_i}(\overline{L_i}^{j_{L_i}-1}(w)) \\
&=_{\tau(\mathcal{N})} \tau(1, w) \quad \text{since } \tau(L_i^{j_{L_i}}, x) = \tau(1, x) \in \tau(\mathcal{N}).
\end{aligned}
$$

$$
\begin{aligned}
\tau((u_1 \cdot u_2) \cdot \mathrm{inv}(u_1 \cdot u_2), w) &=_{def} \tau(u_1, \tau(u_2, \tau(\mathrm{inv}(u_2), \tau(\mathrm{inv}(u_1), w)))) \\
&=_{def} \tau(u_1, \tau(u_2 \cdot \mathrm{inv}(u_2), \tau(\mathrm{inv}(u_1), w))) \\
&=_{\tau(\mathcal{N})} \tau(u_1, \tau(1, \tau(\mathrm{inv}(u_1), w)))
\end{aligned}
$$

since by induction hypothesis $\tau(u_2 \cdot \mathrm{inv}(u_2), w') =_{\tau(\mathcal{N})} \tau(1, w')$ for all term, w'.

$$
\begin{aligned}
&=_{def} \tau(u_1, \tau(\mathrm{inv}(u_1), w)) \\
&=_{def} \tau(u_1 \cdot \mathrm{inv}(u_1), w) \\
&=_{\tau(\mathcal{N})} \tau(1, w)
\end{aligned}
$$

since by induction hypothesis $\tau(u_1 \cdot \mathrm{inv}(u_1), w) =_{\tau(\mathcal{N})} \tau(1, w)$.

  - Let us assume that the equation is $L_i^{j_{L_i}} = 1$.

$$\tau(L_i^{j_{L_i}}, w) =_{\tau(\mathcal{N})} \tau(1, w) \quad \text{since } \tau(L_i^{j_{L_i}}, x) = \tau(1, x) \in \tau(\mathcal{N}).$$

– Let us assume that the first equational step occurs strictly under the top of the terms, at a position p: $t_0 \equiv t_0[l\sigma]_p \longleftrightarrow_{\mathcal{G}_{min} \cup \mathcal{N}} t_0[r\sigma]_p \equiv t_1$, with $p > \epsilon$. We reason by cases over the top of t_0 and p, there are four cases, the top of t_0 is equal to $\cdot$, and $p = 1 \cdot p'$ or $p = 2 \cdot p'$, the top of t_0 is equal to inv and $p = 1$ or $p = 1 \cdot p'$ with $p' \neq \epsilon$. We will only consider into details the last two cases.

• If the top of t_0 is equal to inv and $p = 1$, we reason by cases over the equation that has been used. If the equation is $(x \cdot y) \cdot z = x \cdot (y \cdot z)$ or $x \cdot 1 = x$, we get the desired result by applying the definition of τ.

  ∗ Let us assume that the equation is $x \cdot \mathrm{inv}(x) = 1$.

$$
\begin{aligned}
\tau(t_0, w) = \quad & \tau(\mathrm{inv}(t_0' \cdot \mathrm{inv}(t_0')), w) \\
=_{def} \quad & \tau(\mathrm{inv}(\mathrm{inv}(t_0')), \tau(\mathrm{inv}(t_0'), w)) \\
=_{def} \quad & \tau(t_0', \tau(\mathrm{inv}(t_0'), w)) \\
=_{def} \quad & \tau(t_0' \cdot \mathrm{inv}(t_0'), w) \\
=_{\tau(\mathcal{N})} \quad & \tau(1, w) \text{ induction hypothesis with the term } w \text{ of } \mathcal{T}(\overline{\mathcal{F}}).
\end{aligned}
$$

  ∗ Let us assume that the equation is $L_i^{j_{L_i}} = 1$.

$$
\begin{aligned}
\tau(t_0, w) = \quad & \tau(\mathrm{inv}(L_i^{j_{L_i}}), w) \\
=_{def} \quad & \underbrace{\tau(\mathrm{inv}(L_i), \tau(\mathrm{inv}(L_i), \ldots, \tau(\mathrm{inv}(L_i), w)\ldots))}_{j_{L_i} \ times} \\
=_{def} \quad & \overline{L_i}^{j_{L_i} \times (j_{L_i}-1)}(w) \\
=_{\tau(\mathcal{N})} \quad & w \quad \text{since } \tau(L_i^{j_{L_i}}, x) = \tau(1, x) \in \tau(\mathcal{N}). \\
=_{def} \quad & \tau(\mathrm{inv}(1), w) \\
= \quad & \tau(t_1, w)
\end{aligned}
$$

• If the top of t_0 is equal to inv, and $p = 1 \cdot p'$ with $p' \neq \epsilon$, there are three cases, $t_0 = \mathrm{inv}(\mathrm{inv}(t_0'))$ with $p = 1 \cdot 1 \cdot p''$, $t_0 = \mathrm{inv}(t_0' \cdot t_0'')$ with $p = 1 \cdot 1 \cdot p''$, or $p = 1 \cdot 2 \cdot p''$. We will treat into details only the second case, the others are similar. $t_0 = \mathrm{inv}(t_0' \cdot t_0'')$, $t_1 = \mathrm{inv}(t_1' \cdot t_0'')$ and $t_0' \longleftrightarrow_{\mathcal{G}_{min} \cup \mathcal{N}} t_1'$. Hence $\mathrm{inv}(t_0') \longleftrightarrow_{\mathcal{G}_{min} \cup \mathcal{N}} \mathrm{inv}(t_1')$. By induction hypothesis, since the size of the equational proof $\mathrm{inv}(t_0') \longleftrightarrow_{\mathcal{G}_{min} \cup \mathcal{N}} \mathrm{inv}(t_1')$ is strictly smaller than $t_0 \longleftrightarrow_{\mathcal{G}_{min} \cup \mathcal{N}} t_1$, we get that for all term w in $\mathcal{T}(\overline{\mathcal{F}})$,

$$\tau(\mathrm{inv}(t_0'), w) =_{\tau(\mathcal{N})} \tau(\mathrm{inv}(t_1'), w)$$

$$
\begin{aligned}
\tau(t_0, w) = \quad & \tau(\mathrm{inv}(t_0' \cdot t_0''), w) \\
=_{def} \quad & \tau(\mathrm{inv}(t_0''), \tau(\mathrm{inv}(t_0'), w)) \\
=_{\tau(\mathcal{N})} \quad & \tau(\mathrm{inv}(t_0''), \tau(\mathrm{inv}(t_1'), w))
\end{aligned}
$$
by induction hypothesis with the term w of $\mathcal{T}(\overline{\mathcal{F}})$, under the context $\tau(\mathrm{inv}(t_0''), \bullet)$.
$$
\begin{aligned}
=_{def} \quad & \tau(\mathrm{inv}(t_1' \cdot t_0''), w) \\
= \quad & \tau(t_1, w)
\end{aligned}
$$

Proposition 2. *Let $l = r$ be any equation between ground terms in $\mathcal{T}(\mathcal{F})$. Then for all ground terms s, t in $\mathcal{T}(\mathcal{F})$ such that $s \longleftrightarrow_{l=r}^p t$, for all term w in $\mathcal{T}(\overline{\mathcal{F}}, \mathcal{X})$,*

$$\tau(s, w) =_{\tau(\mathcal{N} \cup \{l=r\})} \tau(t, w)$$

Sketch of the proof. By induction on the length of p. There are several cases, according to p and the top symbol of s, $p = \epsilon$, $p = 1 \cdot p'$ or $p = 2 \cdot p'$ and the top symbol of s is equal to $\cdot$, $p = 1$ or $p = 1 \cdot p'$ (with $p' \neq \epsilon$) and the top symbol is equal to inv. We shall detail only two cases, the first and the third one. This last case explains why we need $\tau(\mathcal{N} \cup \{l = r\})$, and not only $\tau(l = r)$.

- If $p = \epsilon$, then since l and r are closed terms, $s = l$ and $t = r$, hence $\tau(s, w) \longleftrightarrow \tau(t, w)$ with the equation $\tau(l, x) = \tau(r, x)$, at the top, and with the substitution $\{x \mapsto w\}$.
- If $s = \mathrm{inv}(s_1)$, $p = 1$, then $t = \mathrm{inv}(t_1)$, and $s_1 \longleftrightarrow_{l=r}^{\epsilon} t_1$. Since s and t are ground terms, so are s_1 and t_1, hence $s_1 = l$, $t_1 = r$. By proposition 1, since $l \cdot \mathrm{inv}(l)$ (resp. $r \cdot \mathrm{inv}(r)$) and 1 are ground terms in $\mathcal{T}(\mathcal{F})$, and $l \cdot \mathrm{inv}(l) =_{\mathcal{G}_c} 1$ (resp. $r \cdot \mathrm{inv}(r) =_{\mathcal{G}_c} 1$), for all term w in $\mathcal{T}(\overline{\mathcal{F}}, \mathcal{X})$,

$$\tau(l \cdot \mathrm{inv}(l), w) =_{\tau(\mathcal{N})} \tau(1, w) \qquad \tau(r \cdot \mathrm{inv}(r), w) =_{\tau(\mathcal{N})} \tau(1, w)$$

Hence, we get

$$
\begin{aligned}
\tau(s, w) = \quad & \tau(\mathrm{inv}(l), w) \\
=_{\tau(\mathcal{N})} \quad & \tau(\mathrm{inv}(l), \tau(r \cdot \mathrm{inv}(r), w)) \\
& \text{Proposition 1 with the term } w, \text{ under the context } \tau(\mathrm{inv}(l), \bullet). \\
=_{def} \quad & \tau(\mathrm{inv}(l), \tau(r, \tau(\mathrm{inv}(r), w))) \\
=_{\tau(l=r)} \quad & \tau(\mathrm{inv}(l), \tau(l, \tau(\mathrm{inv}(r), w))) \\
& \text{under the context } \tau(\mathrm{inv}(l), \bullet), \text{ with the substitution } x \mapsto \tau(\mathrm{inv}(r), w). \\
=_{def} \quad & \tau(\mathrm{inv}(l) \cdot l, \tau(\mathrm{inv}(r), w)) \\
=_{\tau(\mathcal{N})} \quad & \tau(\mathrm{inv}(r), w) \\
& \text{Proposition 1 with the term } \tau(\mathrm{inv}(r), w), \text{ at the top.} \\
= \quad & \tau(t, w)
\end{aligned}
$$

Definition 3. *Let $W[\bullet]$ be any ground context of $\mathcal{T}(\overline{\mathcal{F}})$. The canonical term associated with $W[\bullet]$, $\mathcal{C}(W[\bullet])$ is defined as follows:*
$$\mathcal{C}(\bullet) = 1 \qquad \mathcal{C}(\overline{L_i}(W[\bullet])) = L_i \cdot \mathcal{C}(W[\bullet])$$
.

Lemma 2. *Let $W[\bullet]$ be any ground context in $\mathcal{T}(\overline{\mathcal{F}})$, the following identity holds: $C_{\mathcal{C}(W[\bullet])}[\bullet] = W[\bullet]$ ($C_t[\bullet]$ is defined by lemma 1). Moreover $\|W[\bullet]\| = \|\mathcal{C}(W[\bullet])\|$.*

The proof follows immediately from the definition.

Proposition 3. *Let t and t' be two ground terms in $\mathcal{T}(\mathcal{F})$ such that $\tau(t, \overline{1}) =_{\tau(\mathcal{N})} \tau(t', \overline{1})$. The equality $t =_{\mathcal{G}_c \cup \mathcal{N}} t'$ holds.*

Sketch of the proof. By induction on the pair $(m, \{\|t\|, \|t'\|\})$, with a lexicographic ordering, where m is the length of the equational proof $\tau(t, \overline{1}) =_{\tau(\mathcal{N})} \tau(t', \overline{1})$.

- We assume first that m is equal to 0, hence that $\tau(t, \overline{1}) = \tau(t', \overline{1})$. We reason by cases over the term structure of t and t'. We shall only detail the case when $t = t_1 \cdot t_2$ and $t' = t'_1 \cdot t'_2$, which is the most interesting one. In this case we get that $C_{t_1}[C_{t_2}[\overline{1}]] = C_{t_1'}[C_{t_2'}[\overline{1}]]$. This means that there exist three contexts in $\mathcal{T}(\overline{\mathcal{F}})$, $U_1[\bullet], U_2[\bullet]$ and $U_3[\bullet]$, such that

$$
\begin{array}{lll}
C_{t_1}[\bullet] = U_1[\bullet] & & C_{t_1}[\bullet] = U_1[U_2[\bullet]] \\
C_{t_2}[\bullet] = U_2[U_3[\bullet]] & \text{or such that} & C_{t_2}[\bullet] = U_3[\bullet] \\
C_{t_1'}[\bullet] = U_1[U_2[\bullet]] & & C_{t_1'}[\bullet] = U_1[\bullet] \\
C_{t_2'}[\bullet] = U_3[\bullet] & & C_{t_2'}[\bullet] = U_2[U_3[\bullet]]
\end{array}
$$

We only consider the first case, since the second one is symmetrical. Let u_j be the canonical term associated with the context $U_j[\bullet]$, for $j = 1, 2, 3$. $\tau(u_j, \bullet) = U_j[\bullet]$ holds and moreover $\|U_j[\bullet]\| = \|u_j\|$. We can now rewrite the above equalities as

$$
\begin{array}{ll}
\tau(t_1, \overline{1}) = \tau(u_1, \overline{1}) & \tau(t_1', \overline{1}) = \tau(u_1 \cdot u_2, \overline{1}) \\
\tau(t_2, \overline{1}) = \tau(u_2 \cdot u_3, \overline{1}) & \tau(t_2', \overline{1}) = \tau(u_3, \overline{1})
\end{array}
$$

$$
\|u_1\| = \|U_1[\bullet]\| = \|C_{t_1}[\bullet]\| \le \|t_1\| < \|t\|
$$

$$
\|u_2 \cdot u_3\| = \|u_2\| + \|u_3\| = \|U_2[\bullet]\| + \|U_3[\bullet]\| \le
$$
$$
\|U_1[U_2[\bullet]]\| + \|U_3[\bullet]\| = \|C_{t_1'}[\bullet]\| + \|C_{t_2'}[\bullet]\| \le
$$
$$
\|t_1'\| + \|t_2'\| = \|t'\|
$$

Similarly, $\|u_1 \cdot u_2\| \le \|t\|$ and $\|u_3\| \le \|t_2'\| < \|t'\|$. We can apply the induction hypothesis on (t_1, u_1), $(t_2, u_2 \cdot u_3)$, $(t_1', u_1 \cdot u_2)$, and on (t_2', u_3). We get

$$
\begin{array}{ll}
t_1 =_{\mathcal{G}_c \cup \mathcal{N}} u_1 & t_1' =_{\mathcal{G}_c \cup \mathcal{N}} u_1 \cdot u_2 \\
t_2 =_{\mathcal{G}_c \cup \mathcal{N}} u_2 \cdot u_3 & t_2' =_{\mathcal{G}_c \cup \mathcal{N}} u_3
\end{array}
$$

Hence $t = t_1 \cdot t_2 =_{\mathcal{G}_c \cup \mathcal{N}} u_1 \cdot (u_2 \cdot u_3) =_{\mathcal{G}_c \cup \mathcal{N}} (u_1 \cdot u_2) \cdot u_3 =_{\mathcal{G}_c \cup \mathcal{N}} t_1' \cdot t_2' = t'$.

– Let us now consider the case when there is at least one step in the equational proof $\tau(t, \overline{1}) =_{\tau(\mathcal{N})} \tau(t', \overline{1})$, i.e. $m \ge 1$. We consider the first step of this proof; it uses an equation of the form $\tau(l, x) = \tau(r, x)$, where $l = r$ is an equation of $\mathcal{N}$. Hence $\tau(t, \overline{1}) = W_1[\tau(l, W_2[\overline{1}])] \longleftrightarrow_{\tau(\mathcal{N})} W_1[\tau(r, W_2[\overline{1}])]$. Let u_1 and u_2 be respectively the canonical terms associated with the ground contexts of $\mathcal{T}(\overline{\mathcal{F}})$, $W_1[\bullet]$ and $W_2[\bullet]$. We get that

$$
\tau(t, \overline{1}) = W_1[\tau(l, W_2[\overline{1}])] = \tau(u_1, \tau(l, \tau(u_2, \overline{1}))) = \tau(u_1 \cdot (l \cdot u_2), \overline{1})
$$

Hence by induction hypothesis, we get that $t =_{\mathcal{G}_c \cup \mathcal{N}} u_1 \cdot (l \cdot u_2)$. Since $l = r$ is an equation of $\mathcal{N}$, we also have obviously that $u_1 \cdot (l \cdot u_2) =_{\mathcal{G}_c \cup \mathcal{N}} u_1 \cdot (r \cdot u_2)$. We denote by t'' the term $u_1 \cdot (r \cdot u_2)$, and we have that $\tau(t'', \overline{1}) = W_1[\tau(r, W_2[\overline{1}])]$ is equal to $\tau(t', \overline{1})$ with an equational proof of length $m - 1$. We can apply the induction hypothesis to (t'', t'), and we get that $t'' =_{\mathcal{G}_c \cup \mathcal{N}} t'$. Eventually we conclude that $t =_{\mathcal{G}_c \cup \mathcal{N}} t'$.

Proposition 4. *Let $l = r$ be a ground equation over $\mathcal{T}(\mathcal{F})$, and s and t two ground terms of $\mathcal{T}(\mathcal{F})$. If $\tau(s, \overline{1}) \longleftrightarrow^p_{\tau(l=r)} \tau(t, \overline{1})$, then $s =_{\mathcal{G}_c \cup \mathcal{N} \cup \{l=r\}} t$.*

Sketch of the proof. The proof is by induction over the length of p, the position where the equational step occurs.

From the above propositions, we get the following

Theorem 1. *Let s and t be two ground terms of $\mathcal{T}(\mathcal{F})$. Then*
$$
s =_{\mathcal{G}_c \cup \mathcal{N} \cup \mathcal{E}} t \quad \text{if and only if} \quad \tau(s, \overline{1}) =_{\tau(\mathcal{N} \cup \mathcal{E})} \tau(t, \overline{1}).
$$

Proof. – If $s =_{\mathcal{G}_c \cup \mathcal{N} \cup \mathcal{E}} t$, then by proposition 1 and proposition 2, we get that $\tau(s, w) =_{\tau(\mathcal{N} \cup \mathcal{E})} \tau(t, w)$, for all term w of $\mathcal{T}(\overline{\mathcal{F}}, \mathcal{X})$, in particular for $w = \overline{1}$.
– If $\tau(s, \overline{1}) =_{\tau(\mathcal{N} \cup \mathcal{E})} \tau(t, \overline{1})$, then by proposition 3 and proposition 4, we get that $s =_{\mathcal{G}_c \cup \mathcal{N} \cup \mathcal{E}} t$.

Corollary 1. *The quotient algebras $\mathcal{T}(\mathcal{F})/\mathcal{G}_c \cup \mathcal{N} \cup \mathcal{E}$ and $\mathcal{T}(\overline{\mathcal{F}})/\tau(\mathcal{N} \cup \mathcal{E})$ are isomorphic, in particular, they have the same cardinality.*

Proof. From the above theorem, $s \mapsto \tau(s, \overline{1})$ is a one-to-one mapping from $\mathcal{T}(\mathcal{F})/\mathcal{G}_c \cup \mathcal{N} \cup \mathcal{E}$ into $\mathcal{T}(\overline{\mathcal{F}})/\tau(\mathcal{N} \cup \mathcal{E})$, and if w is a ground term of $\mathcal{T}(\overline{\mathcal{F}})$, then it has the form $C[\overline{1}]$, hence $w = \tau(\mathscr{C}(C[\bullet]), \overline{1})$, and the mapping $s \mapsto \tau(s, \overline{1})$ is surjective.

3   Computing Normal Forms Automata

Our aim is to count the elements of a group presented as in section 2. Assuming that the completion of the "unary" presentation terminates, we build an automaton that recognizes the set of ground normal forms for the convergent rewriting system R defining this group. If the number m of accepting paths is finite then the group we study is finite and has m elements. Otherwise, there is a loop which represents an infinite set of distinct elements in the group. Let w be the word of function symbols corresponding to the labels in this loop, the language of normal forms contains all the w^ps. In order to fold the group into a finite one, w has to have a finite order p, therefore we shall add an equation of the form $w^p = 1$. Note that we have no information on p, which is left to the choice of the user.

A standard method to build a normal forms automaton for R is to build one automaton for each left hand side of the rules in R, to compute the union of these automata, to determinize this automaton and then to compute its complement. This clearly leads to a normal forms automaton. The main problem of such an algorithm is the use of very costly operations on automata: union, determinization and complementation. This algorithm is therefore untractable with large automata. We address this problem with a new direct algorithm to compute the normal forms automaton. This algorithm avoids the use of union, determinization and complementation on automata. In addition it builds an automaton with labeled states. The labels are terms which are interpreted as formulae describing the set of elements of the group recognized by this state. We present in this paper a very specialized version of the algorithm described in [12]. The later applies to general signatures with arbitrary arities, AC symbols and sets of left-linear rules. Note that thanks to the translation of section 2 we translate a non left-linear rewrite system into a left-linear one. Therefore we avoid the use of the conditional automata techniques[1].

Definition 4 (Bottom-Up Tree Automaton). *A bottom-up tree automaton is a quadruple $(\mathcal{F}, Q, Q_f, \mathcal{R})$ where*

- $\mathcal{F}$ *is a finite alphabet of function symbols equipped with an arity,*
- Q *is a set of states,*
- $Q_f \subseteq Q$ *is a set of final states ,*
- $\mathcal{R}$ *is a rewriting system on $\mathcal{F} \uplus Q$, where the symbols in Q are unary and where the rule are of the form $f(q_1(x_1), \ldots, q_n(x_n)) \longrightarrow q(f(x_1, \ldots, x_n))$ where $f \in \mathcal{F}$ has arity n, $q_1, \ldots, q_n, q$ are in Q and $x_1, \ldots, x_n$ are distinct variables. Such a rule will be denoted by $(q_1, \ldots, q_n) \xrightarrow{f} q$ and is called a transition.*

Definition 5 (Run of a Bottom-Up Tree Automaton). *A run of an automaton $(\mathcal{F}, Q, Q_f, \mathcal{R})$ on a ground term $t \in T(\mathcal{F})$ is a rewriting sequence of $\mathcal{R}$ from t to $q(t)$ where $q \in Q$. If such a run exists and $q \in Q_f$ then t is accepted by the automaton else t is rejected.*

The bottom-up tree automata we actually use are very particular ones, since the function symbols are of arity ≤ 1. Therefore, they are very similar to standard finite state automata, except that terms are read bottom-up, that is from right to left, when they are considered as strings.

Example 1. Consider the automaton given in section 1. A run of this automaton on the term $\overline{T}(\overline{T}(\overline{S}(\overline{1})))$ ends up in the final state $T^2 S$. Therefore this term is accepted by the automaton. The term $\overline{T}^5(\overline{1})$ is rejected because $(\overline{T}^4(\overline{1}))$ is accepted by the state T^4 and no transition sources from T^4 with the symbol $\overline{T}$.

From now on the states will be uniquely labeled with terms. Therefore we will speak of *the state-term t* instead of the state labeled with the term t. We will use the usual vocabulary for terms and states on state-terms.

Definition 6 (Minimal Generalization of a Term). *Let t be a term and let $\mathcal{T}$ be a set of terms. A minimal generalization of t in $\mathcal{T}$ is a term u such that*

- $u \in \mathcal{T}$,
- *t is an instance of u,*
- *if $v \in \mathcal{T}$ and t is an instance of v then u is an instance of v.*

Lemma 3. *In the case of a signature where all symbols have an arity equal to 0 or 1, a minimal generalization of a term in a set is either unique up to renaming or does not exist.*

In the following, we denote by R a set of rules defined over the algebra $T(\overline{\mathcal{F}}, \mathcal{X})$.

Theorem 2. *The algorithm presented in figure 2 computes a normal forms automaton for the rewriting system R. A state-term t of this automaton recognizes exactly the set of terms in ground normal forms $\mathcal{T}$ such that for all $u \in \mathcal{T}$, t is the minimal generalization of u in the set of all state-terms.*

We split the proof into several lemmas.

- Construction:
  - States:
    - * Start with an automaton with one final state-term $\overline{1}$
    - * For each integer $0 \leq i \leq n$ add a final state-term $\overline{L_i}(x)$
    - * For all strict non-variable subterm t of the left-hand side of a rule in R add a final state-term t
  - Transitions:
    - * Add the transition $\xrightarrow{\overline{1}} \overline{1}$
    - * For every state-term t and every symbol $\overline{L_i}$ if $\overline{L_i}(t)$ is not reducible by R then add a transition $t \xrightarrow{\overline{L_i}} s$ where s is the minimal generalization of $\overline{L_i}(t)$ in the set of all state-terms
- Cleaning: Clean the automaton, removing the inaccessible states.

Fig. 2. Algorithm to compute the normal forms automaton

Lemma 4. *If a term u is accepted by the automaton in a state-term s then u is an instance of s.*

Proof. The proof is by induction on the size of the recognized terms. If $u = \overline{1}$, the only accepting state for u is $\overline{1}$. If $u = \overline{L_i}(v)$ is a term accepted by a state-term s, then let us consider the last transition of an accepting run; it necessarily of the form $t \xrightarrow{\overline{L_i}} s$. By definition of a run, v is accepted by the state-term t, and by induction hypothesis v is an instance of t, hence $u = \overline{L_i}(v)$ is an instance of $\overline{L_i}(t)$. Moreover, by construction of the automaton, $\overline{L_i}(t)$ is an instance of s. Eventually, we get that u is an instance of s.

Lemma 5. *If a term is in normal form then it is accepted by the automaton.*

Proof. The proof is by induction on the size of the normal term u. If $u = \overline{1}$ then it is accepted by the state-term $\overline{1}$. Let $u = \overline{L_i}(v)$. Since u is normal, v is normal. By the induction hypothesis v is accepted by a state t and the preceding lemma proves that v is an instance of t. Since $u = \overline{L_i}(v)$ is an instance of $\overline{L_i}(t)$, $\overline{L_i}(t)$ is in normal form. By construction there is a transition $t \xrightarrow{\overline{L_i}} s$ where s is a final state-term. Thus $u = \overline{L_i}(v)$ is accepted by the automaton in the state s.

Lemma 6. *If u is a normal form accepted by the state-term s then s is the minimal generalization of u.*

Proof. The proof is by induction on the size of u.

- Case $u = \overline{1}$: the only state accepting u is $\overline{1}$. It is the minimal generalization state-term of the term $\overline{1}$ in the set of state-terms.
- Case $u = \overline{L_i}(v)$: let $t \xrightarrow{\overline{L_i}} s$ be the last transition of an accepting sequence of u in the state-term s. Clearly v is accepted in t and by induction hypothesis, t is the minimal generalization of v. In particular $v = t\sigma_0$, $u = \overline{L_i}(v) = \overline{L_i}(t)\sigma_0$. Let G_u be the set of the generalized state-terms of u and $G_{\overline{L_i}(t)}$ be the one of $\overline{L_i}(t)$. The inclusion $\{s\} \subset G_{\overline{L_i}(t)} \subset G_u$ clearly holds since u is an instance

of $\overline{L_i}(t)$ which is itself an instance of s. Since all symbols have an arity ≤ 1, the set G_u is totally ordered with respect to the instantiation ordering. Let $g \in G_u$, we shall prove that $g \in G_{\overline{L_i}(t)}$. Let us consider the three cases described in the following scheme:

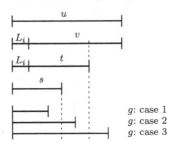

g: case 1
g: case 2
g: case 3

- Cases 1 and 2: $g \in G_{\overline{L_i}(t)}$ by definition.
- Case 3: $g = \overline{L_i}(t)\sigma = L_i(t\sigma)$. Let $g' = t\sigma$. Since t is a state-term, it is not a variable. Hence g' is not a variable. Therefore g is a state-term which is a strict subterm of the left hand side of one rule. Consequently g' is also a strict subterm of the left hand side of one rule. g' is a generalized term of v therefore $g' = t$ (because t is the minimal generalized term). We conclude $g \in G_{\overline{L_i}(t)}$.

Therefore $G_{\overline{L_i}(t)} = G_u$ and they have the same minimal state-term s which is the minimal generalized term of u.

Lemma 7. *If a term is reducible then it is rejected by the automaton.*

Proof. Let u be a reducible term. Without loss of generality, we may assume that $u = \overline{L_i}(u')$ and u' is in normal form (by taking the minimal reducible subterm of u). u' is accepted by its minimal generalized state-term t'. Let $l = \overline{L_i}(t'')$ be the left hand side of a rule reducing u. u' is an instance of t'' and t'' is a state-term: by minimality t' is an instance of t''. Therefore $\overline{L_i}(t')$ is an instance of l i.e. is reducible. The construction of the automaton will not create a transition labeled $\overline{L_i}$ coming out from t'. Thus u is rejected.

Example 2. We detail the computation of the automaton given in section 1 by exhibiting the transitions added from two particular state-terms.

Transitions from the state-term $\overline{T}(x)$. Note that in section 1, the state $\overline{T}(x)$ is denoted by T in the automaton.

- Transition by $\overline{S}$: the term $\overline{S}(\overline{T}(x))$ is in normal form. The set of generalized state-terms of $\overline{S}(\overline{T}(x))$ is $\{\overline{S}(x); \overline{S}(\overline{T}(x))\}$, and the minimal one is $\overline{S}(\overline{T}(x))$. Therefore the transition $\overline{T}(x) \xrightarrow{\overline{S}} \overline{S}(\overline{T}(x))$ is added.
- Transition by $\overline{T}$: $\overline{T}(\overline{T}(x))$ is in normal form. The set of generalized state-terms of $\overline{T}(\overline{T}(x))$ is $\{\overline{T}(x); \overline{T}^2(x)\}$, and the minimal one is $\overline{T}^2(x)$. Therefore the transition $\overline{T}(x) \xrightarrow{\overline{T}} \overline{T}^2(x)$ is added.

Transitions from the state-term $\overline{T}^2(\overline{S}(\overline{T}(x)))$.

- Transition by $\overline{S}$: the term $\overline{S}(\overline{T}^2(\overline{S}(\overline{T}(x))))$ is reducible (by the 6th rule of the rewriting system R given in section 1). Therefore no transition is added.
- Transition by $\overline{T}$: the term $\overline{T}^3(\overline{S}(\overline{T}(x)))$ is in normal form. Its set of generalized state-terms is $\{\overline{T}(x); \overline{T}^2(x); \overline{T}^3(x); \overline{T}^3(\overline{S}(x))\}$, and the minimal state-term is $\overline{T}^3(\overline{S}(x))$. Thus the transition $\overline{T}^2(\overline{S}(\overline{T}(x))) \xrightarrow{\overline{T}} \overline{T}^3(\overline{S}(x))$ is added.

Note that the computation of the automaton does not rely on the convergence of the rewriting system: this computation is possible for any rewrite system, and when the system is not confluent, the number of paths in the automaton yields an upper bound on the number of distinct elements in the quotient algebra. When there are some loops in the automaton, they can be used to guess the kind of equations that have to be added in order to "fold" the algebra, in the same way as in the convergent case (see the beginning of the section).

4   Conclusion

We have shown how to apply rewriting and automata techniques to the study of quotients of $SL_2(\mathbb{Z})$. The main technical issues are, first the correspondence between a particular class of finitely generated groups and a quotient algebra defined over a signature containing only function symbols of arity ≤ 1, and second the direct construction of a normal forms automaton for a convergent rewrite system built over such a "unary" signature.

Concerning the first point, we have defined an isomorphism on the ground terms of the quotient algebras, which is compatible with the equational theory. This translation is possible not only for the equational theory, but also step by step for the completion process (when the strategy synchronizes the processing of one equation or rule and its extension, in the binary case). The unary presentation allows to get rid of the convergent rewriting system for groups and of the extensions of the other rules. If the quotient algebra is finite, the completion will always terminate, but in case we suspect the completion to diverge, it would be interesting to see whether a completion using SOUR graphs [11] could help, giving some hints on what equations can be added to the equational theory in order to get a finite group.

When the completion terminates, and yields a convergent rewrite system, the automaton may have loops which represent infinite sets of distinct elements in the group. Such a loop can be easily represented as a regular language $\{w^p \mid p \in \mathbb{N}\}$, thanks to the pumping lemma and may help to understand how to fold the group into a finite group.

A more prospective research is to try to do parameterized completion for some particular sets of equations, where there are some integers, which are actually parameters; for example some powers of given words occurring in the equations defining a class of groups, such as the above integer p.

We have been able to treat the example of $\mathbb{P}SL_2(\mathbb{Z}/5\mathbb{Z})$, which is small, but not trivial, and it seems to us that this shows that our method is suitable for this kind of study. From a practical point of view, we used CiME [3], in which completion (and also completion modulo the group theory) is implemented. CiME also has the general version of the direct construction of normal forms automaton, with loop detection and path counting. This means that we are able to treat our motivating example in a fully automatic way.

Our ultimate goal is to be able to treat some examples with several million elements, hence for efficiency sake, the construction of normal forms automata in the case when there are only symbols of arity ≤ 1 will be implemented.

References

[1] Bruno Bogaert and Sophie Tison. Equality and disequality constraints on brother terms in tree automata. In P. Enjalbert, A. Finkel, and K. W. Wagner, editors, *9th Annual Symposium on Theoretical Aspects of Computer Science*, volume 577 of *Lecture Notes in Computer Science*, pages 161–171, Paris, France, 1992. Springer-Verlag.

[2] Reinhard Bündgen. *Term Completion versus Algebraic Completion*. PhD thesis, Universität Tübingen, 1991.

[3] Evelyne Contejean and Claude Marché. CiME: Completion Modulo E. In Harald Ganzinger, editor, *7th International Conference on Rewriting Techniques and Applications*, volume 1103 of *Lecture Notes in Computer Science*, pages 416–419, New Brunswick, NJ, USA, July 1996. Springer-Verlag. System Description available at `http://www.lri.fr/~demons/cime.html`.

[4] A. Coste and T. Gannon. Congruence subgroups. Submitted, Electronic version available at `http://xxx.lpthe.jussieu.fr/abs/math/9909080`.

[5] Antoine Coste. Quotients de $SL_2(\mathbb{Z})$ pour les caractères. Preprint, Institut des Hautes Études Scientifiques, November 1997.

[6] Nachum Dershowitz and Jean-Pierre Jouannaud. Rewrite systems. In J. van Leeuwen, editor, *Handbook of Theoretical Computer Science*, volume B, pages 243–309. North-Holland, 1990.

[7] Robert H. Gilman. Presentations of groups and monoids. *Journal of Algebra*, 57:544–554, 1979.

[8] Derek F. Holt. Decision problems in finitely presented groups. In *Proceedings of the Euroconference "Computational Methods for Representations of Groups and Algebras"*, Essen, April 1997.

[9] C. Itzykson and J.-M. Douffe. *Théorie statistique des champs*. Éditions du CNRS, 1989.

[10] Donald E. Knuth and Peter B. Bendix. Simple word problems in universal algebras. In J. Leech, editor, *Computational Problems in Abstract Algebra*, pages 263–297. Pergamon Press, 1970.

[11] Christopher Lynch and Polina Strogova. Sour graphs for efficient completion. *Discrete Mathematics and Theoretical Computer Science*, 2:1–25, 1998.

[12] Benjamin Monate. Automates de formes normales et réductibilité inductive. In *Journées du pôle Contraintes et programmation logique*, pages 21–31, Rennes, nov 1997. PRC/GDR Programmation du CNRS.

[13] John Stillwell. The word problem and the isomorphism problem for groups. *Bulletin of the American Mathematical Society*, 6(1):33–56, January 1982.

Normal Forms and Reduction for Theories of Binary Relations

Dan Dougherty and Claudio Gutiérrez

Computer Science Group, Wesleyan University
Middletown CT 06459 USA
{ddougherty,cgutierrez}@wesleyan.edu

Abstract. We consider equational theories of binary relations, in a language expressing composition, converse, and lattice operations. We treat the equations valid in the standard model of sets and also define a hierarchy of equational axiomatisations stratifying the standard theory. By working directly with a presentation of relation-expressions as *graphs* we are able to define a notion of reduction which is confluent and strongly normalising, in sharp contrast to traditional treatments based on first-order terms. As consequences we obtain unique normal forms, decidability of the decision problem for equality for each theory. In particular we show a non-deterministic polynomial-time upper bound for the complexity of the decision problems.

1 Introduction

The theory of binary relations is a fundamental conceptual and methodological tool in computer science. The formal study of relations was central to early investigations of logic and the foundations of mathematics [11, 20, 24, 25, 26] and has more recently found application in program specification and derivation, [2, 6, 4, 18] denotational and axiomatic semantics of programs, [8, 10, 22, 19] and hardware design and verification [7, 16].

The collection of binary relations on a set has rich algebraic structure: it forms a monoid under composition, each relation has a converse, and it forms a Boolean algebra under the usual set-theoretic operations. In fact the equational theory in this language is undecidable, since it is possible to encode set theory [26]. Here we eliminate complementation as an operation, and investigate the set $E_\mathcal{R}$ of equations between relation-expressions valid when interpreted over sets, as well as certain equational axiomatic theories naturally derived from $E_\mathcal{R}$.

Now, the most popular framework for foundations and for implementations of theorem provers, proof-checkers, and programming languages remains the λ-calculus. It seems reasonable to say that this is due at least in part to the fact that the equational theory of λ-terms admits a computational treatment which is well-behaved: 'b-reduction is confluent, and terminating in typed calculi, so that the notion of *normal form* is central to the theory.

To our knowledge, no analogous notion of normal form for terms in $E_\mathcal{R}$ is known. In fact the calculus of relations has a reputation for being complex.

L. Bachmair (Ed.): RTA 2000, LNCS 1833, pp. 95–109, 2000.
© Springer-Verlag Berlin Heidelberg 2000

Bertrand Russell (quoted in [21]) viewed the classical results of Peirce and Schröder on relational calculus as being "difficult and complicated to so great a degree as to doubt their utility." And in their recent monograph [4, page 81] Bird and de Moor observe that "the calculus of relations has gained a good deal of notoriety for the apparently enormous number of operators and laws one has to memorise in order to do proofs effectively."

But in this paper we suggest that a rather attractive syntactic/computational treatment of the theory of relations is indeed available, at least for the fragment of the theory not including complementation.

The essential novelty derives from the idea of taking certain graphs as the representation of relations. These graphs, called here "diagrams," arise very naturally and have been used since Peirce by researchers in the relation community (e.g. Tarski, Lyndon, Jónsson, Maddux, etc.); recent formalisations appear in [12, 1, 7]. What we do here is to take graphs seriously as a notation alternative to first-order terms, i.e., to treat diagrams as first-class *syntactic* entities, and specifically as candidates for *rewriting*.

One can see diagram rewriting as an instance of a standard technique in automated deduction. It is well-known that certain equations inhibit classical term-rewriting techniques — the typical examples are associativity and commutativity — and that a useful response can be to pass to computing *modulo* these equations. In Table 1 we exhibit a set E_D of equations such that diagrams are the natural data structure for representing terms modulo E_D.

Summary of Results

It is not hard to see that in the absence of complementation equality between relation-expressions can be reduced to equality between expressions not involving union, essentially because union distributes over the other operations. So we ultimately restrict attention to the complement- and union-free fragment of the full signature (see Definition 1). It is known [1, 12] that the set of equations true in set-relation algebras in this signature is decidable.

We clarify the relationship between terms and diagrams by showing that the algebra of diagrams is precisely the free algebra for the set E_D of equations between terms. It is rather surprising that a finite set of equations accounts for precisely the identifications between terms induced by compiling them into diagrams.

Freyd and Scedrov [12] isolated the theory of *allegories*, a finitely axiomatisable subtheory of the theory of relations which corresponds to a certain geometrically-motivated restricted class of morphisms between diagrams. We refine this by constructing a proper hierarchy of equational theories, beginning with the theory of allegories, which stratifies the equational theory of set-relations.

Our main result is a computational treatment of diagrams via a notion of reduction. Actually each of the equational theories in the hierarchy induces its own reduction relation; but we prove *uniformly* that each reduction satisfies strong normalisation and Church-Rosser properties. Therefore each theory enjoys unique (diagram-) normal forms and decidability. In fact the decision problem for each theory is in NP, non-deterministic polynomial time.

We feel that the existence of computable unique normal forms is our most striking result. The virtue of treating diagrams as syntax is highlighted by the observation that $E_\mathcal{R}$ is not finitely axiomatisable [15], so no finite *term* rewriting system can even claim to correctly present the theory, much less be a convergent presentation.

In light of the characterisation of the set of diagrams as the free algebra for the set $E_\mathcal{D}$ of equations, these results can be seen — if one insists — as results about rewriting of terms modulo $E_\mathcal{D}$. But for us the diagram presentation is the primary one and is ultimately the closest to our intuition.

Related Work

The case for using a calculus of relations as a framework for concepts and methods in mathematics and computer science is compellingly made by Freyd and Scedrov in [12]. They define allegories as certain *categories*; the structures modeled in this paper are in that sense one-object allegories. One may view this as the distinction between typed and untyped calculi.

Bird and de Moor's book [4] is an extended presentation of the application of relational calculus to program specification and derivation, building explicitly on the theory of allegories. There, terms in relation calculus are not programs *per se*, but the authors do raise the question of how one might *execute* relation-expressions [4, page 110]. As noted there, a promising proposal is made by Lipton and Chapman in [18], where a notion of rewriting terms using the allegory axioms is presented. It should be very interesting to explore the relationship between the Lipton-Chapman model and the one presented here.

Brown and Hutton [7] apply relational methods to the problems of designing and verifying hardware circuits. They observe that people naturally reason informally about pictures of circuits and seek to provide formal basis, again based on allegories, for such reasoning; their vehicle is the relational language RUBY used to design hardware circuits. To our knowledge they do not claim decidability or normal forms for the theory they implement. An implementation of their method is distributed at [16].

Two other investigations of graphical relation-calculi are the work of Kahl [17] and that of Curtis and Lowe [9].

The general topic of diagrammatic reasoning has been attracting interest in several areas lately (see for example [3]). The present research might be viewed as a case-study in reasoning with diagrams in the general sense.

Further indication of the range of current investigations into relations and relation-calculi may be found in, for example, the books [23] or [5] or the proceedings of the roughly annual RelMiCS conferences.

2   Preliminaries

Definition 1. *The signature Σ is composed of the binary operations* intersection $\cap$ *and* composition ; *(usually written as concatenation), two unary operations* converse $(\ )^{\circ}$ *and* domain dom, *and a constant* 1.

When we exhibit terms, the composition is to be interpreted in "diagrammatic order" so that xy means "x first then y. The operation dom has a natural interpretation as the domain of a relation, and under this interpretation it is definable from the other operations as $\mathrm{dom}\,x = 1 \cap xx^{\circ}$. The inclusion of dom in the signature is non-traditional, but there is a very good technical reason for its inclusion, made clear in the remark following Theorem 2.

The standard models are sets and binary relations; the following definition is from [1].

Definition 2. *A* (subpositive) set relation algebra *is an algebra of the form* $\langle A, \cap, ; , (\ )^{\circ}, \mathrm{dom}, 1 \rangle$ *where A is a set of binary relations on some base set closed under the operations, which have the standard relational meaning (1 being the identity relation).*

By $\mathcal{R}$ we will denote the class of algebras isomorphic to set relation algebras and by $E_{\mathcal{R}}$ the set of equations valid in $\mathcal{R}$.

Definition 3. *A* undirected graph g *is a pair* (V_g, E_g) *of sets (vertices and edges) together with a map* $E_g \longrightarrow [V_g]^2$, *where the elements of* $[V_g]^2$ *are the 2-element multisets from V. Such a graph is* connected *if there is a path between any two vertices. An undirected graph h is a* minor *of g if h can be obtained from a subgraph g' of g by a sequence of contractions of vertices of g'.*

The notion of directed graph *is obtained by replacing* $[V]^2$ *by* $V \times V$ *in the definition above; note that any directed graph g obviously has an undirected graph underlying it. A directed graph g is* labelled *by a set X if there is a function* $l(g) : E_g \longrightarrow X$. *We will be interested in this paper in directed labelled graphs g with a distinguished* start *vertex s_g and a distinguished* finish *vertex f_g. We allow these to be the same vertex.*

For the sake of brevity the term *graph* will always mean: directed, labelled graph with distinguished start and finish vertices, whose underlying undirected graph is connected.

Let $\mathcal{G}$ denote the set of such graphs. Strictly speaking the set $\mathcal{G}$ depends on the particular set of labels chosen, but this set will never change in the course of our work, so we suppress mention of the label-set in the notation. We do assume that the set of labels is infinite.

A *morphism* φ between graphs g and h is a pair of functions $\varphi_V : V_g \longrightarrow V_h$ and $\varphi_E : E_g \longrightarrow E_h$ which

- preserves edges and direction, *i.e.*, for all $v, w \in V_g$, if e is an edge in g between v and w, then $\varphi_E(e)$ is an edge in h between $\varphi_V(v)$ and $\varphi_V(w)$,
- preserves labels, *i.e.*, for all $e \in E_g$, $l(e) = l(\varphi_E(e))$, and
- preserves start and finish vertices, *i.e.*, $\varphi_V(s_g) = s_h$ and $\varphi_V(f_g) = f_h$.

If it is clear from context, we will simply write φ instead of φ_V or φ_E.

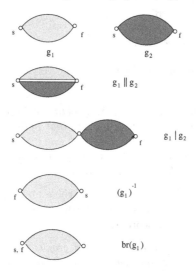

Fig. 1. The distinguished graphs 1, 2_a, and 2_a^{-1}.

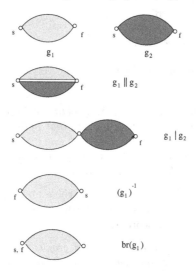

Fig. 2. Operations on graphs

2.1   Diagrams for Binary Relations

Here we introduce the central notion of *diagram* for a relational term. The set of diagrams supports an algebraic structure reflecting that of relations themselves, and a categorical structure in which morphisms between diagrams correspond to equations between relational terms valid in all set relation algebras. This material is standard and provides a foundation for our work in the rest of the paper.

There are some distinguished graphs in $\mathcal{G}$. The graph with only one vertex which is at the same time the start and finish, and no edges, will be denoted by 1. The graph with edge labelled a from the start vertex to the (distinct) finish vertex is denoted 2_a; the graph obtained by reversing the sense of the edge is $2_{a^{-1}}$. (See Figure 1.)

Definition 4. *Let g, g_1, g_2 be graphs in $\mathcal{G}$. We define the following operations in $\mathcal{G}$ (see Figure 2 for a graphical presentation.)*

1. *The* parallel composition, *$g_1 \| g_2$, is the graph obtained by (1) identifying the start vertices of the graphs g_1, g_2 (this is the new start), and (2) identifying the finish vertices of the graphs g_1, g_2 (the new finish).*
2. *The* sequential composition, *$g_1 | g_2$, is the graph obtained by identifying the finish of g_1 with the start of g_2, and defining the new start to be s_{g_1} and new finish to be f_{g_2}.*

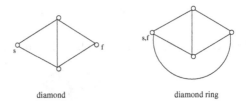

Fig. 3. Some graphs not in $\mathcal{D}$. Note the vertices s, f in each case.

3. *The converse of g, denoted by g^{-1}, is obtained from g by interchanging its start and finish. It is important to note that neither labels nor direction of edges changes.*

4. *The branching of g, denoted by $\mathrm{br}(g)$, is the graph obtained from g by redefining its finish to be the same as the start.*

Not every graph in $\mathcal{G}$ can be built using these operations. Figure 3 gives two examples (the edges in these pictures can be directed at will). Further significance of these graphs is given in Theorem 5.

Definition 5. *Let $\mathcal{D}$ denote the the set of graphs generated by 1, the 2_a, and the operations of branching and sequential and parallel composition.*

The set $\mathcal{D}$ is a Σ-algebra in a natural way; Theorem 2 below says more. $\mathcal{D}$ will play a key role in the normalisation process and has interesting properties in its own right.

Let $T_\Sigma(X)$ be the set of first-order terms over Σ with the labels X as variables. Then there is a surjective homomorphism

$$T_\Sigma(X) \longrightarrow \mathcal{D}, \quad t \mapsto g_t$$

defined recursively by $g_1 = 1$, $g_a = 2_a$ for $a \in X$, $g_{t_1;t_2} = g_{t_1} \mid g_{t_2}$, $g_{t_1 \cap t_2} = g_{t_1} \| g_{t_2}$, $g_{t^\circ} = (g_t)^{-1}$, and $g_{\mathrm{dom}\, t} = \mathrm{br}(g_t)$.

We can see the power of diagrams in the following important representation theorem. Recall that $\mathcal{R}$ denotes the class of algebras isomorphic to subpositive set relation algebras.

Theorem 1 ((Freyd-Scedrov, Andreka-Bredikhin)). *Let r, t be terms in the signature Σ. Then the equation $r = t$ is valid in $\mathcal{R}$ if and only if there are morphisms $g_r \longrightarrow g_t$ and $g_t \longrightarrow g_r$.*

Proof. The relationship between graphs in $\mathcal{D}$ and set relation algebras goes as follows. For an algebra $\mathcal{A}$ with base A and relations $R_1^\mathcal{A}, \ldots, R_n^\mathcal{A}$ of $\mathcal{A}$ define the graph $g_{\mathcal{A},\bar{R}}$ to have the set of vertices A and an edge (a, b) with label j for each $(a, b) \in R_j^\mathcal{A}$. Observe that $g_{\mathcal{A},\bar{R}}$ is not necessarily in $\mathcal{D}$. By an induction on terms it can be proved that for each term t in $T_\Sigma(R_1, \ldots, R_n)$ and elements $a, b \in A$ it holds $(a, b) \in t^\mathcal{A}[\bar{R}]$ if and only if there is a $\mathcal{G}$-morphism $g_t \longrightarrow g_{\mathcal{A},\bar{R}}$ which takes s to a and f to b.

The statement of the theorems follows now easily.   ///

The strength of Theorem 1 is to reduce equational reasoning in binary relations to reasoning about graph theoretical morphisms. In particular since diagrams are finite one can check whether or not there are morphisms between two given ones, so we have a decision procedure for equality (in $E_\mathcal{R}$).

This result can be improved in at least two directions from a computational point of view:

- Refine the morphisms in order to stratify the equations, hence possibly getting better computational tools for interesting fragments.
- Investigate rewrite systems and normal forms in this new representation.

We pursue these directions in the following two sections. In fact the developments are independent of one another, so that the reader interested primarily in diagram-rewriting can on a first reading proceed directly to Section 4.

3   Terms and Equations as Diagrams and Morphisms

Theorem 1 shows that morphisms between diagrams reflect equations valid in the theory of binary relations. Unfortunately it puts all these valid equations in one sack. Experience shows that certain equations appear more often than others in practice and are in some sense are more fundamental. Our program in this section is to classify equations by their operational meaning.

3.1   Equational Characterisation of $\mathcal{D}$

$$x1 = x$$
$$x(yz) = (xy)z$$
$$x \cap y = y \cap x$$
$$x \cap (y \cap z) = (x \cap y) \cap z$$
$$x^{\circ\circ} = x$$
$$(xy)^\circ = y^\circ x^\circ$$
$$(x \cap y)^\circ = x^\circ \cap y^\circ$$

$$1^\circ = 1$$
$$1 \cap 1 = 1$$
$$(1 \cap x)(1 \cap y) = (1 \cap x \cap y)$$
$$x \cap y(1 \cap z) = (x \cap y)(1 \cap z)$$
$$1 \cap x(y \cap z) = 1 \cap (x \cap y^\circ)z$$

$$\mathrm{dom}\, 1 = 1$$
$$(\mathrm{dom}\, x)^\circ = \mathrm{dom}\, x$$
$$\mathrm{dom}((x \cap y)z) = 1 \cap x(\mathrm{dom}\, z)y^\circ$$
$$\mathrm{dom}((\mathrm{dom}\, x)y) = (\mathrm{dom}\, x) \cap (\mathrm{dom}\, y)$$

Table 1. The equations $E_\mathcal{D}$.

To start with, the class $\mathcal{D}$ itself embodies certain equations in the sense that each graph in $\mathcal{D}$ can come from several different terms. It is interesting that these identifications can be axiomatised by a finite set of equations, $E_\mathcal{D}$, shown in Table 1. The equations in the left column capture the essential properties of

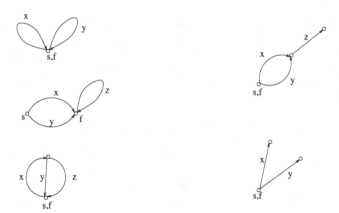

Fig. 4. Graphs representing some terms in the equations in Table 1. The ones in the left correspond to the last three equations about $1 \cap x$. The ones on the right to the last two equations about dom. Observe that in each of these equations the left- and right- hand side terms are represented by the same graph.

the operators (associativity, commutativity, and the involutive laws for converse), those in the upper right hand deal with identifications among terms of the form $1 \cap x$, and the rest take care of the identification of terms which contain the operator dom. All of these equations are trivially valid in $\mathcal{D}$: their left- and right-hand sides compile to the same diagram.

Theorem 2. $\mathcal{D}$ *is the free algebra over the set of labels for the set of equations* $E_{\mathcal{D}}$.

Proof. For the non-trivial direction we have to show that if r, t are not provably equal under $E_{\mathcal{D}}$, then $g_r \neq g_t$. This is done by proving that $E_{\mathcal{D}}$ can be completed by a finite number of equations into a confluent and terminating rewrite system modulo the equations for associativity of composition, AC of intersection and $1 \cap x(y \cap z) = 1 \cap (x \cap y^\circ)z$. A complete proof is in [14].   ///

3.2   Equations Capturing Morphisms

Freyd and Scedrov in [12] made the observation (without proof) that $\mathcal{D}$-morphisms which collapse at most two vertices at a time correspond to a simple and natural equational theory, an abstract theory of relations, the theory of *allegories*. Motivated by this idea we introduce a proper hierarchy of equational theories, stratifying $E_{\mathcal{R}}$ in terms of complexity of the morphisms acting on the data, each of which has a geometric as well as algebraic aspect.

Definition 6. *Let* $\varphi : g_1 \longrightarrow g_2$ *be an arrow in* $\mathcal{D}$. *We call* φ *an* n-*arrow if* $|V_{g_1}| \leq |V\varphi(g_1)| + n$.

$$x \cap x = x$$
$$x(y \cap z) = x(y \cap z) \cap xy$$
$$xy \cap z = (x \cap zy^{\circ})y \cap z.$$

Table 2. The operational equations

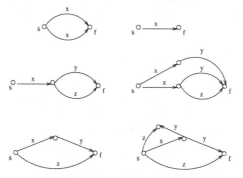

Fig. 5. The graph representations of the equations in Table 2.

Note that in general the composition of n-arrows is not an n-arrow. This motivates the following definition.

Definition 7. *Let $n \geq 0$ be a natural number.*

1. *The category $\mathcal{D}^n$ is the category whose objects are the graphs in $\mathcal{D}$, and whose arrows are finite compositions of n-arrows in $\mathcal{D}$.*
2. *The theory $E_{\mathcal{R}}^n$ is the set of equations $r = t$ between Σ-terms such that $g_r \longrightarrow g_t$ and $g_t \longrightarrow g_r$ in $\mathcal{D}^n$.*

We have the following chain of inclusions of categories: $\mathcal{D}^0 \subseteq \mathcal{D}^1 \subseteq \cdots \subseteq \mathcal{D}$. It can be shown that if $n \geq 1$ then $E_{\mathcal{R}}^n$ is closed under deduction. So we have a hierarchy of equational theories $E_{\mathcal{R}}^0 \subseteq E_{\mathcal{R}}^1 \subseteq \cdots \subseteq \bigcup_i E_{\mathcal{R}}^i$.

Theorem 1 can now be rephrased as $E_{\mathcal{R}} = \bigcup_i E_{\mathcal{R}}^i$.

The equational theory of *allegories* is presented by the axioms in the left column of Table 1 plus the ones shown in Table 2. These last three equations correspond (in the sense of Theorem 1) to 1-arrows over graphs in $\mathcal{D}$, as can be checked from the graphical representation in Figure 5. The next theorem formalises the converse statement i.e., that they are sufficient to axiomatise the equations obtained from 1-arrows.

Theorem 3. $E_{\mathcal{R}}^1$ *is exactly the equational theory of allegories.*

Proof. The proof is delicate and we do not present it here; a complete proof can be found in [14].   ///

Theorem 4. *For each $n \geq 2$ it holds $E_{\mathcal{R}}^{2n-1} \subset E_{\mathcal{R}}^{2n}$ (the inclusion is proper).*

Proof. An adaptation of the technique in [12, 2.158] showing that $E_{\mathcal{R}}$ is not finitely axiomatisable.   ///

Unfortunately we do not know yet know much more about each of the steps in the hierarchy. But we conjecture that $E_{\mathcal{R}}^1 = E_{\mathcal{R}}^2 = E_{\mathcal{R}}^3$. We also conjecture that for every $n \geq 2$, the equational theory $E_{\mathcal{R}}^n$ is finitely axiomatisable. We do not know the answer to the question: "is $E_{\mathcal{R}}^{2n} = E_{\mathcal{R}}^{2n+1}$ for $n \geq 1$?"

4   Normalisation

This section presents a very general combinatorial lemma concerning the set of functions over finite structures viewed as an abstract reduction system. It is most conveniently presented using the language of categories, but no more than rudimentary category theory is required for the presentation. The material in this section is condensed from [13].

In this section juxtaposition denotes composition of arrows in a category, and is to be read in the standard way, so that fg means "g first." We use $A \cong B$ to indicate that A and B are isomorphic.

Definition 8. *Let C be any category, and A, B objects of C. Define the relation $\rightleftarrows$ between objects of C as follows:*

$A \rightleftarrows B$ *if and only if there are arrows $A \longrightarrow B$ and $B \longrightarrow A$.*

Clearly $\rightleftarrows$ is an equivalence relation. Our goal is to find simple conditions which make the relationship $\rightleftarrows$ decidable.

The following notion is motivated by the observation that a (set-theoretic) function f between sets A and B can be seen as an map onto its image $f(A)$ followed by the inclusion of $f(A)$ into B.

Definition 9. *An arrow m is* mono *if whenever $ma = mb$ then $a = b$. An arrow e is* epi *if whenever $ae = be$ then $a = b$.*

An arrow $A \xrightarrow{f} B$ has an epi-mono factorisation if there exist arrows e epi and m mono such that $f = me$. A category C has epi-mono factorisation if every arrow in $\mathcal{A}$ has such a factorisation.

Definition 10. *An object A is* hom-finite *if the set $Hom(A, A)$ of maps from A to A is finite. A category C is* hom-finite *if each object of C is hom-finite.*

Of course any concrete category of sets of finite objects will be hom-finite. In particular, the categories $\mathcal{G}$, $\mathcal{D}$ and $\mathcal{D}^n$ for each n are each hom-finite.

Lemma 1. *Suppose that A is hom-finite. If $m : A \longrightarrow A$ is a monomorphism, then m is an isomorphism. Also, If there are monomorphisms $m_1 : A \longrightarrow B$ and $m_2 : B \longrightarrow A$, then A and B are isomorphic.*

Proof. For the first assertion: consider the monomorphisms $m, m^2, \cdots : A \longrightarrow A$. Because A is hom-finite, there are integers $i < j$ such that $m^i = m^j$. Now, using the fact that m is mono, $1_A = m^k$ for $k = j - i$. Thus m is a monomorphism with a right inverse; this implies that m is an isomorphism.

For the second claim: we have that $m_2 m_1 : A \longrightarrow A$ is mono, hence, by the first part it is an isomorphism. Hence there exists f with $m_2 m_1 f = 1_A$. Thus m_2 is a monomorphism with a right inverse; this implies that it is an isomorphism.

$///$

Definition 11. *Let A, B be objects in a category $\mathcal{C}$. Define $A \Longrightarrow B$ if and only if B is both (the target of) a quotient- and (the source of) a sub-object of A; that is, there is an epimorphism e and a monomorphism m such that*

$$A \xrightarrow{\ e\ } B \xrightarrow{\ m\ } A.$$

We also require that e not be an isomorphism.

By $\stackrel{*}{\Longrightarrow}$ we will denote the reflexive-transitive closure of $\Longrightarrow$, where reflexivity is defined up to isomorphism. Thus $A \stackrel{*}{\Longrightarrow} B$ means either $A \cong B$ or there is a finite sequence $A \Longrightarrow C_1 \Longrightarrow \cdots \Longrightarrow C_n \Longrightarrow B$.

Lemma 2. *If A is hom-finite then there are no infinite $\Longrightarrow$-reductions out of A.*

Proof. For sake of contradiction suppose there were such a sequence. For each i we have maps $A_i \xrightarrow{e_i} A_{i+1} \xrightarrow{m_i} A_i$, and so we may define $a_i = m_1 \cdots m_i e_i \cdots e_1 : A \to A$. Since A is hom-finite there are $i < j$ with $a_i = a_j$. Cancelling the monos on the left and the epis on the right, we have $1_A = m_{i+1} \cdots m_j e_j \cdots e_{i+1}$. This implies that e_{i+1} is iso, a contradiction. $///$

Observe that $A \stackrel{*}{\Longleftrightarrow} B$ implies that $A \rightleftarrows B$. The converse need not be true in general, but the next result provides a strong converse in certain categories.

Proposition 1. *Suppose $\mathcal{C}$ is a hom-finite category with epi-mono factorisation. Then if $A \rightleftarrows B$, then there exists C such that $A \stackrel{*}{\Longrightarrow} C \stackrel{*}{\Longleftarrow} B$.*

Proof. The proof is by Noetherian induction over $\Longrightarrow$, out of the multiset $\{A, B\}$. We are given $A \xrightarrow{f} B \xrightarrow{g} A$. If both f and g are mono then by Lemma 1 A and B are isomorphic and we may take C to be A. Otherwise, by symmetry we may suppose f is not mono without loss of generality. Factor the arrow gf as epi-mono, obtaining: $A \xrightarrow{e} X \xrightarrow{m} A$. Now, e is not mono, otherwise gf would be, contradicting the assumption that f is not mono. In particular e is not iso, and so $A \Longrightarrow X$. Since $X \rightleftarrows B$ we may apply the induction hypothesis to $\{X, B\}$, obtaining C with $A \Longrightarrow X \stackrel{*}{\Longrightarrow} C \stackrel{*}{\Longleftarrow} B$ as desired. $///$

The previous results imply that the relation $\Longrightarrow$ is a terminating and confluent abstract reduction system capturing the equivalence relation $\rightleftarrows$:

Corollary 1 ((Normal Forms for $\rightleftarrows$)). *Suppose $\mathcal{C}$ is a hom-finite category with epi-mono factorisation.*

If $A \rightleftarrows B$, then there is a C, unique up to isomorphism, such that $A \overset{*}{\Longrightarrow} C$ and $B \overset{*}{\Longrightarrow} C$ and C is $\Longrightarrow$-irreducible.

Proof. By Lemma 2 we may let C and C' be any $\Longrightarrow$-irreducible objects such that $A \overset{*}{\Longrightarrow} C$ and $B \overset{*}{\Longrightarrow} C'$ respectively. Then $C \rightleftarrows C'$. But Proposition 1 and the irreducibility of C and C' imply that C and C' are isomorphic.   ///

Observe that in the preceding Corollary we have $A \rightleftarrows C$ and $B \rightleftarrows C$. Also note that by taking B to be A we may conclude that for each A there is a C, unique up to isomorphism such that $A \overset{*}{\Longrightarrow} C$ and C is $\Longrightarrow$-irreducible. We refer to such a C as a "$\Longrightarrow$-normal form for A".

5   Normal Forms for Diagrams

We want to apply the results of the previous section to the categories $\mathcal{D}$ and $\mathcal{D}^n$. The following facts about $\mathcal{D}$ and each $\mathcal{D}^n$ are easy to check: (i) a map is epi if and only if it is surjective on vertices and on edges, and (ii) a map is an isomorphism if it is bijective on vertices and on edges. The next result is deeper; a proof can be found in [14].

Theorem 5. *Let $g \in \mathcal{G}$. Then g is in $\mathcal{D}$ if and only if the underlying undirected graph of g does not have diamond and diamond ring (see Figure 3) as minors.*

In particular, the set of graphs $\mathcal{D}$ is closed under the formation of subgraphs, i.e., if $g \in \mathcal{D}$ and h is a connected subgraph containing s_g and f_g then $h \in \mathcal{D}$.

Observe that the categories $\mathcal{D}^n$ have the same objects as $\mathcal{D}$ so the above theorem applies immediately to the $\mathcal{D}^n$. Theorem 5 is crucial in verifying that $\mathcal{D}^n$ supports the techniques of the previous section.

Proposition 2. *The categories $\mathcal{D}$ and each $\mathcal{D}^n$ are hom-finite and has epi-mono factorisation.*

Proof. The first assertion is easy to see. For the second, let $\varphi : g_1 \longrightarrow g_2$ be an arrow in $\mathcal{D}^n$. Then the graph $\varphi(g_1)$ is a subgraph of $g_2 \in \mathcal{D}$, hence by Theorem 5 it is also in $\mathcal{D}$. So we have $g_1 \overset{\varphi'}{\longrightarrow} \varphi(g_1) \overset{i}{\longrightarrow} g_2$, where φ' and i are epi and mono respectively.   ///

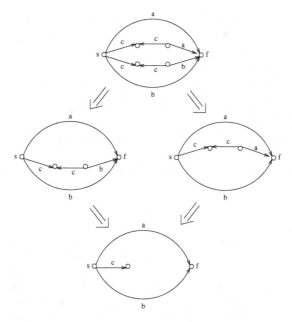

Fig. 6. A graph in $\mathcal{D}$ (top) and possible reductions. Observe that the reduction of the graphs in the middle to a common graph (bottom) depends on the existence of the operator br (dom).

Remark. It is precisely here that we see the benefit of our extended signature. The class of diagrams built over the traditional signature — without dom — does **not** have epi-mono factorisation, due to the fact that it is not closed under subgraphs. Figure 6 shows this by example.

We can now present our main results.

Theorem 6. *Let $\mathcal{C}$ be either $\mathcal{D}$ or one of the $\mathcal{D}^n$.*

If $g \rightleftarrows h$ then there is a k, unique up to isomorphism, such that $g \Longrightarrow k$ and $h \Longrightarrow k$ and k is $\Longrightarrow$-irreducible.

For each graph g, there is a graph $\mathrm{nf}(g)$ such that $\mathrm{nf}(g)$ is $\Longrightarrow$-irreducible and such that for any h, $g \rightleftarrows h$ in $\mathcal{C}$ if and only if $\mathrm{nf}(g) \cong \mathrm{nf}(h)$. The graph $\mathrm{nf}(g)$ is unique up to isomorphism.

Proof. This is an immediate consequence of Corollary 1 and Proposition 2. ///

It is important to note that the notions $\rightleftarrows$ and $\Longrightarrow$ in the previous Theorem are taken relative to the category ($\mathcal{D}$ or $\mathcal{D}^n$) one has chosen to work in.

Theorem 7. *For $\mathcal{D}$ and for each $\mathcal{D}^n$ the relation $\rightleftarrows$ is decidable in non-deterministic polynomial time. Each theory $E_{\mathcal{R}}^n$ is decidable in non-deterministic polynomial time.*

Proof. Decidability follows from the previous results and the fact that $\Longrightarrow$ is computable. To get the NP upper bound we examine the complexity of reduction to normal form. If $A \Longrightarrow A'$ then the sum of the number of vertices and edges of A must exceed that of A', since epimorphisms are surjections and bijections are isomorphisms (and we know the map from A to A' is not an isomorphism by definition). So any reduction of a diagram A to normal form takes a number of steps bounded by the size of A. So to test whether $A \rightleftarrows B$ we can generate sequences of morphisms reducing each of A and B — not necessarily to normal form — and test that the results are isomorphic. The latter test is of course itself in NP.

The second assertion follows immediately from the definition of $E_{\mathcal{R}}^n$.   ///

6   Conclusion

We have examined the equational theory $E_{\mathcal{R}}$ of binary relations over sets and a family $E_{\mathcal{R}}^n$ of approximations to this theory. The theory $E_{\mathcal{R}}^1$ is Freyd and Scedrov's theory of allegories. By working with a natural notion of diagram for a relation-expression we have defined a notion of reduction of a diagram which yields an analysis of the theories above. A surprisingly important aspect was the inclusion of a "domain" operator in the signature: the corresponding operation is definable in terms of the traditional operations, but the class of diagrams for the enriched signature has better closure properties.

Since each notion of reduction of diagrams is terminating and confluent, we may compute unique normal forms for each of the theories. Each theory is therefore decidable and in fact normal forms can be computed in non-deterministic polynomial time.

The decidability of $E_{\mathcal{R}}^n$ is reminiscent of the decidability of $E_{\mathcal{R}}$, but has been more difficult to establish. This is because although equality in $E_{\mathcal{R}}$ is witnessed by any pair of graph morphisms between diagrams, equality in $E_{\mathcal{R}}^n$ is witnessed by a *sequence* of restricted morphisms. The length of this sequence could be bounded only after our work relating $\rightleftarrows$ and $\Longrightarrow$.

References

[1] H. Andréka and D.A. Bredikhin. The equational theory of union-free algebras of relations. *Algebra Universalis*, 33:516–532, 1995.

[2] R. C. Backhouse and P. F. Hoogendijk. Elements of a relational theory of datatypes. In B. Möller, H. Partsch, and S. Schuman, editors, *Formal Program Development*, volume 755 of *Lecture Notes in Computer Science*, pages 7–42. Springer-Verlag, 1993.

[3] J. Barwise and G. Allwein, editors. *Logical Reasoning with Diagrams*. Oxford University Press, 1996.

[4] R. Bird and O. de Moor. *Algebra of Programming*. Prentice Hall, 1997.

[5] C. Brink, W. Kahl, and G. Schmidt, editors. *Relational Methods in Computer Science*. Advances in Computing. Springer-Verlag, Wien, New York, 1997. ISBN 3-211-82971-7.

[6] P. Broome and J. Lipton. Combinatory logic programming: computing in relation calculi. In M. Bruynooghe, editor, *Logic Programming*. MIT Press, 1994.

[7] C. Brown and G. Hutton. Categories, allegories and circuit design. In *Logic in Computer Science*, pages 372–381. IEEE Computer Society Press, 1994.

[8] C.A.R.Hoare. An axiomatic basis for computer programming. *CACM*, 12(10):576–583, 1969.

[9] S Curtis and G Lowe. Proofs with graphs. *Science of Computer Programming*, 26:197–216, 1996.

[10] J. W. De Bakker and W. P. De Roever. A calculus for recursive program schemes. In M. Nivat, editor, *Automata, Languages and Programming*, pages 167–196. North-Holland, 1973.

[11] A. De Morgan. On the syllogism, no. IV, and on the logic of relations. *Transactions of the Cambridge Philosophical Society*, 10:331–358, 1860.

[12] P. Freyd and A. Scedrov. *Categories and Allegories*. North-Holland Mathematical Library, Vol. 39. North-Holland, 1990.

[13] C. Gutiérrez. Normal forms for connectedness in categories. *Annals of pure and applied logic*. Special issue devoted to the XI Simposio Latinoamericano de Logica Matematica, Venezuela, July 1998. To appear.

[14] C. Gutiérrez. *The Arithmetic and Geometry of Allegories: normal forms and complexity of a fragment of the theory of relations*. PhD thesis, Wesleyan University, 1999.

[15] M. Haiman. Arguesian lattics which are not linear. *Bull. Amer. Math. Soc.*, 16:121–123, 1987.

[16] G. Hutton, E. Meijer, and E. Voermans. A tool for relational programmers. Distributed on the mathematics of programming electronic mailing list, January 1994. http://www.cs.nott.ac.uk/~gmh/allegories.gs, 1994.

[17] Wolfram Kahl. Relational matching for graphical calculi of relations. *Journal of Information Science*, 119(3–4):253–273, December 1999.

[18] J. Lipton and E. Chapman. Some notes on logic programming with a relational machine. In Ali Jaoua, Peter Kempf, and Gunther Schmidt, editors, *Relational Methods in Computer Science*, Technical Report Nr. 1998-03, pages 1–34. Fakultät für Informatik, Universität der Bundeswehr München, July 1998.

[19] R.D. Maddux. Relation-algebraic semantics. *Theoretical Computer Science*, 160:1–85, 1996.

[20] C. S. Peirce. *Collected Papers*. Harvard University Press, 1933.

[21] V.R. Pratt. *Origins of the calculus of binary relations*, pages 248–254. IEEE Computer Soc. Press, 1992.

[22] J. G. Sanderson. *A Relational Theory of Computing*, volume 82 of *Lecture Notes in Computer Science*. Springer-Verlag, 1980.

[23] G. W. Schmidt and T. Ströhlein. *Relations and Graphs: Discrete Mathematics for Computer Scientists*. EATCS Monographs on Theoretical Computer Science. Springer-Verlag, 1993.

[24] E. Schröder. *Vorlesungen über der Algebra der Logik, Vol. 3, "Algebra und Logik der Relative"*. 1895.

[25] A. Tarski. On the calculus of relations. *Journal of Symbolic Logic*, 6(3):73–89, 1941.

[26] A. Tarski and S. Givant. *A formalization of set theory without variables*. AMS Colloquium Publications, Vol. 41. American Mathematical Society, 1988.

Parallelism Constraints

Katrin Erk* and Joachim Niehren**

Programming Systems Lab, Universität des Saarlandes, Saarbrücken, Germany
http://www.ps.uni-sb.de/~{erk,niehren}

Abstract. Parallelism constraints are logical descriptions of trees. They
are as expressive as context unification, i.e. second-order linear unification. We present a semi-decision procedure enumerating all "most general
unifiers" of a parallelism constraint and prove it sound and complete.
In contrast to all known procedures for context unification, the presented procedure terminates for the important fragment of dominance
constraints and performs reasonably well in a recent application to underspecified natural language semantics.

1 Introduction

Parallelism constraints [7, 16] are logical descriptions of trees. They are equal
in expressive power to context unification [4], a variant of linear second-order
unification [13, 18]. The decidability of context unification is a prominent open
problem [20] even though several fragments are known decidable [22, 21, 4].

Parallelism constraints state relations between the nodes of a tree: mother-of, sibling-
of and labeling, dominance (ancestor-of), disjointness, inequality, and parallelism. Parallelism $\pi_1/\pi_2{\sim}\pi_3/\pi_4$, as illustrated in Figure
1, holds in a tree if the structure of the tree
between the nodes π_1 and π_2 — i.e., the tree
below π_1 minus the tree below π_2 — is isomorphic to that between π_3 and π_4.

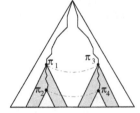

Fig. 1. Parallelism
$\pi_1/\pi_2{\sim}\pi_3/\pi_4$

Parallelism constraints differ from context
unification in their perspective on trees. They
view trees from inside, talking about the *nodes* of a single tree, rather than from
the outside, talking about relations between several *trees*. This difference has important consequences. First, it is not only a difference of nodes versus trees but
also one of occurrences versus structure. Second, different decidable fragments
can be distinguished for parallelism constraints and context unification. Third,
different algorithms can be devised. For instance, the language of *dominance
constraints* [15, 24, 1, 9] is a decidable fragment of parallelism constraints for

---

* Supported by the DFG through the Graduiertenkolleg Kognition in Saarbrücken.
** Supported through the Collaborative Research Center (SFB) 378 of the DFG, the
Esprit Working Group CCL II (EP 22457), and the Procope project of the DAAD.

L. Bachmair (Ed.): RTA 2000, LNCS 1833, pp. 110–126, 2000.

which powerful solver exist [6, 5, 16]. But when encoded into context unification, dominance constraints are not subsumed by any of the decidable fragments mentioned above, not even by subtree constraints [23], although they look similar. The difference is again that dominance constraints speak about occurrences of subtrees whereas subtree constraints speak about their structure.

Parallelism constraints form the backbone of a recent underspecified analysis of natural language semantics [7, 11]. This analysis uses the fragment of *dominance constraints* to describe scope ambiguities in a similar fashion as [19, 2], while the full expressivity of parallelism is needed for modeling ellipsis. An earlier treatment of semantic underspecification [17] was based directly on context unification. The implementation used an incomplete procedure [10] which guesses trees top-down by imitation and projection, leaving out flex-flex. This procedure performs well on the parallelism phenomena encountered in ellipsis resolution, but when dealing with scope ambiguities, it consistently runs into combinatoric explosion. To put it differently, this procedure does not perform well enough on the context unification equivalent of dominance constraints.

In this paper, we propose a new semi-decision procedure for parallelism constraints built on top of a powerful, terminating solver for dominance constraints. We prove our procedure sound and complete: We define the notion of a *minimal solved form* for parallelism constraints, which plays the same role as *most general unifiers* in unification theory. We then show that our procedure enumerates all minimal solved forms of a given parallelism constraint.

Plan of the Paper. In the following section, we describe the syntax and semantics of dominance and parallelism constraints. Section 3 presents an algorithm for dominance constraints which in section 4 is extended to a semi-decision procedure for parallelism constraints. In sections 5 and 6 we sketch a proof of soundness and completeness. Section 7 further extends the semi-decision procedure by optional rules, which yield stronger propagation speeding up a first prototype implementation. Section 8 concludes. Many proofs are omitted for lack of space but can be found in an extended version [8].

2   Syntax and Semantics

Semantics. We assume a signature Σ of function symbols ranged over by $f, g, \ldots$, each of which is equipped with an arity $\mathsf{ar}(f) \geq 0$. Constants are function symbols of arity 0 denoted by a, b. We further assume that Σ contains at least one constant and a symbol of arity at least 2.

A (finite) *tree* τ is a ground term over Σ, for instance $f(g(a,a))$. A *node* of a tree can be identified with its *path* from the root down, expressed by a word over $\mathbb{N}_+$, the set of natural numbers excluding 0. We write ε for the empty path and $\pi_1\pi_2$ for the concatenation of π_1 and π_2. A path π is a prefix of a path π' if there exists some (possibly empty) π'' such that $\pi\pi'' = \pi'$.

Fig. 2.
$f(g(a,a))$

A tree can be characterized uniquely by a tree domain (the set of its paths) and a labeling function. A *tree domain* D is a finite nonempty prefix-closed set of paths. A path $\pi i \in D$ is the i-th child of the node/path $\pi \in D$. A *labeling function* is a function $L : D \to \Sigma$ fulfilling the condition that for every $\pi \in D$ and $k \geq 1$, $\pi k \in D$ iff $k \leq \mathsf{ar}(L(\pi))$. We write D_τ for the domain of a tree τ and L_τ for its labeling function. For instance, the tree $\tau = f(g(a,a))$ displayed in Fig. 2 satisfies $D_\tau = \{\varepsilon, 1, 11, 12\}$, $L_\tau(\varepsilon) = f$, $L_\tau(1) = g$, and $L_\tau(11) = a = L_\tau(12)$.

Definition 1. *The tree structure $\mathcal{M}^\tau$ of a tree τ is a first-order structure with domain D_τ. It provides a labeling relation $:f^\tau \subseteq D_\tau^{\mathsf{ar}(f)+1}$ for each $f \in \Sigma$:*

$$:f^\tau = \{(\pi, \pi 1, \ldots, \pi n) \mid L_\tau(\pi) = f, \mathsf{ar}(f) = n\}$$

We write $\mathcal{M}^\tau \models \pi{:}f(\pi_1, \ldots, \pi_n)$ for $(\pi, \pi_1, \ldots, \pi_n) \in :f^\tau$; this relation states that node π of τ is labeled by f and has π_i as its i-th child (for $1 \leq i \leq n$). Every tree structure $\mathcal{M}^\tau$ can be extended conservatively by relations for dominance, disjointness, and parallelism. *Dominance* is the prefix relation between paths $\pi \triangleleft^* \pi'$; restricted to D_τ, it is the ancestor relation of τ; we write $\pi \triangleleft^+ \pi'$ if $\pi \triangleleft^* \pi'$ and $\pi \neq \pi'$. *Disjointness* $\pi \perp \pi'$ holds if neither $\pi \triangleleft^* \pi'$ nor $\pi' \triangleleft^* \pi$. Concerning parallelism, let $\mathsf{betw}_\tau(\pi_1, \pi_2)$ be the set of nodes in the substructure of τ between π_1 and π_2: If $\pi_1 \triangleleft^* \pi_2$ holds in $\mathcal{M}^\tau$, we define

$$\mathsf{betw}_\tau(\pi_1, \pi_2) = \{\pi \in D_\tau \mid \pi_1 \triangleleft^* \pi \text{ but not } \pi_2 \triangleleft^+ \pi\}.$$

The node π_2 plays a special role: it is part of the substructure of τ between π_1 and π_2, but its label is not. This is expressed in Def. 2, which is illustrated in Fig. 1.

Definition 2. *Parallelism* $\mathcal{M}^\tau \models \pi_1/\pi_2 {\sim} \pi_3/\pi_4$ *holds iff* $\pi_1 \triangleleft^* \pi_2$ *and* $\pi_3 \triangleleft^* \pi_4$ *are valid in $\mathcal{M}^\tau$ and there exists a correspondence function* $c : \mathsf{betw}_\tau(\pi_1, \pi_2) \to \mathsf{betw}_\tau(\pi_3, \pi_4)$, *a bijective function which satisfies* $c(\pi_1) = \pi_3$ *and* $c(\pi_2) = \pi_4$ *and preserves the tree structure of $\mathcal{M}^\tau$, i.e. for all* $\pi \in \mathsf{betw}_\tau(\pi_1, \pi_2) - \{\pi_2\}$, $f \in \Sigma$, *and* $n = \mathsf{ar}(f)$:

$$\mathcal{M}^\tau \models \pi{:}f(\pi 1, \ldots, \pi n) \quad \textit{iff} \quad \mathcal{M}^\tau \models c(\pi){:}f(c(\pi 1), \ldots, c(\pi n))$$

Lemma 1. *If* $c : \mathsf{betw}_\tau(\pi_1, \pi_2) \to \mathsf{betw}_\tau(\pi_3, \pi_4)$ *is a correspondence function, then* $c(\pi_1\pi) = \pi_3\pi$ *for all* $\pi_1\pi \in \mathsf{betw}_\tau(\pi_1, \pi_2)$.

Syntax. We assume an infinite set $\mathcal{V}$ of (node) variables ranged over by X, Y, Z, U, V, W. A *(parallelism) constraint* ϕ is a conjunction of *atomic constraints* or *literals* for parallelism, dominance, labeling, disjointness, and inequality. A *dominance constraint* is a constraint without parallelism literals. The abstract syntax of parallelism constraints is defined as follows:

$$\varphi, \psi ::= X_1/X_2{\sim}Y_1/Y_2 \mid X{\lhd}^*Y \mid X{:}f(X_1, \ldots, X_n) \qquad (\mathrm{ar}(f) = n)$$
$$\mid \quad X{\perp}Y \mid X{\neq}Y \mid \textbf{false} \mid \varphi \wedge \psi$$

Abbreviations: $X{=}Y$ for $X{\lhd}^*Y \wedge Y{\lhd}^*X$ and $X{\lhd}^+Y$ for $X{\lhd}^*Y \wedge X{\neq}Y$

For simplicity, we view parallelism, inequality, and disjointness literals as symmetric. We also write XRY, where $R \in \{{\lhd}^*, {\lhd}^+, \perp, \neq, =\}$. A richer set of relations could be used, as proposed in [6], but this would complicate matters slightly. For a comparison to context unification, we refer to [16]. An example for the simpler case of string unification is given below (see Figure 4).

First order formulas Φ built from constraints and the usual logical connectives are interpreted over the class of tree structures in the usual Tarskian way. We write $\mathcal{V}(\Phi)$ for the set of variables occurring in Φ. If a pair $(\mathcal{M}^\tau, \alpha)$ of a tree structure $\mathcal{M}^\tau$ and a variable assignment $\alpha : \mathcal{G} \to D_\tau$, for some set $\mathcal{G} \supseteq \mathcal{V}(\Phi)$, satisfies Φ, we write this as $(\mathcal{M}^\tau, \alpha) \models \Phi$ and say that $(\mathcal{M}^\tau, \alpha)$ is a *solution* of Φ. We say that Φ is *satisfiable* iff it possesses a solution. Entailment $\Phi \models \Phi'$ means that all solutions of Φ are also solutions of Φ'.

We often draw constraints as graphs with the nodes representing variables; a labeled variable is connected to its children by solid lines, while a dotted line represents dominance. For example, the graph for $X{:}f(X_1, X_2) \wedge X_1{\lhd}^*Y \wedge X_2{\lhd}^*Y$ is displayed in Fig. 3. As trees do not branch upwards, this constraint is unsatisfiable.

Fig. 3. An unsatisfiable constraint

Parallelism literals are shown graphically as well as textually: the square brackets in Fig. 4 illustrate the parallelism literal written beside the graph. This graph encodes the string unification [14] problem $gx = xg$; the two brackets represent the two occurrences of x. Disjointness and inequality literals are not represented graphically.

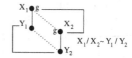

Fig. 4. String unification

3 Solving Dominance Constraints

Our semi-decision procedure for parallelism constraints consists of two parts: a terminating dominance constraint solver, and a part dealing with parallelism proper. Having our procedure terminate for general dominance constraints and perform well for dominance constraints in linguistic applications was an important design requirement for us.

In this section, we present the first part of our procedure, the solver for dominance constraints. This solver, which is similar to the algorithms in [12, 6] and could in principle be replaced by them, terminates in non-deterministic polynomial time. Actually, satisfiability of dominance constraints is NP-complete [12]. Boolean satisfiability is encoded by forcing graph fragments to "overlap" and making the algorithm choose between different possible overlappings. For instance, the constraint to the right entails $X=Y \lor X=Y_1$. The solver is intended to perform well in cases without overlap, where distinct variables denote distinct values. This can typically be assumed in linguistic applications.

Fig. 5. Overlap

We organize all procedures in this paper as *saturation algorithms*. A saturation algorithm consists of a set of *saturation rules*, each of which has the form $\varphi \to \bigvee_{i=1}^n \varphi_i$ for some $n \geq 1$. A rule is a *propagation rule* if $n = 1$, and a *distribution rule* otherwise. The only critical rules with respect to termination are those which introduce fresh variables on their right hand side. A rule $\varphi \to \Phi$ is *correct* if $\varphi \models \exists V \Phi$ where $V = \mathcal{V}(\Phi) - \mathcal{V}(\varphi)$.

By a slight abuse of notation, we identify a constraint with the *set* of its literals. This way, subset inclusion defines a partial ordering $\subseteq$ on constraints; we also write $=^{set}$ for the corresponding equality $\subseteq \cap \supseteq$, and $\subset$ for the strict variant $\subseteq \cap \neq^{set}$. This way, we can define saturation for a set S of saturation rules as follows: We assume that each rule $\rho \in S$ comes with an application condition $C_\rho(\varphi)$ deciding whether ρ can be applied to φ or not. A *saturation step* $\to_S$ consists of one application of a rule in S:

$$\frac{\varphi' \subseteq \varphi \qquad \rho \in S}{\varphi \to_S \varphi \land \varphi_i} \text{ if } C_\rho(\varphi) \text{ where } \rho \text{ is } \varphi' \to \bigvee_{i=1}^n \varphi_i$$

For this section, we let $C_{\varphi' \to \bigvee_{i=1}^n \varphi_i}(\varphi)$ be true iff $\varphi_i \not\subseteq \varphi$ for all $1 \leq i \leq n$. We call a constraint *S-saturated* if it is irreducible with respect to $\to_S$ and *clash-free* if it does not contain **false**. We also say that a constraint is in *S-solved form* if it is *S*-saturated and clash-free.

Figure 6 contains schemata for saturation rules that together solve dominance constraints. Let D be the (infinite) set of instances of these schemata. Both clash schemata are obvious. Next, there are standard schemata for reflexivity, transitivity, decomposition, and inequality. Schema (D.Lab.Dom) declares that a parent dominates its children.

We illustrate the remaining schemata of propagation rules by an example: We reconsider the unsatisfiable constraint $X{:}f(X_1, X_2) \land X_1 \vartriangleleft^* Y \land X_2 \vartriangleleft^* Y$ of Fig. 3. By (D.Lab.Disj), we infer $X_1 \bot X_2$, from which (D.Prop.Disj) yields $Y \bot Y$, which then clashes by (D.Clash.Disj).

Propagation rules:

(D.Clash.Ineq) $X{=}Y \land X{\neq}Y \to$ **false**

(D.Clash.Disj) $X \perp X \to$ **false**

(D.Dom.Refl) $\varphi \to X \triangleleft^* X$ where $X \in \mathcal{V}(\varphi)$

(D.Dom.Trans) $X \triangleleft^* Y \land Y \triangleleft^* Z \to X \triangleleft^* Z$

(D.Eq.Decom) $X{:}f(X_1, \ldots, X_n) \land Y{:}f(Y_1, \ldots, Y_n) \land X{=}Y \to \wedge_{i=1}^n X_i{=}Y_i$

(D.Lab.Ineq) $X{:}f(\ldots) \land Y{:}g(\ldots) \to X{\neq}Y$ where $f \neq g$

(D.Lab.Disj) $X{:}f(\ldots X_i, \ldots, X_j, \ldots) \to X_i \perp X_j$ for $1 \leq i < j \leq n$

(D.Prop.Disj) $X \perp Y \land X \triangleleft^* X' \land Y \triangleleft^* Y' \to Y' \perp X'$

(D.Lab.Dom) $X{:}f(\ldots, Y, \ldots) \to X \triangleleft^+ Y$

Distribution rules:

(D.Distr.NotDisj) $X \triangleleft^* Z \land Y \triangleleft^* Z \to X \triangleleft^* Y \lor Y \triangleleft^* X$

(D.Distr.Child) $X \triangleleft^* Y \land X{:}f(X_1, \ldots, X_n) \to Y{=}X \lor \bigvee_{i=1}^n X_i \triangleleft^* Y$

Fig. 6. Solving dominance constraints: rule set D

There are only two situations where distribution is necessary. The situation shown in Fig. 7 is handled by
(D.Distr.NotDisj): the tree nodes denoted by X and Y cannot be at disjoint positions because they both dominate Z. **Fig. 7.**
The distribution rule (D.Distr.Children) is applicable to the Nondisjointness
constraint in Fig. 5: As the constraint contains $Y{:}f(Y_1, Y_2) \land$
$Y \triangleleft^* X$, we must have either $Y{=}X$ or $Y_1 \triangleleft^* X$ or $Y_2 \triangleleft^* X$. Propagation proves that
the third choice results in a clash, while the others lead to satisfiable constraints.

Proposition 1 (Soundness). *Any dominance constraint in D-solved form is
satisfiable.*

Along the lines of [12]. On the other hand, the saturation algorithm for D is
complete in the sense that it computes every *minimal solved form* of a dominance
constraint.

Definition 3. *Let φ, φ' be constraints, S a set of saturation rules and $\preceq$ an
partial order on constraints. Then φ' is a $\preceq$-minimal S-solved form for φ iff φ'
is an S-solved form that is $\prec$-minimal satisfying $\varphi \preceq \varphi'$.*

For dominance constraints, we can simply use set inclusion. As an example, a $\subseteq$-minimal D-solved form for the con-
straint in Fig. 8 is $X \triangleleft^* Y \land X \triangleleft^* Z \land X \triangleleft^* X \land Y \triangleleft^* Y \land Z \triangleleft^* Z$.
(Note that X does not need to be labeled.)

Fig. 8. A solved
Lemma 2 (Completeness). *Let φ be a dominance* form
*constraint and φ' a $\subseteq$-minimal D-solved form for φ. Then
$\varphi \to_D^* \varphi'$.*

Proof. By well-founded induction on the strict partial order $\supset$ on the set $\{\psi \mid \psi \subseteq \varphi'\}$. If φ is D-solved then $\varphi =^{set} \varphi'$ by minimality and we are done. Otherwise, there is a rule $\psi \rightarrow \vee_{i=1}^{n}\psi_i$ in D which applies to φ. Since $\varphi \subseteq \varphi'$ and φ' is in D-solved form, there exists an i such that $\psi_i \subseteq \varphi'$. By the inductive hypothesis, $\varphi \wedge \psi_i \rightarrow_D^* \varphi'$ and thus $\varphi \rightarrow_D^* \varphi'$.

4 Processing Parallelism Constraints

We extend the dominance constraint solver of the previous section to a semi-decision procedure for parallelism constraints. The main idea is to compute the correspondence functions for all parallelism literals in the input constraint (compare Def. 2). We use a new kind of literals, *path equalities*, to accomplish this with as much propagation and as little case distinction as possible.

We define the set of variables $\mathsf{betw}_\varphi(X_1, X_2)$ *between* X_1 and X_2 as the syntactic counterpart of the set of nodes $\mathsf{betw}_\tau(\pi_1, \pi_2)$: If $X_1 \triangleleft^* X_2 \in \varphi$, then

$$\mathsf{betw}_\varphi(X_1, X_2) = \{X \in \mathcal{V}(\varphi) \mid X_1 \triangleleft^* X \in \varphi \text{ and } (X \triangleleft^* X_2! \in \varphi \text{ or } X \perp X_2 \in \varphi)\}.$$

Given a parallelism literal $X_1/X_2 \sim Y_1/Y_2$, we need to establish a syntactic correspondence function $c : \mathsf{betw}_\varphi(X_1, X_2) \rightarrow \mathsf{betw}_\varphi(Y_1, Y_2)$. In doing this, we may have to add new local variables to φ. In the following, we always consider a constraint φ together with a set $\mathcal{G} \subseteq \mathcal{V}$ of *global* variables; all other variables are *local*. For an input constraint φ, we assume $\mathcal{V}(\varphi) \subseteq \mathcal{G}$.

We record syntactic correspondences by use of a new, auxiliary kind of constraints: a *path equality* $\mathrm{p}\left(\begin{smallmatrix}X_1 & Y_1 \\ X & Y\end{smallmatrix}\right)$ states, informally speaking, that X below X_1 corresponds to Y below Y_1. More precisely, a path equality relation $\mathcal{M}^\tau \models \mathrm{p}\left(\begin{smallmatrix}\pi_1 & \pi_3 \\ \pi_2 & \pi_4\end{smallmatrix}\right)$ is true iff there exists a path π such that $\pi_2 = \pi_1\pi$ and $\pi_4 = \pi_3\pi$, and for each $\pi' \triangleleft^+ \pi$, $L_\tau(\pi_1\pi') = L_\tau(\pi_3\pi')$. We are only interested in "generated" constraints where each path equality establishes a correspondence for some parallelism literal.

Definition 4. *Let φ be a constraint. φ is called* generated *iff for any* $\mathrm{p}\left(\begin{smallmatrix}U_1 & V_1 \\ U_2 & V_2\end{smallmatrix}\right) \in \varphi$ *there exists some atomic parallelism constraint $X_1/X_2 \sim Y_1/Y_2 \in \varphi$ such that $U_1 = X_1 \wedge V_1 = Y_1$ is in φ, and $U_2 \in \mathsf{betw}_\varphi(X_1, X_2)$ or $V_2 \in \mathsf{betw}_\varphi(Y_1, Y_2)$.*

If $U_2 \in \mathsf{betw}_\varphi(X_1, X_2)$, then it must correspond to V_2 and inference will determine that V_2 must be between Y_1 and Y_2, and vice versa.

Figure 9 shows the schemata of the sets P and N of saturation rules handling parallelism. The rule set $D \cup P \cup N$ forms a sound and complete semi-decision procedure for parallelism constraints, which we abbreviate by DPN (and accordingly for other rule set combinations). In section 7 we extend DPN by a set T of optional rules affording stronger propagation on path equality literals. DPN is more succinct and thus forms the basis of our proofs. $DPNT$ is more interesting for practical purposes.

Propagation Rules:

(P.Root) $\quad X_1/X_2 {\sim} Y_1/Y_2 \rightarrow \text{p}\left(\begin{smallmatrix}X_1 & Y_1 \\ X_1 & Y_1\end{smallmatrix}\right) \wedge \text{p}\left(\begin{smallmatrix}X_1 & Y_1 \\ X_2 & Y_2\end{smallmatrix}\right)$

(P.Copy.Dom) $\quad U_1 R U_2 \wedge \bigwedge_{i=1}^{2} \text{p}\left(\begin{smallmatrix}X_1 & Y_1 \\ U_i & V_i\end{smallmatrix}\right) \rightarrow V_1 R V_2 \quad$ where $R \in \{\lhd^*, \perp, \neq\}$

(P.Copy.Lab) $\quad U_0 {:} f(U_1, \dots, U_n) \wedge \bigwedge_{i=0}^{n} \text{p}\left(\begin{smallmatrix}X_1 & Y_1 \\ U_i & V_i\end{smallmatrix}\right) \wedge X_1/X_2 {\sim} Y_1/Y_2 \rightarrow$
$\qquad V_0 {:} f(V_1, \dots, V_n) \quad$ where $U_0 \perp X_2 \in \varphi$ or $U_0 \lhd^+ X_2 \in \varphi$

(P.Path.Sym) $\quad \text{p}\left(\begin{smallmatrix}X & Y \\ U & V\end{smallmatrix}\right) \rightarrow \text{p}\left(\begin{smallmatrix}Y & X \\ V & U\end{smallmatrix}\right)$

(P.Path.Dom) $\quad \text{p}\left(\begin{smallmatrix}X & Y \\ U & V\end{smallmatrix}\right) \rightarrow X \lhd^* U \wedge Y \lhd^* V$

(P.Path.Eq.1) $\quad \text{p}\left(\begin{smallmatrix}X_1 & X_3 \\ X_2 & X_4\end{smallmatrix}\right) \wedge \bigwedge_{i=1}^{4} X_i {=} Y_i \rightarrow \text{p}\left(\begin{smallmatrix}Y_1 & Y_3 \\ Y_2 & Y_4\end{smallmatrix}\right)$

(P.Path.Eq.2) $\quad \text{p}\left(\begin{smallmatrix}X & X \\ U & V\end{smallmatrix}\right) \rightarrow U {=} V$

Distribution Rules:

(P.Distr.Crown) $\quad X_1 \lhd^* X \wedge X_1/X_2 {\sim} Y_1/Y_2 \rightarrow X \lhd^* X_2 \vee X \perp X_2 \vee X_2 \lhd^+ X$

(P.Distr.Project) $\quad \varphi \rightarrow X {=} Y \vee X {\neq} Y \quad$ where $X, Y \in \mathcal{V}(\varphi)$

Introduction of local variables:

(N.New) $\qquad \varphi \wedge X_1/X_2 {\sim} Y_1/Y_2 \rightarrow \text{p}\left(\begin{smallmatrix}X_1 & Y_1 \\ X & X'\end{smallmatrix}\right) \quad$ where $X \in$
$\qquad \text{betw}_\varphi(X_1, X_2)$; X' new and local

Fig. 9. Schemata of rule sets P and N for handling parallelism

Schema (P.Root) states, with respect to a parallelism literal $X_1/X_2{\sim}Y_1/Y_2$, that X_1 corresponds to Y_1 and X_2 corresponds to Y_2. To see how to go on from there, consider the constraint in Fig. 10. Variable X is between X_1 and X_2, and Y is between Y_1 and Y_2. But they are just *dominated* by X_1 and Y_1, respectively, their position is not fixed. So it would be precipitous to assume that X and Y correspond — there is nothing in the constraint which would force us to do that. Schema (N.New) acts on this idea as follows: Given a literal $X_1/X_2{\sim}Y_1/Y_2$ and a variable

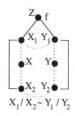

Fig. 10.
Correspondence

$X \in \text{betw}_\varphi(X_1, X_2)$, correspondence $\text{p}\left(\begin{smallmatrix}X_1 & Y_1 \\ X & X'\end{smallmatrix}\right)$ is stated between X and a variable $X' \notin \mathcal{V}(\varphi) \cup \mathcal{G}$. If the structure of the constraint enforces correspondence between X and some other variable $Y \in \text{betw}_\varphi(Y_1, Y_2)$, then this will be inferred by saturation. (N.New) need only be applied if X does not yet possess a correspondent within $X_1/X_2{\sim}Y_1/Y_2$. We adapt the application condition for (N.New)-rules accordingly:

$C_{\varphi' \rightarrow \text{p}\left(\begin{smallmatrix}X_1 & Y_1 \\ X & X'\end{smallmatrix}\right)}(\varphi)$ is true iff $X' \notin \mathcal{V}(\varphi) \cup \mathcal{G}$ and $\text{p}\left(\begin{smallmatrix}X_1 & Y_1 \\ X & Y\end{smallmatrix}\right) \notin \varphi$ for all variables Y

Recall that $\mathcal{G}$ is the set of global variables with respect to which we saturate our constraint. Given $X_1/X_2{\sim}Y_1/Y_2 \in \varphi$, (P.Copy.Dom) and (P.Copy.Lab) copy dominance, disjointness, inequality, and labeling literals from $\text{betw}_\varphi(X_1, X_2)$ to

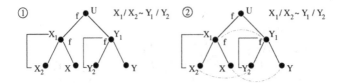

Fig. 11. Resolving an atomic parallelism constraint

betw$_\varphi(Y_1, Y_2)$ and vice versa. The condition on the position of U_0 in (P.Copy.Lab) makes sure that the labels of X_2 and Y_2 are not copied.

P contains two additional distribution rule schemata. (P.Distr.Crown) deals with situations like that in Fig. 12: We have to decide whether X is in betw$_\varphi(X_1, X_2)$ or not. Only then do we know whether we need to apply (N.New) to X. (P.Distr.Project), on the other hand, guesses whether two variables should be identified or not. It is a very powerful schema, so

Fig. 12. X "inside" or "outside"?

we do not want to use it too often in practice. Much of what it does by case distinction is accomplished through propagation by the optional rule schemata we present in section 7.

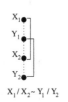

How does syntactic correspondence as established by *DPN* relate to semantic correspondence functions as defined in Def. 2? (P.Root) implements the first property of correspondence functions, the "preservation of tree structure" property remains to be examined. Consider Fig. 11. Constraint 1 constitutes the input to the procedure, while constraint 2 shows, as grey arcs, the correspondences that must hold by Def. 2. These correspondences are computed by *DPN*: We infer $p \begin{pmatrix} X_1 & Y_1 \\ X_1 & Y_1 \end{pmatrix} \wedge p \begin{pmatrix} X_1 & Y_1 \\ X_2 & Y_2 \end{pmatrix}$ by (P.Root).

Fig. 13. Self-overlap

(N.New) is applicable to X and yields $p \begin{pmatrix} X_1 & Y_1 \\ X & X' \end{pmatrix}$ for a new local variable X'. We have $X_1 \triangleleft^+ X_2$ by (D.Lab.Dom), so we may apply (P.Copy.Lab) to $X_1{:}f(X_2, X)$ and get $Y_1{:}f(Y_2, X')$. But since the constraint also contains $Y_1{:}f(Y_2, Y)$, (D.Eq.Decom) gives us $X'{=}Y$, from which (P.Path.Eq.1) infers $p \begin{pmatrix} X_1 & Y_1 \\ X & Y \end{pmatrix}$. We see that the structure of the constraint has enforced correspondence between X and Y, and saturation has made the correct inferences.

While *DPN* computes only finitely many solved forms for the constraint in Fig. 11, the constraint in Fig. 13 possesses infinitely many different solved forms. One solved form contains $X_1{=}X_2{=}Y_1{=}Y_2$. Another contains $X_1 \triangleleft^+ X_2 {=} Y_1 \triangleleft^+ Y_2$. For the case of $X_1 \triangleleft^+ Y_1 \triangleleft^+ X_2 \triangleleft^+ Y_2$, there is one solved form with one local variable, two with two, one with three, two with four, and so on ad infinitum.

5   Soundness

Clearly, all rules in *DPN* are correct. For the soundness of *DPN*-saturation is remains to show that generated *DPN*-solved forms are satisfiable. First, we show that a special class of generated *DPN*-solved forms, called "simple", are satisfiable. Then we lift the result to arbitrary generated *DPN*-solved forms. We can safely restrict our attention to generated constraints:

Lemma 3. *Let φ be a constraint without path equalities and let $\varphi \rightarrow_{\text{DPN}} \varphi'$. Then φ' is generated.*

Definition 5. *Let φ be a constraint. A variable $X \in \mathcal{V}(\varphi)$ is called* labeled *in φ iff $\exists X' \in \mathcal{V}(\varphi)$ such that $X=X'$ and $X':f(X_1,\ldots,X_n)$ are in φ for some term $f(X_1,\ldots,X_n)$. We call φ* simple *if all its variables are labeled and there exists some root variable $Z \in \mathcal{V}(\varphi)$ such that $Z\triangleleft^* X$ is in φ for all $X \in \mathcal{V}(\varphi)$.*

Lemma 4. *A simple generated constraint in DPN-solved form is satisfiable.*

Proof. The constraint graph of a simple generated constraint φ in *DPN*-solved form can be seen as a tree (plus redundant dominance edges and parallelism literals). So we can transform φ into a tree τ by a standard construction. For every parallelism literal in φ, the corresponding parallelism holds in $\mathcal{M}^\tau$: As suggested by the examples in the previous section, *DPN* enforces that the computed path equalities encode valid correspondence functions in $\mathcal{M}^\tau$.

Now suppose we have a generated non-simple constraint φ in *DPN*-solved form. Take for instance the constraint in Fig. 14. We want to show that there is an *extension* $\varphi \wedge \varphi'$ of it that is simple, generated, and in *DPN*-solved form. We proceed by successively labeling unlabeled variables. Suppose we want to label

Fig. 14. Extension

X first. The main idea is to make all variables minimally dominated by X into X's children, i.e. all variables V with $X\triangleleft^+V$ such that there is no intervening W with $X\triangleleft^+W\triangleleft^+V$.

So in the constraint in Fig. 14, Y, Z, U are minimally dominated. However, we choose only one of Z, U as we have $Z=U$. Hence, we would like to label X by some function symbol of arity 2, extending the constraint, for instance, by $X:f(Y,Z)$. (If there is no symbol of suitable arity in Σ, we can always simulate it by a constant symbol and a symbol of arity ≥ 2.) However, we have to make sure that we preserve solvedness during extension. For example, when adding $X:f(Y,Z)$ to the constraint in Fig. 14, we also add $Y \perp Z$ so as not to make (D.Lab.Disj) applicable. Specifically, we have to be careful when labeling a variable like X_1 in Fig. 15 (where grey arcs stand for path equality literals): X_1 is in

Fig. 15.
Extension
and parallelism

(1) Eliminating/introducing a local variable

$$X{=}Z \;\land\; \varphi =_{\mathcal{G}}^{loc} \varphi \quad \text{if } X \notin \mathcal{G},\, X \notin \mathcal{V}(\varphi),\, Z \in \mathcal{V}(\varphi)$$

(2) Renaming a local variable

$$\varphi =_{\mathcal{G}}^{loc} \varphi[Y/X] \qquad \text{if } X \notin \mathcal{G},\, Y \notin \mathcal{V}(\varphi) \cup \mathcal{G}$$

(3) Exchanging representatives of an equivalence class in a constraint

$$X{=}Y \;\land\; \varphi =_{\mathcal{G}}^{loc} X{=}Y \;\land\; \varphi[Y/X]$$

(4) Set equivalence (associativity, commutativity, idempotency)

$$\varphi =_{\mathcal{G}}^{loc} \psi \qquad\qquad \text{if } \varphi =^{set} \psi$$

Fig. 17. The equivalence relation $=_{\mathcal{G}}^{loc}$ on constraints handling local variables

betw$_\varphi(X_1, X_2)$, and when we add $X_1{:}g(X)$ for some unary g, we also have to add $X_2{:}g(X')$, otherwise (P.Copy.Lab) would be applicable.

So, by adding a finite number of atomic constraints and without adding any new local variables, we can label at least one further unlabeled variable in the constraint, while keeping it in *DPN*-solved form. Thus, if we repeat this process a finite number of times, we can extend our generated constraint in *DPN*-solved form to a simple generated constraint in *DPN*-solved form, from which we can then read off a solution right away.

Theorem 1 (Soundness). *A generated constraint in DPN-solved form is satisfiable.*

6 Completeness

DPN-saturation is complete in the sense that it computes every *minimal solved form* of a parallelism constraint. For parallelism constraints, the set inclusion order we have used previously is not sufficient; we adapt it such that it takes local variables into account.

Consider Fig. 16. If (N.New) is applied to X first, this yields p $\left(\begin{smallmatrix} X_1 & Y_1 \\ X & X' \end{smallmatrix}\right)$ for a new local variable X', plus $Y_1{:}g(X')$ and $X'{=}Y$ by (P.Copy.Lab) and (D.Eq.Decom). Accordingly, if (N.New) is applied to Y first, we get p $\left(\begin{smallmatrix} X_1 & Y_1 \\ Y' & Y \end{smallmatrix}\right) \land X_1{:}g(Y') \land Y'{=}X$ for a new local variable Y'. The nondeterministic choice in applying (N.New) leads to two *DPN*-solved forms incomparable by $\subseteq$ which, however, we do not want to distinguish.

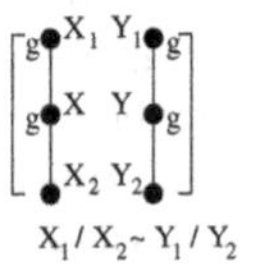

To solve this problem, we use an equivalence relation handling local variables: Let $\mathcal{G} \subseteq \mathcal{V}$, then $=_{\mathcal{G}}^{loc}$ is the smallest equivalence relation on constraints satisfying the axioms in Fig. 17. From this equivalence and subset inclusion, we define the new partial order $\leq_{\mathcal{G}}$.

Fig. 16. Local variables?

Definition 6. *For* $\mathcal{G} \subseteq \mathcal{V}$ *let* $\leq_{\mathcal{G}}$ *be the reflexive and transitive closure* $(\subseteq \cup =_{\mathcal{G}}^{loc})^*$.

We also write $=_{\mathcal{G}}$ for $\leq_{\mathcal{G}} \cap \geq_{\mathcal{G}}$. We return to our above example concerning Fig. 16. Let $\mathcal{G} = \{X_1, X_2, Y_1, Y_2, X, Y\}$. Then $X_1{:}g(X) \wedge Y_1{:}g(Y) \wedge Y_1{:}g(X') \wedge X'{=}Y =_{\mathcal{G}}^{loc} X_1{:}g(X) \wedge Y_1{:}g(Y) \wedge X'{=}Y$ by axioms (3) and (4). This, in turn, is $=_{\mathcal{G}}^{loc}$ equivalent to $X_1{:}g(X) \wedge Y{:}g(Y)$ by axiom (1). Again by axiom (1), this is $=_{\mathcal{G}}^{loc}$ equivalent to $X_1{:}g(X) \wedge Y_1{:}g(Y) \wedge Y'{=}X$, which equals $X_1{:}g(X) \wedge X_1{:}g(Y') \wedge Y_1{:}g(Y) \wedge Y'{=}X$ by axioms (4) and (3).

Lemma 5. *The partial order $\leq_{\mathcal{G}}$ can be factored out into the relational composition of its components, i.e., $\leq_{\mathcal{G}}$ is $\subseteq \circ =_{\mathcal{G}}^{loc}$.*

Lemma 6. *If $\varphi \leq_{\mathcal{G}} \psi$ and ψ is a DPN-solved form, then there exists a DPN-solved form φ' such that $\varphi \subseteq \varphi' =_{\mathcal{G}}^{loc} \psi$.*

Lemma 7. *Let φ be a constraint, $\mathcal{G} \subseteq \mathcal{V}$, and ψ a DPN-solved form with $\varphi \leq_{\mathcal{G}} \psi$. If a rule $\rho \in DPN$ is applicable to φ, then there exists a constraint φ' satisfying $\varphi \rightarrow_{\{\rho\}} \varphi'$ and $\varphi' \leq_{\mathcal{G}} \psi$.*

Proof. By Lemma 6 there exists a DPN-solved form ψ' with $\varphi \subseteq \psi' =_{\mathcal{G}}^{loc} \psi$. First, suppose ρ is a rule $\overline{\varphi} \rightarrow \vee_{i=1}^{n} \overline{\varphi}_i$ in DP. Then there exists an i such that $\overline{\varphi}_i \subseteq \psi'$, hence $\varphi \wedge \overline{\varphi}_i \subseteq \psi'$. Now suppose that $\rho \in N$: Let ρ be $\overline{\varphi} \rightarrow \text{p}\,(\begin{smallmatrix} X_1 & Y_1 \\ X & X' \end{smallmatrix})$ with $X' \notin \mathcal{G} \cup \mathcal{V}(\varphi)$. Then $\text{p}\,(\begin{smallmatrix} X_1 & Y_1 \\ X & Y \end{smallmatrix}) \in \psi$ for some variable Y. But then by axiom (2) of Fig. 17, we have $\psi' =_{\mathcal{G}}^{loc} \psi'[Z'/X']$ for some $Z' \notin \mathcal{G} \cup \mathcal{V}(\psi') \cup \mathcal{V}(\varphi)$, which by axiom (1) is $=_{\mathcal{G}}^{loc}$ equivalent to $\psi'[Z'/X'] \wedge Y{=}X'$, which in turn equals $\psi'[Z'/X'] \wedge Y{=}X' \wedge \text{p}\,(\begin{smallmatrix} X_1 & Y_1 \\ X & X' \end{smallmatrix})$ by axiom (3). Call this last constraint ψ'', then $\varphi \wedge \text{p}\,(\begin{smallmatrix} X_1 & Y_1 \\ X & X' \end{smallmatrix}) \subseteq \psi'' =_{\mathcal{G}}^{loc} \psi$.

It remains to show that there exists a DPN-branch of finite length from φ to each of its minimal solved forms. If saturation rules can be applied in any order, N can speculatively generate an arbitrary number of local variables. For example, for the constraint in Fig. 18, it could successively postulate $\text{p}\,(\begin{smallmatrix} X_1 & Y_1 \\ Y_1 & Y_1' \end{smallmatrix})$, $\text{p}\,(\begin{smallmatrix} X_1 & Y_1 \\ Y_1' & Y_1'' \end{smallmatrix})$, ... We solve this problem by choosing a special rule application order in our completeness proof: After each $\rightarrow_N$ step, we first form a DP-saturation before considering another rule from N. We use a *distance measure* between a smaller and a larger constraint to prove completeness for DPN saturation obeying this application order. The two elements of the measure are: the number of distinct

Fig. 18. Termination?

variables in the larger constraint not present in the smaller one; and the minimum number of correspondences still to be computed for a constraint.

Definition 7. *We define the number $\text{lc}(\mathcal{S}, \varphi)$ of lacking correspondents in φ for a set $\mathcal{S} \subseteq \mathcal{V}(\varphi)$ by*

$$\text{lc}(\mathcal{S}, \varphi) = \sum \left\{ \text{lc}_{X_2 Y_2}^{X_1 Y_1}(X, \varphi) + \text{lc}_{Y_2 X_2}^{Y_1 X_1}(X, \varphi) \mid X \in \mathcal{S} \text{ and } X_1/X_2 {\sim} Y_1/Y_2 \in \varphi \right\}$$

where we fix the values of the auxiliary terms be setting for all $W, U, U', V, V' \in \mathcal{V}(\varphi)$:

$$\mathsf{lc}^{UU'}_{VV'}(W, \varphi) = \begin{cases} 1 \text{ if } W \in \mathsf{betw}_\varphi(U, V) \text{ and } p\left(\begin{smallmatrix} U & U' \\ W & W' \end{smallmatrix}\right) \text{ is not in } \varphi \text{ for any } W' \\ 0 \text{ otherwise} \end{cases}$$

Definition 8. *For constraints $\varphi_1 \subseteq \varphi_2$, let $\mathsf{diff}(\varphi_1, \varphi_2)$ be the size of the set $\{X \in \mathcal{V}(\varphi_2) \mid X \neq Y \in \varphi_2 \text{ for all } Y \in \mathcal{V}(\varphi_1)\}$.*

We call a set $\mathcal{S} \subseteq \mathcal{V}(\varphi)$ of variables an inequality set *for φ iff $X \neq Y \in \varphi$ for any distinct $X, Y \in \mathcal{S}$.*

For constraints φ_2 that are saturated with respect to (P.Distr.Project), $\mathsf{diff}(\varphi_1, \varphi_2)$ is the number of variables X in φ_2 such that $X = Y \notin \varphi_2$ for all $Y \in \mathcal{V}(\varphi_1)$.

Definition 9. *Let φ, ψ be constraints and $\mathcal{G} \subseteq \mathcal{V}$ with $\varphi \leq_\mathcal{G} \psi$. Then the $\mathcal{G}$-measure $\mu_\mathcal{G}(\varphi, \psi)$ for φ and ψ is the sequence $\left(\mu^1_\mathcal{G}(\varphi, \psi), \mu^2(\varphi)\right)$, where:*

- $\mu^1_\mathcal{G}(\varphi, \psi) = \min\{\mathsf{diff}(\varphi, \psi') \mid \varphi \subseteq \psi' =^{loc}_\mathcal{G} \psi \text{ and } \psi' \text{ is DPN-solved }\}$
- $\mu^2(\varphi) = \min\{\mathsf{lc}(\mathcal{S}, \varphi) \mid \mathcal{S} \text{ is a maximal inequality set for } \varphi\}$.

We order $\mathcal{G}$-measures by the lexicographic ordering $<$ on sequences of natural numbers, which is well-founded. The main idea of the following proof is that after each $\rightarrow_N$ step and subsequent DP-saturation, the $\mathcal{G}$-measure between a constraint and its solved form has strictly decreased.

Theorem 2 (Completeness). *Let φ be a constraint, $\mathcal{G} \subseteq \mathcal{V}$, and ψ a $\leq_\mathcal{G}$-minimal DPN-solved form for φ. Then there exists a DPN-solved form $\psi' =_\mathcal{G} \psi$ which can be reached from φ, i.e. $\varphi \rightarrow^*_{\mathrm{DPN}} \psi'$.*

Proof. W.l.o.g. let φ be DP-closed. If no rule from N is applicable to φ then $\varphi =_\mathcal{G} \psi$ by the minimality of ψ. If a rule $\rho \in N$ is applicable to φ, then by Lemma 7 there exist φ', φ'' such that $\varphi \rightarrow_{\{\rho\}} \varphi'' \rightarrow^*_{\mathrm{DP}} \varphi' \leq_\mathcal{G} \psi$, and φ' is DP-saturated. By induction, it is sufficient to show that $\mu_\mathcal{G}(\varphi', \psi) < \mu_\mathcal{G}(\varphi, \psi)$. Note that because φ is DP-closed, a maximal inequality set within φ contains exactly one variable from each syntactic variable equivalence class represented in φ; and $\mathsf{lc}(\{X\}, \varphi) = \mathsf{lc}(\{Y\}, \varphi)$ whenever $X = Y \in \varphi$ because of saturation under (P.Path.Eq.1). The value of $\mathsf{diff}(\varphi, \psi')$ is minimal for ψ' if for any $Y \in \mathcal{V}(\psi')$ that is inequal to all $X \in \mathcal{V}(\varphi)$, Y is local[1] and there is no other variable $Z \in \mathcal{V}(\psi')$ distinct from Y with $Y = Z \in \psi'$.

Let φ'' be $\varphi \wedge p\left(\begin{smallmatrix} X_1 & Y_1 \\ X & X' \end{smallmatrix}\right)$. In φ', (P.Distr.Project) has been applied to X' and all variables in $\mathcal{V}(\varphi)$. Let $\psi' =^{loc}_\mathcal{G} \psi$ with $\varphi \subseteq \psi'$ and minimal $\mathsf{diff}(\varphi, \psi')$. The constraint ψ' contains $p\left(\begin{smallmatrix} X_1 & Y_1 \\ X & Z \end{smallmatrix}\right)$ for some Z. W.l.o.g. we pick a ψ' that does not contain X'.

---

[1] The variable Y is local because $\mathcal{V}(\psi') \cap \mathcal{G} = \mathcal{V}(\psi) \cap \mathcal{G} = \mathcal{V}(\varphi) \cap \mathcal{G}$, otherwise the value of $\mathsf{d}\blacksquare(\varphi, \psi')$ would not be minimal for ψ'.

(T.Trans.H) $p\left(\begin{smallmatrix}X & Y\\ U & V\end{smallmatrix}\right) \wedge p\left(\begin{smallmatrix}Y & Z\\ V & W\end{smallmatrix}\right) \rightarrow p\left(\begin{smallmatrix}X & Z\\ U & W\end{smallmatrix}\right)$

(T.Trans.V) $p\left(\begin{smallmatrix}X_1 & Y_1\\ X_2 & Y_2\end{smallmatrix}\right) \wedge p\left(\begin{smallmatrix}X_2 & Y_2\\ X_3 & Y_3\end{smallmatrix}\right) \rightarrow p\left(\begin{smallmatrix}X_1 & Y_1\\ X_3 & Y_3\end{smallmatrix}\right)$

(T.Diff.1) $p\left(\begin{smallmatrix}X_1 & Y_1\\ X_2 & Y_2\end{smallmatrix}\right) \wedge p\left(\begin{smallmatrix}X_1 & Y_1\\ X_3 & Y_3\end{smallmatrix}\right) \wedge X_2 \vartriangleleft^* X_3 \wedge Y_2 \vartriangleleft^* Y_3 \rightarrow p\left(\begin{smallmatrix}X_2 & Y_2\\ X_3 & Y_3\end{smallmatrix}\right)$

(T.Diff.2) $p\left(\begin{smallmatrix}X_1 & Y_1\\ X_3 & Y_3\end{smallmatrix}\right) \wedge p\left(\begin{smallmatrix}X_2 & Y_2\\ X_3 & Y_3\end{smallmatrix}\right) \wedge X_1 \vartriangleleft^* X_2 \wedge Y_1 \vartriangleleft^* Y_2 \rightarrow p\left(\begin{smallmatrix}X_1 & Y_1\\ X_2 & Y_2\end{smallmatrix}\right)$

Fig. 19. Optional propagation by rules T: transitivity of path equalities

- If $X'{=}Y \in \varphi'$ for some $Y \in \mathcal{V}(\varphi)$, then $\mu^2(\varphi') < \mu^2(\varphi)$ and $\mu_{\mathcal{G}}^1(\varphi', \psi) = \mu_{\mathcal{G}}^1(\varphi', \psi)$: $\mathsf{lc}(\{V\}, \varphi') < \mathsf{lc}(\{X\}, \varphi)$ whenever $V{=}X \in \varphi'$, and either X or some other member of its equivalence class must be in each maximal inequality set. At the same time, a maximal inequality set within φ' can contain only one of X' and Y, so X' contributes nothing additional to $\mu^2(\varphi')$.
  Let ψ'' be $\psi' \wedge X'{=}Z \wedge \psi'[X'/Z]$. Then ψ'' is DPN-solved, and $\varphi' \subseteq \psi''$. We have $\mathsf{diff}(\varphi', \psi'') = \mathsf{diff}(\varphi, \psi')$ because for any $V{\neq}Y \in \psi'$, ψ'' contains $V{\neq}Y \wedge V{\neq}X'$. Furthermore, $\mathsf{diff}(\varphi', \psi'')$ is minimal because the only variable in ψ'' not in ψ' is X'.
- If $X'{\neq}Y \in \varphi'$ for all $Y \in \mathcal{V}(\varphi)$, then $\mu_{\mathcal{G}}^1(\varphi', \psi) < \mu_{\mathcal{G}}^1(\varphi, \psi)$: Let ψ'' be $\psi'[X'/Z]$. Thus, $\psi' =_{\mathcal{G}}^{loc} \psi''$ by axiom (2) and because Z must be local, and $Z{=}Z'$ is not in ψ'' for any distinct Z' because of the minimality of $\mathsf{diff}(\varphi, \psi')$, as pointed out above. Obviously ψ'' is a DPN-solved form with $\varphi' \subseteq \psi''$. Furthermore, $\mathsf{diff}(\varphi', \psi'') = \mathsf{diff}(\varphi, \psi) - 1$ because we must have had $Z{\neq}V \in \psi'$ for all $V \in \mathcal{V}(\varphi)$.

7   Improving Propagation

In this section, we extend DPN by the set of optional propagation rules T, which states transitivity properties of path equalities. Using these rules, much (but not all) of what (P.Distr.Project) achieves by case distinction can be gained through propagation. Interestingly, though, we have not yet encountered a linguistically relevant example requiring (P.Distr.Project).

The set of saturation rules T comprises all instances of the schemata in Fig. 19. Scheme (T.Trans.H) describes horizontal transitivity of path equality constraints, while (T.Trans.V), (T.Diff.1) and (T.Diff.2) all deal with vertical transitivity. The correctness of these rules is obvious.

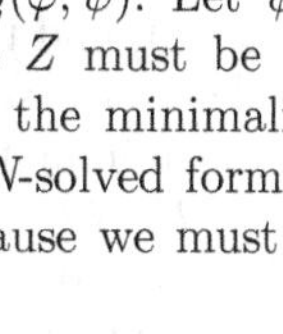

By generatedness, each path equality infered by DPN saturation describes a correspondence for some parallelism literal. With T, this is different. Consider, for example, Fig. 20 where DPN saturation can infer the correspondence $p\left(\begin{smallmatrix}X_1 & Y_1\\ U_1 & V_1\end{smallmatrix}\right)$. (P.Root) yields $p\left(\begin{smallmatrix}U_1 & V_1\\ U_2 & V_2\end{smallmatrix}\right)$. Now (T.Trans.V) can add $p\left(\begin{smallmatrix}X_1 & X_2\\ V_1 & V_2\end{smallmatrix}\right)$, a path equality that does not describe any syntactic correspondence for any of the two parallelism literals present. Actually, the reason why we record correspondence

Fig. 20. Vertical transitivity

by path equalities, as quadruples of variables, is that they support transitivity rules and thus allow for stronger propagation.

Fig. 21 shows an example where T-saturation can apply a propagation rule to make an inference that would need case distinction by (P.Distr.Project) with DPN-saturation alone. (N.New) can be applied to $U \in \text{betw}_\varphi(X_1, Y_1)$, yielding p $\left(\begin{smallmatrix} X_1 & X_2 \\ U & U' \end{smallmatrix}\right)$ for a new local variable U', and (P.Copy.Dom) adds $X_2 \triangleleft^* U' \triangleleft^* U''$. Applying (N.New) and (P.Copy.Dom) two more times each, we get first p $\left(\begin{smallmatrix} X_2 & X_3 \\ U' & U'' \end{smallmatrix}\right)$ and $X_3 \triangleleft^* U'' \triangleleft^* Y_3$ and then p $\left(\begin{smallmatrix} X_3 & X_4 \\ U'' & U''' \end{smallmatrix}\right)$ and $X_4 \triangleleft^* U''' \triangleleft^* Y_4$. Do we now need to apply (N.New) to U''' in $X_4/Y_4 \sim X_1/Y_1$? In $DPNT$, we do not: We get p $\left(\begin{smallmatrix} X_1 & X_3 \\ U & U'' \end{smallmatrix}\right)$ and then p $\left(\begin{smallmatrix} X_1 & X_4 \\ U & U''' \end{smallmatrix}\right)$ by (T.Trans.H). Note in passing that

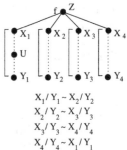

$X_1/Y_1 \sim X_2/Y_2$
$X_2/Y_2 \sim X_3/Y_3$
$X_3/Y_3 \sim X_4/Y_4$
$X_4/Y_4 \sim X_1/Y_1$

Fig. 21. Horizontal transitivity

p $\left(\begin{smallmatrix} X_1 & X_3 \\ U & U'' \end{smallmatrix}\right)$ does not describe a correspondence for any parallelism literal in φ. In DPN, on the other hand, we would need to apply (N.New) to U''', yielding p $\left(\begin{smallmatrix} X_4 & X_1 \\ U''' & U^{\text{iv}} \end{smallmatrix}\right)$, and then use (P.Distr.Project) to guess whether $U = U^{\text{iv}}$ or $U \neq U^{\text{iv}}$. Only the first case leads to a terminating saturation process.

A $\leq_{\mathcal{G}}$-minimal $DPNT$-solved form for a constraint φ may differ from a $\leq_{\mathcal{G}}$-minimal DPN-solved one — there may be additional path equalities. To ensure that they do not differ in more aspects, we impose an additional minor condition (which could possibly be omitted): We apply (P.Copy.Dom) only if the path equalities p $\left(\begin{smallmatrix} X_1 & Y_1 \\ U_i & V_i \end{smallmatrix}\right)$, $i = 1, 2$, describe a correspondence for some parallelism literal.[2]

Lemma 8. *Let φ be a constraint, $\mathcal{G} \subseteq \mathcal{V}$, and ψ a $\leq_{\mathcal{G}}$-minimal DPN-solved form for φ. Then there exists a conjuction ψ' of path equalities such that $\psi \wedge \psi'$ is a $\leq_{\mathcal{G}}$-minimal DPNT-solved form for φ.*

Proposition 2 (Soundness). *If φ is generated and $\varphi \rightarrow^*_{DPNT} \psi$ for a DPNT-solved form ψ, then ψ is satisfiable.*

Proposition 3 (Completeness). *Let φ be a constraint, $\mathcal{G} \subseteq \mathcal{V}$, and ψ a $\leq_{\mathcal{G}}$-minimal DPNT-solved form for φ. Then there exists a constraint $\psi' =_{\mathcal{G}} \psi$ in DPNT-solved form with $\varphi \rightarrow^*_{DPNT} \psi'$.*

Implementation. A first prototype implementation of $DPNT$ is available as an applet on the Internet [3]. Saturation rules are applied in an order refining the order mentioned above: A distribution rule is only applied to a constraint saturated under the propagation rules from DPT. A rule from N is only applied to a constraint saturated under DPT. This implementation handles ellipses in

---

[2] I.e. we now have (P.Copy.Dom) $U_1 R U_2 \wedge \bigwedge_{i=1}^{2}$ p $\left(\begin{smallmatrix} X_1 & Y_1 \\ U_i & V_i \end{smallmatrix}\right) \wedge X_1/X_2 \sim Y_1/Y_2 \rightarrow$ $V_1 R V_2$, where $R \in \{\triangleleft^*, \perp, \neq\}$ and $U_1, U_2 \in \text{betw}_\varphi(X_1, X_2)$.

natural language equally well as the previously mentioned implementation based on context unification [17]. But the two implementations differ with respect to scope ambiguities, i.e. dominance constraint solving: While the context unification based program could handle scope ambiguities with at most 3 quantifiers, the parallelism constraint procedure resolves scope ambiguities of 5 quantifiers in only 6 seconds and can even deal with more quantifiers.

8   Conclusion

We have presented a semi-decision procedure for parallelism constraints which terminates for the important fragment of dominance constraints. It uses path equality constraints to record correspondence, allowing for strong propagation. We have proved the procedure sound and complete. In the process, we have introduced the concept of a minimal solved form for parallelism constraints.

Many things remain to be done. One important problem is to describe the linguistically relevant fragment of parallelism constraints and see whether it is decidable. Then, the prototype implementation we have is not optimized in any way. We would like to replace it by one using constraint technology and to see how that scales up to large examples from linguistics. Also, we would like to apply parallelism constraints to a broader range of linguistic phenomena.

References

[1] R. Backofen, J. Rogers, and K. Vijay-Shanker. A first-order axiomatization of the theory of finite trees. *J. Logic, Language, and Information*, 4:5–39, 1995.

[2] J. Bos. Predicate logic unplugged. In *10th Amsterdam Colloquium*, pages 133–143, 1996.

[3] The CHORUS demo system. http://www.coli.uni-sb.de/cl/projects/chorus/demo.html, 1999.

[4] H. Comon. Completion of rewrite systems with membership constraints. In *Coll. on Automata, Languages and Programming*, volume 623 of *LNCS*, 1992.

[5] D. Duchier and C. Gardent. A constraint-based treatment of descriptions. In *Proc. of the Third International Workshop on Computational Semantics*, 1999.

[6] D. Duchier and J. Niehren. Dominance constraints with set operators. Submitted, 2000.

[7] M. Egg, J. Niehren, P. Ruhrberg, and F. Xu. Constraints over Lambda-Structures in Semantic Underspecification. In *Proc. COLING/ACL'98*, Montreal, 1998.

[8] K. Erk and J. Niehren. Parallelism constraints. Technical report, Universität des Saarlandes, Programming Systems Lab, 2000. Extended version of RTA 2000 paper. http://www.ps.uni-sb.de/Papers/abstracts/parallelism.html.

[9] C. Gardent and B. Webber. Describing discourse semantics. In *Proc. 4th TAG+ Workshop*, Philadelphia, 1998. University of Pennsylvania.

[10] A. Koller. Evaluating context unification for semantic underspecification. In *Proc. Third ESSLLI Student Session*, pages 188–199, 1998.

[11] A. Koller, J. Niehren, and K. Striegnitz. Relaxing underspecified semantic representations for reinterpretation. In *Proc. Sixth Meeting on Mathematics of Language*, 1999.

[12] A. Koller, J. Niehren, and R. Treinen. Dominance constraints: Algorithms and complexity. In *Proc. Third Conf. on Logical Aspects of Computational Linguistics*, Grenoble, 1998.

[13] J. Lévy. Linear second order unification. In *7th Int. Conference on Rewriting Techniques and Applications*, volume 1103 of *LNCS*, pages 332–346, 1996.

[14] G. Makanin. The problem of solvability of equations in a free semigroup. *Soviet Akad. Nauk SSSR*, 223(2), 1977.

[15] M. P. Marcus, D. Hindle, and M. M. Fleck. D-theory: Talking about talking about trees. In *Proc. 21st ACL*, pages 129–136, 1983.

[16] J. Niehren and A. Koller. Dominance Constraints in Context Unification. In *Proc. Third Conf. on Logical Aspects of Computational Linguistics*, Grenoble, 1998. To appear in LNCS.

[17] J. Niehren, M. Pinkal, and P. Ruhrberg. A uniform approach to underspecification and parallelism. In *Proc. ACL'97*, pages 410–417, Madrid, 1997.

[18] M. Pinkal. Radical underspecification. In *Proc. 10th Amsterdam Colloquium*, pages 587–606, 1996.

[19] U. Reyle. Dealing with ambiguities by underspecification: construction, representation, and deduction. *Journal of Semantics*, 10:123–179, 1993.

[20] Decidability of context unification. The RTA list of open problems, number 90, http://www.lri.fr/~rtaloop/, 1998.

[21] M. Schmidt-Schauß. Unification of stratified second-order terms. Technical Report 12/99, J. W. Goethe-Universität Frankfurt, Fachbereich Informatik, 1999.

[22] M. Schmidt-Schauß and K. Schulz. On the exponent of periodicity of minimal solutions of context equations. In *RTA*, volume 1379 of *LNCS*, 1998.

[23] K. N. Venkatamaran. Decidability of the purely existential fragment of the theory of term algebra. *J. ACM*, 34(2):492–510, 1987.

[24] K. Vijay-Shanker. Using descriptions of trees in a tree adjoining grammar. *Computational Linguistics*, 18:481–518, 1992.

Linear Higher-Order Matching Is NP-Complete

Philippe de Groote

LORIA UMR n° 7503 – INRIA
Campus Scientifique, B.P. 239
54506 Vandœuvre lès Nancy Cedex – France
degroote@loria.fr

Abstract. We consider the problem of higher-order matching restricted to the set of linear λ-terms (i.e., λ-terms where each abstraction $\lambda x. M$ is such that there is exactly one free occurrence of x in M). We prove that this problem is decidable by showing that it belongs to NP. Then we prove that this problem is in fact NP-complete. Finally, we discuss some heuristics for a practical algorithm.

1 Introduction

Higher-order unification, which is the problem of solving a syntactic equation (modulo β or $\beta\eta$) between two simply typed λ-terms is known to be undecidable [8], even in the second-order case [7]. On the other hand, if one of the two λ-terms does not contain any unknown, the problem (which is called, in this case, higher-order matching) becomes simpler. Indeed, second-order [10], third-order [5], and fourth-order [17] matching have been proved to be decidable (See also [21] for a survey, including lower and upper bounds on the complexity). In the general case, however, the decidability of higher-order matching is still open.

Since higher-order unification is known to be undecidable, it is natural to investigate whether one can recover decidability for some restricted form of it. For instance, Miller's higher-order patterns form a class of λ-terms for which unification is decidable [14]. Another restricted form of higher-order unification (in fact, in this case, second-order unification) is context unification [3, 4], which may be seen as a generalisation of word unification. While the decidability of context unification is open in general, some subcases of it have been shown to be decidable [3, 4, 11, 12, 19].

In this paper, we study higher-order matching for a quite restricted class of λ-terms, namely, the λ-terms that correspond to the proofs of the implicative fragment of multiplicative linear logic [6]. These λ-terms are *linear* in the sense that each abstraction $\lambda x. M$ is such that there is exactly one free occurrence of x in M. Our main result is the decidability of *linear higher-order matching*. Whether linear higher-order unification is decidable is an open problem related, in the second-order case, to context unification [11].

Linear higher-order unification (in the sense of this paper) has been investigated by Cervesato and Pfenning [2], and by Levy [11]. Cervesato and Pfenning

L. Bachmair (Ed.): RTA 2000, LNCS 1833, pp. 127–140, 2000.
© Springer-Verlag Berlin Heidelberg 2000

consider the problem of higher-order unification for a λ-calculus whose type system corresponds to full intuitionistic linear logic. They are not interested in decidability results—they know that the problem they study is undecidable because the simply-typed λ-calculus may be embedded in the calculus they consider—but in giving a semi-decision procedure in the spirit of Huet's pre-unification algorithm [9]. Levy, on the other hand, is interested in decidability results but only for the second-order case, whose decidability implies the decidability of context unification.

Both in the case of Levy and in the case of Cervesato and Pfenning, there was no reason for considering matching rather than unification. In the first case, the matching variant of the problem is subsumed by second-order matching, which is known to be decidable. In the second case, the matching variant of the problem subsumes full higher-order matching, which is known to be a hard open problem.

Our own motivations in studying linear higher-order matching come from the use of categorial grammars in computational linguistics. Natural language parsing using modern categorial grammars [15, 16] amounts to automated deduction in logics akin to the multiplicative fragment of linear logic. Consequently, the syntactic structures that result from categorial parsing may be seen, through the Curry-Howard correspondence, as linear λ-terms. As a consequence, higher-order unification and matching restricted to the set of linear λ-terms have applications in this categorial setting. In [13], for instance, Morrill and Merenciano show how to use linear higher-order matching to generate a syntactic form (i.e., a sentence in natural language) from a given logical form.

The paper is organised as follows. In the next section, we review several basic definitions and define precisely what is a linear higher-order matching problem. In Section 3, we prove that linear higher-order matching is decidable while, in Section 4, we prove its NP-completeness. Finally, in Section 5, we specify a more practical algorithm and we discuss some implementation issues.

2   Basic Definitions

Definition 1. *Let $\mathcal{A}$ be a finite set, the elements of which are called* atomic types. *The set $\mathcal{F}$ of* linear functional types *is defined according to the following grammar:*

$$\mathcal{F} \ ::= \ \mathcal{A} \ | \ (\mathcal{F} \multimap \mathcal{F}).$$

∎

We let the lowercase Greek letters $(\alpha, \beta, \gamma, \ldots)$ range over $\mathcal{F}$.

Definition 2. *Let $(\Sigma_\alpha)_{\alpha \in \mathcal{F}}$ be a family of pairwise disjoint finite sets indexed by $\mathcal{F}$, whose almost every member is empty. Let $(\mathcal{X}_\alpha)_{\alpha \in \mathcal{F}}$ and $(\mathcal{Y}_\alpha)_{\alpha \in \mathcal{F}}$ be two families of pairwise disjoint countably infinite sets indexed by $\mathcal{F}$, such that $(\bigcup_{\alpha \in \mathcal{F}} \mathcal{X}_\alpha) \cap (\bigcup_{\alpha \in \mathcal{F}} \mathcal{Y}_\alpha) = \varnothing$. The set $\mathcal{T}$ of* raw λ-terms *is defined according to the following grammar:*

$$\mathcal{T} \ ::= \ \Sigma \ | \ \mathcal{X} \ | \ \mathcal{Y} \ | \ \lambda \mathcal{X}.\mathcal{T} \ | \ (\mathcal{T}\,\mathcal{T}),$$

where $\Sigma = \bigcup_{\alpha \in \mathcal{F}} \Sigma_\alpha$, $\mathcal{X} = \bigcup_{\alpha \in \mathcal{F}} \mathcal{X}_\alpha$, *and* $\mathcal{Y} = \bigcup_{\alpha \in \mathcal{F}} \mathcal{Y}_\alpha$. ∎

In the above definition, the elements of Σ are called the *constants*, the elements of $\mathcal{X}$ are called the λ-*variables*, and the elements of $\mathcal{Y}$ are called the *meta-variables* or the *unknowns*. We let the lowercase Roman letters $(a, b, c, \ldots)$ range over the constants, the lowercase italic letters $(x, y, z, \ldots)$ range over the λ-variables, the uppercase bold letters $(\mathbf{X}, \mathbf{Y}, \mathbf{Z}, \ldots)$ range over the unknowns, and the uppercase italic letters $(M, N, O, \ldots)$ range over the λ-terms. The notions of free and bound occurrences of a λ-variable are defined as usual, and we write $\mathrm{FV}(M)$ for the set of λ-variables that occur free in a λ-term M. Finally, a λ-term that does not contain any unknown is called a *pure* λ-*term*.

If $\mathrm{P} = \{i_0, i_1, \ldots, i_n\}$ is a (linearly ordered) set of indices, we write $\lambda \overline{x}_\mathrm{P}. M$ for the λ-term $\lambda x_{i_0}. \lambda x_{i_1}. \ldots \lambda x_{i_n}. M$. Similarly, we write $M (\overline{N}_\mathrm{P})$ or $M (N_i)_{i \in \mathrm{P}}$ for $(\ldots ((M N_{i_0}) N_{i_1}) \ldots N_{i_n})$. As in set theory, we let $n = \{0, 1, \ldots, n - 1\}$. These notations will be extensively used in Section 5.

We then define the notion of typed linear λ-term.

Definition 3. *The family* $(\mathcal{T}_\alpha)_{\alpha \in \mathcal{F}}$ *of sets of typed linear* λ-*terms is inductively defined as follows:*

1. *if* $a \in \Sigma_\alpha$ *then* $a \in \mathcal{T}_\alpha$;
2. *if* $\mathbf{X} \in \mathcal{Y}_\alpha$ *then* $\mathbf{X} \in \mathcal{T}_\alpha$;
3. *if* $x \in \mathcal{X}_\alpha$ *then* $x \in \mathcal{T}_\alpha$;
4. *if* $x \in \mathcal{X}_\alpha$, $M \in \mathcal{T}_\beta$, *and* $x \in \mathrm{FV}(M)$, *then* $\lambda x. M \in \mathcal{T}_{(\alpha \multimap \beta)}$;
5. *if* $M \in \mathcal{T}_{(\alpha \multimap \beta)}$, $N \in \mathcal{T}_\alpha$, *and* $\mathrm{FV}(M) \cap \mathrm{FV}(N) = \varnothing$, *then* $(M N) \in \mathcal{T}_\beta$. ∎

Clauses 4 and 5 imply that any typed linear λ-term $\lambda x. M$ is such that there is exactly one free occurrence of x in M. Remark, on the other hand, that constants and unknowns may occur several times in the same linear λ-term.

¿From now on, we define $\mathcal{T}$ to be $\bigcup_{\alpha \in \mathcal{F}} \mathcal{T}_\alpha$ (which is a proper subset of the set of raw λ-terms). It is easy to prove that the sets $(\mathcal{T}_\alpha)_{\alpha \in \mathcal{F}}$ are pairwise disjoint. Consequently, we may define the type of a typed linear λ-term M to be the unique linear type α such that $M \in \mathcal{T}_\alpha$.

We take for granted the usual notions of α-conversion, η-expansion, β-redex, one step β-reduction $(\rightarrow_\beta)$, n step β-reduction $(\overset{n}{\rightarrow}_\beta)$, many step β-reduction $(\twoheadrightarrow_\beta)$, and β-conversion $(=_\beta)$. We use Δ (possibly with a subscript) to range over β-redexes, and we write $\Delta \subset M$ to say that Δ is a β-redex occurring in a λ-term M.

We let $M[x:=N]$ denote the usual capture-avoiding substitution of a λ-variable by a λ-term. Similarly, $M[\mathbf{X}:=N]$ denotes the capture-avoiding substitution of an unknown by a λ-term. Note that any β-redex $(\lambda x. M) N$ occurring in a linear λ-term is such that $\mathrm{FV}(\lambda x. M) \cap \mathrm{FV}(N) = \varnothing$. Moreover we may suppose, by α-conversion, that $x \notin \mathrm{FV}(N)$. Consequently, when writing $M[x:=N]$ (or $M[\mathbf{X}:=N]$) we always assume that $\mathrm{FV}(M) \cap \mathrm{FV}(N) = \varnothing$. Finally we abbreviate $M[x_0:=N_0] \cdots [x_{n-1}:=N_{n-1}]$ as $M[x_i:=N_i]_{i \in n}$.

In Section 5, we will also consider a more semantic version of substitution, i.e., a function $\sigma : \mathcal{Y} \rightarrow \mathcal{T}$ that is the identity almost everywhere. The finite subset of $\mathcal{Y}$ where $\sigma(\mathbf{X}) \neq \mathbf{X}$ is called the domain of the substitution (in notation, $dom(\sigma)$). It is clear that such a substitution σ determines a unique syntactic substitution $(\mathbf{X} := \sigma(\mathbf{X}))_{\mathbf{X} \in dom(\sigma)}$, and conversely.

We now give a precise definition of the matching problem with which we are concerned.

Definition 4. *A linear higher-order matching problem modulo β (respectively, modulo $\beta\eta$) is a pair of typed linear λ-terms $\langle M, N \rangle$ of the same type such that N is pure (i.e., does not contain any unknown). Such a problem admits a solution if and only if there exists a substitution $(\mathbf{X}_i := O_i)_{i \in n}$ such that $M[\mathbf{X}_i := O_i]_{i \in n} =_\beta N$ (respectively, $M[\mathbf{X}_i := O_i]_{i \in n} =_{\beta\eta} N$).* ∎

In the sequel of this paper, a pair of λ-terms $\langle M, N \rangle$ obeying the conditions of the above definition will also be called a *syntactic equation*. Moreover, we will assume that the right-hand side of such an equation (namely, N) is a pure *closed* λ-term. There is no loss of generality in this assumption because, for $x \in \mathrm{FV}(N)$, $\langle M, N \rangle$ admits a solution if and only if $\langle \lambda x.\, M, \lambda x.\, N \rangle$ admits a solution.

3   Decidability

We first define the size of a term.

Definition 5. *The size $|M|$ of a linear λ-term M is inductively defined as follows:*

1. $|\mathsf{a}| = 1$
2. $|\mathbf{X}| = 0$
3. $|x| = 1$
4. $|\lambda x.\, M| = |M|$
5. $|M\,N| = |M| + |N|$ ∎

The set of linear typed λ-term is a subset of the simply typed λ-terms that is closed under β-reduction. Consequently it inherits the properties of confluence, subject reduction, and strong normalisation. This last property is also an obvious consequence of the following easy lemma.

Lemma 1. *Let M and N be two linear typed λ-terms such that $M \rightarrow_\beta N$. Then $|M| = |N| + 1$.*

Proof. A direct consequence of the fact that any λ-abstraction $\lambda x.\, M$ is such that there is one and only one free occurrence of x in M. □

As a consequence of this lemma, we obtain that all the reduction sequences from one term to another one have the same length.

Lemma 2. *Suppose that M and N are two linear typed λ-terms such that $M \xrightarrow{n}_\beta N$ and $M \xrightarrow{m}_\beta N$. Then $m = n$.*

Proof. By iterating Lemma 1, we have $n = |M| - |N| = m$. □

The above lemma allows the following definition to be introduced.

Definition 6. *The reducibility degree $\nu(M)$ of a linear typed λ-term M is defined to be the unique natural number n such that $M \xrightarrow{n}_\beta N$, where N is the β-normal form of M.* ∎

We also introduce the following notions of complexity.

Definition 7. *The complexity $\rho(\alpha)$ of a linear type α is defined to be the number of "$\multimap$" it contains:*

1. $\rho(a) = 0$
2. $\rho(\alpha \multimap \beta) = \rho(\alpha) + \rho(\beta) + 1$

The complexity $\rho(\Delta)$ of a β-redex $\Delta = (\lambda x.\, M)\, N$ is defined to be the complexity of the type of the abstraction $\lambda x.\, M$. ∎

Lemma 3. *Let $M \in T_{\alpha \multimap \beta}$, and $N \in T_\alpha$. Then*

$$\sum_{\Delta \subset MN} \rho(\Delta) \le \sum_{\Delta \subset M} \rho(\Delta) + \sum_{\Delta \subset N} \rho(\Delta) + \rho(\alpha \multimap \beta).$$

Proof. In case M is not an abstraction, we have $\sum_{\Delta \subset MN} \rho(\Delta) = \sum_{\Delta \subset M} \rho(\Delta) + \sum_{\Delta \subset N} \rho(\Delta)$. Otherwise we have $\sum_{\Delta \subset MN} \rho(\Delta) = \sum_{\Delta \subset M} \rho(\Delta) + \sum_{\Delta \subset N} \rho(\Delta) + \rho(\alpha \multimap \beta)$, because $M\,N$ itself is a β-redex whose complexity is $\rho(\alpha \multimap \beta)$. □

Lemma 4. *Let $M \in T$, and $N, x \in T_\alpha$ be such that $x \in \mathrm{FV}(M)$. Then*

$$\sum_{\Delta \subset M[x:=N]} \rho(\Delta) \le \sum_{\Delta \subset M} \rho(\Delta) + \sum_{\Delta \subset N} \rho(\Delta) + \rho(\alpha).$$

Proof. By induction on the structure of M. The only case that is not straightforward is when $M \equiv x\,O$ and N is an abstraction. In this case, one additional redex of complexity $\rho(\alpha)$ is created. □

We now prove the key lemma of this section.

Lemma 5. *Let M be any linear typed λ-term. The following inequality holds:*

$$\nu(M) \le \sum_{\Delta \subset M} \rho(\Delta) \tag{1}$$

Proof. The proof is done by induction on the length of M. We distinguish between two cases according to the structure of M.

$M \equiv \xi N_0 \dots N_{n-1}$ where $\xi \equiv a$, $\xi \equiv \mathbf{X}$, or $\xi \equiv x$, and where the sequence of terms $(N_i)_{i \in n}$ is possibly empty. We have that $\nu(M) = \sum_{i \in n} \nu(N_i)$ and that $\sum_{\Delta \subset M} \rho(\Delta) = \sum_{i \in n} \sum_{\Delta \subset N_i} \rho(\Delta)$. Consequently, (1) holds by induction hypothesis.

$M \equiv (\lambda x. N) N_0 \ldots N_{n-1}$.   Let the type of $\lambda x. N$ be $\alpha \multimap \beta$. If $n = 0$ (i.e., the sequence of terms $(N_i)_{i \in n}$ is empty) the induction is straightforward. If $n = 1$, we have:

$$\begin{aligned}
\nu(M) &= \nu(N[x{:=}N_0]) + 1 \\
&\leq \sum_{\Delta \subset N[x := N_0]} \rho(\Delta) + 1 && \text{(by induction hypothesis)} \\
&\leq \sum_{\Delta \subset N} \rho(\Delta) + \sum_{\Delta \subset N_0} \rho(\Delta) + \rho(\alpha) + 1 && \text{(by lemma 4)} \\
&\leq \sum_{\Delta \subset N} \rho(\Delta) + \sum_{\Delta \subset N_0} \rho(\Delta) + \rho(\alpha \multimap \beta) \\
&= \sum_{\Delta \subset M} \rho(\Delta)
\end{aligned}$$

Finally, if $n \geq 2$:

$$\begin{aligned}
\nu(M) &= \nu(N[x{:=}N_0] N_1 \ldots N_{n-1}) + 1 \\
&\leq \sum_{\Delta \subset N[x:=N_0] N_1 \ldots N_{n-1}} \rho(\Delta) + 1 && \text{(by induction hypothesis)} \\
&= \sum_{\Delta \subset N[x:=N_0] N_1} \rho(\Delta) + \sum_{i \in n-2} \sum_{\Delta \subset N_{i+2}} \rho(\Delta) + 1 \\
&\leq \sum_{\Delta \subset N[x:=N_0]} \rho(\Delta) + \rho(\beta) + \sum_{i \in n-1} \sum_{\Delta \subset N_{i+1}} \rho(\Delta) + 1 && \text{(by lemma 3)} \\
&\leq \sum_{\Delta \subset N} \rho(\Delta) + \rho(\alpha) + \rho(\beta) + \sum_{i \in n} \sum_{\Delta \subset N_i} \rho(\Delta) + 1 && \text{(by lemma 4)} \\
&= \sum_{\Delta \subset N} \rho(\Delta) + \rho(\alpha \multimap \beta) + \sum_{i \in n} \sum_{\Delta \subset N_i} \rho(\Delta) \\
&= \sum_{\Delta \subset M} \rho(\Delta) && \square
\end{aligned}$$

Linearity plays a central role in the above Lemmas. In fact, Lemmas 1, 2, 4, and 5 do not hold for the simply typed λ-calculus. This is quite clear for Lemmas 1 and 2. Moreover, without the latter, Definition 6 does not make sense. Nevertheless, one might try to adapt this definition to the case of the simply typed λ-calculus by defining the reducibility degree of a λ-term M to be the *maximal* natural number n such that $M \xrightarrow{n}_\beta N$ (where N is the β-normal form of M). Lemma 4 could also be adapted by taking into account the number of free occurrences of x in M. But then, any attempt in adapting Lemma 5 would fail because linearity does not play a part only in the statement of this last lemma, but also in its proof. Indeed, this proof is done by induction on the length of a term M and, in case $M \equiv (\lambda x. N) N_0$, we apply the induction hypothesis to the term $N[x{:=}N_0]$. Now, if x occurs more than once in N, there is no reason why the length of $N[x{:=}N_0]$ should be less than the length of $(\lambda x. N) N_0$.

We now prove the main result of this paper.

Proposition 1. *Linear higher-order matching (modulo β) is decidable.*

Proof. We prove that the length of any possible solution is bounded, which implies that the set of possible solutions is finite.

Let $\langle M, N \rangle$ be a linear higher-order matching problem, and assume, without loss of generality, that M and N are β-normal. Let $(\mathbf{X}_i)_{i \in n}$ be the unknowns that occur in M, and suppose that $(\mathbf{X}_i{:=}O_i)_{i \in n}$ is a solution to the problem where the O_i's are β-normal. By Lemma 1, we have:

$$\nu(M[\mathbf{X}_i{:=}O_i]_{i \in n}) = |M[\mathbf{X}_i{:=}O_i]_{i \in n}| - |N| \tag{1}$$

Let n_i be the number of occurrences of $\mathbf{X}_i$ in M. Equation (1) may be rewritten as follows:

$$\nu(M[\mathbf{X}_i{:=}O_i]_{i \in n}) = \sum_{i \in n} n_i |O_i| + |M| - |N| \tag{2}$$

On the other hand, by Lemma 5, we have:

$$\nu(M[\mathbf{X}_i{:=}O_i]_{i\in n}) \leq \sum_{\Delta \in M[\mathbf{X}_i := O_i]_{i\in n}} \rho(\Delta) \tag{3}$$

Since M and $(O_i)_{i\in n}$ are β-normal, the β-redexes that occur in $M[\mathbf{X}_i{:=}O_i]_{i\in n}$ are created by the substitution, which is the case whenever some O_i is an abstraction and some subterm of the form $\mathbf{X}_i\,P$ occurs in M. Consequently, we have:

$$\sum_{\Delta \in M[\mathbf{X}_i := O_i]_{i\in n}} \rho(\Delta) \leq \sum_{i\in n} n_i\rho(\alpha_i) \tag{4}$$

where α_i is the type of the unknown $\mathbf{X}_i$. ¿From (3) and (4), we have

$$\nu(M[\mathbf{X}_i{:=}O_i]_{i\in n}) \leq \sum_{i\in n} n_i\rho(\alpha_i) \tag{5}$$

Finally, from (2) and (5), we obtain

$$\sum_{i\in n} n_i|O_i| \leq |N| - |M| + \sum_{i\in n} n_i\rho(\alpha_i) \tag{6}$$

which gives an upper bound to the length of the solution.   □

As a corollary to this proposition, we immediately obtain that higher-order linear matching modulo $\eta\beta$ is decidable because the set of η-expanded β-normal forms closed by abstraction and application is provably closed under substitution and β-reduction [10].

4   NP-Completeness

Note that the upper bound given by Proposition 1 is polynomial in the length of the problem. Moreover, β- and $\beta\eta$-conversion between two pure linear λ-terms may be decided in polynomial time since normalisation is linear. Hence, a non deterministic Turing machine may guess a substitution and check that this substitution is indeed a solution in polynomial time. Consequently, linear higher-order matching belongs to NP. In fact, as we show in this section, it is NP-complete.

Let Σ be an alphabet containing at least two symbols, and let $\mathcal{X}$ be a countable set of variables. We write Σ^* (respectively, Σ^+) for the set of words (respectively, non-empty words) generated by Σ. We denote the concatenation of two words u and v by $u \cdot v$. The next proposition, which states the NP-completeness of associative matching, is due to Angluin [1, THEOREM 3.6].

Proposition 2. *Given $v \in (\Sigma \cup \mathcal{X})^*$ and $w \in \Sigma^*$, deciding whether there exists a non-erasing substitution $\sigma : \mathcal{X} \to \Sigma^+$ such that $\sigma(v) = w$ is NP-complete.*   □

We need a slight variant of this proposition in order to take erasing substitutions into account.

Proposition 3. *Given $v \in (\Sigma \cup \mathcal{X})^*$ and $w \in \Sigma^*$, deciding whether there exists a substitution $\sigma : \mathcal{X} \to \Sigma^*$ such that $\sigma(v) = w$ is NP-complete.*

Proof. Consider a symbol, say #, that does not belong to Σ. For any word $u \in (\Sigma \cup \mathcal{X})^*$, let $\underline{u}$ be inductively defined as follows:

1. $\underline{\epsilon} = \epsilon$, where ϵ is the empty word;
2. $\underline{\mathbf{a} \cdot u'} = \# \cdot \mathbf{a} \cdot \underline{u'}$, where $\mathbf{a} \in (\Sigma \cup \mathcal{X})$, and $u' \in (\Sigma \cup \mathcal{X})^*$.

Let $v' \in (\Sigma \cup \mathcal{X})^*$ and $w' \in \Sigma^*$. It is almost immediate that there exists a substitution $\sigma : \mathcal{X} \to \Sigma^+$ such that $\sigma(v') = w'$ if and only if there exists a substitution $\tau : \mathcal{X} \to (\Sigma \cup \{\#\})^*$ such that $\tau(\underline{v'}) = \underline{w'}$. $\qquad\square$

In order to get our NP-completeness result, it remains to reduce the problem of the above proposition to a linear higher-order matching problem. The trick is to encode word concatenation as function composition.

Proposition 4. *Linear higher-order matching is NP-complete.*

Proof. Let $\iota \in \mathcal{A}$ be an atomic type, and let $\Sigma_{\iota - o\iota} = \Sigma$ and $\mathcal{Y}_{\iota - o\iota} = \mathcal{X}$. Finally, let $x \in \mathcal{X}_\iota$. For any word $u \in (\Sigma \cup \mathcal{X})^*$, we inductively define $\underline{u}$ as follows:

1. $\underline{\epsilon} = \lambda x.\, x$, where ϵ is the empty word;
2. $\underline{\mathbf{a} \cdot u'} = \lambda x.\, \mathbf{a}\, (\underline{u'}\, x)$, where $\mathbf{a} \in (\Sigma \cup \mathcal{X})$, and $u' \in (\Sigma \cup \mathcal{X})^*$.

It is easy to show that there exist a substitution $\sigma : \mathcal{X} \to \Sigma^*$ such that $\sigma(v) = w$ if and only if the syntactic equation $\langle \underline{v}, \underline{w} \rangle$ admits a solution (modulo β, or modulo $\beta\eta$). $\qquad\square$

The existence of a set of constants $\Sigma_{\iota - o\iota}$ seems to be crucial in the above proof. Indeed, contrarily to the case of the simply typed λ-calculus, there is an essential difference between constants and free λ-variables. Clause 4 of Definition 3 implies that there is at most one free occurrence of any λ-variable in any linear λ-term. There is no such restriction on the constants. Consequently, a given constant may occur several time in the same linear λ-term. This fact is implicitly used in the above proof.

5   Heuristics for an Implementation

In this section, we give a practical algorithm obtained by specialising Huet's unification procedure [9]. We first specify this algorithm by giving a set of transformations in the spirit of [11, 20]. These transformations obey the following form:

$$e \longrightarrow \langle S_e, \sigma_e \rangle \qquad\qquad (*)$$

where e is a syntactic equation, S_e is a set of syntactic equations, and σ_e is a substitution. Transformations such as $(*)$ may then be applied to pairs $\langle S, \sigma \rangle$ made of a set S of syntactic equations, and a substitution σ:

$$\langle S, \sigma \rangle \longrightarrow \langle \sigma_e((S \setminus \{e\}) \cup S_e), \sigma_e \circ \sigma \rangle \qquad (**)$$

provided that $e \in S$. By iterating $(**)$, one possibly exhausts the set S:

$$\langle S, \sigma \rangle \longrightarrow^* \langle \varnothing, \tau \rangle,$$

in which case τ is intended to be a solution of the system of syntactic equations S.

For the sake of simplicity, we specify an algorithm that solves the problem modulo $\beta\eta$. The set of transformations is given by the three schemes listed below. All the λ-terms occurring in these schemes are considered to be in η-expanded β-normal forms.

1. *Simplification*:

$$e \equiv \langle \lambda \overline{x}_{\mathrm{P}}.\, \mathbf{a}(\overline{M}_{\mathrm{Q}}), \lambda \overline{x}_{\mathrm{P}}.\, \mathbf{a}(\overline{N}_{\mathrm{Q}}) \rangle$$
$$S_e \equiv \{ \langle \lambda \overline{x}_{\mathrm{P}_i}.\, M_i, \lambda \overline{x}_{\mathrm{P}_i}.\, N_i, \rangle \mid i \in Q \}$$
$$\sigma_e \equiv id$$

provided that $\mathrm{FV}(M_i) = \mathrm{FV}(N_i)$, and where $\mathbf{a}$ is either a constant or a bound variable, and the family of sets $(P_i)_{i \in Q}$ is such that $\{ x_j \mid j \in P_i \} = \mathrm{FV}(M_i)$.

2. *Imitation*:

$$e \equiv \langle \lambda \overline{x}_{\mathrm{P}}.\, \mathbf{X}(\overline{M}_{\mathrm{Q}}), \lambda \overline{x}_{\mathrm{P}}.\, a(\overline{N}_{\mathrm{R}}) \rangle$$
$$S_e \equiv \{ e \}$$
$$\sigma_e \equiv (\mathbf{X} := \lambda \overline{y}_{\mathrm{Q}}.\, a(\lambda \overline{z}_{n_i}.\, \mathbf{Y}_i(\overline{y}_{\mathrm{Q}_i})(\overline{z}_{n_i}))_{i \in \mathrm{R}})$$

where $(Q_i)_{i \in \mathrm{R}}$ is a family of disjoint sets such that $\bigcup_{i \in \mathrm{R}} Q_i = Q$, $(\mathbf{Y}_i)_{i \in \mathrm{R}}$ is a family of fresh unknowns whose types may be inferred from the context.

3. *Projection*:

$$e \equiv \langle \lambda \overline{x}_{\mathrm{P}}.\, \mathbf{X}(\overline{M}_{\mathrm{Q}}), \lambda \overline{x}_{\mathrm{P}}.\, \mathbf{a}(\overline{N}_{\mathrm{R}}) \rangle$$
$$S_e \equiv \{ e \}$$
$$\sigma_e \equiv (\mathbf{X} := \lambda \overline{y}_{\mathrm{Q}}.\, y_k(\lambda \overline{z}_{n_i}.\, \mathbf{Y}_i(\overline{y}_{\mathrm{Q}_i})(\overline{z}_{n_i}))_{i \in m})$$

where $k \in Q$, m is the arity of y_k, $(Q_i)_{i \in m}$ is a family of disjoint sets such that $\bigcup_{i \in m} Q_i = Q \setminus \{k\}$, $(\mathbf{Y}_i)_{i \in m}$ is a family of fresh unknowns of the appropriate type.

We now sketch the proof that the above set of transformations is correct and complete.

Lemma 6. *Let* $e \longrightarrow \langle S_e, \sigma_e \rangle$ *be one of the above transformations. Let* S *be a set of syntactic equations such that* $e \in S$. *If* σ *is a substitution that solves all the syntactic equations in* $\sigma_e((S \setminus \{e\}) \cup S_e)$ *then* $\sigma \circ \sigma_e$ *solves all the syntactic equations in* S.

Proof. It suffices to show that $\sigma \circ \sigma_e$ solves e whenever σ solves $\sigma_e(S_e)$, which is trivial for the three transformations. $\qquad\square$

Lemma 7. *Let* S *and* T *be two sets of syntactic equations, and let* τ *be a substitution such that* $\langle S, id \rangle \longrightarrow^* \langle T, \tau \rangle$. *If* σ *is a substitution that solves all the syntactic equations in* T *then* $\sigma \circ \tau$ *solves all the syntactic equations in* S.

Proof. By iterating the previous lemma. $\qquad\square$

As a direct consequence of this lemma, we obtain the correctness of the transformational algorithm.

Proposition 5. *Let* $e = \langle M, N \rangle$ *be a syntactic equation and* σ *be a substitution such that* $\langle \{e\}, id \rangle \longrightarrow^* \langle \varnothing, \sigma \rangle$. *Then* $\sigma(M) = N$. $\qquad\square$

We now prove the completeness of our algorithm.

Proposition 6. *Let* $e = \langle M, N \rangle$ *be a syntactic equation and* σ *be a substitution such that* $\sigma(M) = N$. *Then, there exists a sequence of transitions such that* $\langle \{e\}, id \rangle \longrightarrow^* \langle \varnothing, \sigma' \rangle$, *and* σ' *agrees with* σ *on the set of unknowns occurring in* M.

Proof. Let S be a non-empty set of syntactic equations and let σ be a substitution that solves all the equations in S. One easily shows—see [20, Lemmas 4.16 and 4.17], for details—that there exists $e \in S$ together with a transformation

$$e \longrightarrow \langle S_e, \sigma_e \rangle, \tag{1}$$

and a substitution τ such that:

1. $\sigma = \tau \circ \sigma_e$,
2. τ solves $\sigma_e((S \setminus \{e\}) \cup S_e)$.

Consequently, by iterating (1) on some system R_0 that is solved by some substitution σ_0, one obtains a sequence of transitions:

$$\langle R_0, id \rangle \longrightarrow \langle R_1, \rho_1 \rangle \longrightarrow \langle R_2, \rho_2 \rangle \longrightarrow \cdots \tag{2}$$

together with a sequence of substitutions $(\sigma_0, \sigma_1, \sigma_2, \ldots)$ such that:

1. $\sigma_0 = \sigma_i \circ \rho_i$,
2. σ_i solves R_i.

It remains to prove that (2) eventually terminates and, therefore, exhausts the set R_0. To this end, define the size of a system S (in notation, $|S|$) to be the sum of the sizes of the right-hand sides of the syntactic equations occurring in S. Define also the size of a substitution σ with respect to a system S (in notation, $|\sigma|_S$) to be the sum of the sizes of the terms substituted for the unknowns occurring in S. Transformation (1) and substitution τ are such that

$$|\tau|_T \leq |\sigma|_S,$$

where $T = \sigma_e((S \setminus \{e\}) \cup S_e)$. It is then easy to show that each transition of (2) strictly decreases the pair $(|R_i|, |\sigma_i|_{R_i})$ according to the lexicographic ordering.
$\square$

The three transformations we have given specify a non-deterministic algorithm. Its practical implementation would therefore appeal to some backtracking mechanism. This is not surprising since we have proved linear higher-order matching to be NP-complete. Nevertheless, some source of non-determinism could be avoided. We conclude by discussing this issue.

A naive implementation of the transformational algorithm would give rise to backtracking steps for two reasons:

1. the current non-empty set of syntactic equations is such that no transformation applies;
2. the size of the current substitution is strictly greater than the upper bound given by proposition 1.

It is easy to see that the first case of failure can be detected earlier. Indeed, if no transformation applies to a system S, it means that all the syntactic equations in S have the following form:

$$\langle \lambda \overline{x}_m . \mathbf{a}(\overline{M}_n), \lambda \overline{x}_m . \mathbf{b}(\overline{N}_o) \rangle \tag{3}$$

where either $\mathbf{a} \neq \mathbf{b}$, or $\mathbf{a} = \mathbf{b}$ but there exists $k \in n$ such that $\mathrm{FV}(M_k) \neq \mathrm{FV}(N_k)$. Now, it is clear that such equations cannot be solved. Consequently, one may fail as soon as a system S contains at least one equation like (3). In addition, one may easily prove that any application of the *simplification* rule does not alter the set of possible solutions. Therefore *simplification* may be applied deterministically. These observations give rise to the following heuristic:

Start by Applying repeatedly simplification *until all the heads of the left-hand sides of the equations are unknowns. If this is not possible, fail.*

This heuristic is not proper to our linear matching problem. In fact, it belongs to the folklore of higher-order unification. We end this section by giving some further heuristic principles that are specific to the *linear aspects* of our problem.

The next three lemmas, whose elementary proofs are left to the reader, will allow us to state another condition of possible failure that we may check before applying any transformation. Let $\#_a(M)$ denote the number of occurrences of a given constant "a" in some λ-term M. Similarly, let $\#_{\mathbf{X}}(M)$ denote the number of occurrences in M of some unknown $\mathbf{X}$.

Lemma 8. *Let M and N be two linear λ-terms, and let $x \in \mathrm{FV}(M)$. Then, for every constant* a, $\#_\mathrm{a}(M[x{:=}N]) = \#_\mathrm{a}(M) + \#_\mathrm{a}(N)$. $\qquad\qquad\square$

Lemma 9. *Let M and N be two linear λ-terms such that $M \twoheadrightarrow_\beta N$. Then, for every constant* a, $\#_\mathrm{a}(M) = \#_\mathrm{a}(N)$. $\qquad\qquad\square$

Lemma 10. *Let M and N be two linear λ-terms, and $\mathbf{X}$ be some unknown. Then, for every constant* a, $\#_\mathrm{a}(M[\mathbf{X}{:=}N]) = \#_\mathrm{a}(M) + \#_\mathbf{x}(M) \times \#_\mathrm{a}(N)$. $\quad\square$

As a consequence of these lemmas, we have that the number of occurrences of any constant in the left-hand side of any equation cannot decrease. This allows us to state the following failure condition.

Check, for every constant a, *that each equation $\langle M, N \rangle$ is such that $\#_\mathrm{a}(M) \leq \#_\mathrm{a}(N)$. If this is not the case, fail.*

The above condition may be checked before applying any transformation, and then kept as an invariant. To this end, it must be incorporated as a proviso to the simplification rule. Then, the choice between *imitation* and/or *projection* must obey the following principle:

When considering an equation such as

$$\langle\, \lambda\overline{x}_\mathrm{P}.\, \mathbf{X}(\overline{M}_\mathrm{Q}),\ \lambda\overline{x}_\mathrm{P}.\, a(\overline{N}_\mathrm{R}) \,\rangle$$

check whether there exist some equation $\langle A, B \rangle$ (including the above equation) such that $\#_\mathrm{a}(A) + \#_\mathbf{x}(A) > \#_\mathrm{a}(B)$. If this is the case, projection *is forced. Otherwise, try* imitation *before trying* projection.

The reason for trying imitation first, which is a heuristic used by Paulson in Isabelle [18], is that each application of *imitation* gives rise to a subsequent *simplification*.

When applying *imitation*, we face the problem of guessing the family of sets $(Q_i)_{i \in \mathrm{R}}$. This source of non-determinism is typical of linear higher-order unification.[1] Now, since any application of *imitation* may be immediately followed by a *simplification*, the family $(Q_i)_{i \in \mathrm{R}}$ should be such that the subsequent *simplification* may be applied. We now explain how this constraint may be satisfied.

Consider some linear λ-term A whose η-expanded β-normal form is:

$$\lambda\overline{x}_\mathrm{P}.\, \mathbf{a}(\overline{M}_\mathrm{Q})$$

where **a** is not a bound variable. We define the incidence function of A to be the unique $f_A : \mathrm{P} \to \mathrm{Q}$ such that:

$$f_A(i) = j \quad \text{if and only if} \quad x_i \in \mathrm{FV}(M_j).$$

---

[1] It is due to the multiplicative nature of the connective "$\multimap$" and is reminiscent of the context-splitting problem one has to solve when trying to prove a multiplicative sequent of the form $\Gamma \vdash A \otimes B$ by a backward application of the $\otimes$-introduction rule.

Now, consider the two following λ-terms:

$$A \equiv \lambda \overline{x}_{\mathrm{P}}.\,\mathbf{X}(\overline{M}_{\mathrm{Q}})$$
$$B \equiv \lambda \overline{x}_{\mathrm{Q}}.\,\mathbf{a}(\overline{N}_{\mathrm{R}})$$

It is not difficult to show that the incidence function of $A[\mathbf{X}{:=}B]$ is such that:

$$f_{A[\mathbf{X}:=B]} = f_B \circ f_A \tag{4}$$

We will apply the above identity to the case of the *imitation* rule. Let A, B, and C be the terms involved in the definition of an *imitation* step:

$$A \equiv \lambda \overline{x}_{\mathrm{P}}.\,\mathbf{X}(\overline{M}_{\mathrm{Q}})$$
$$B \equiv \lambda \overline{x}_{\mathrm{P}}.\,\mathbf{a}(\overline{N}_{\mathrm{R}})$$
$$C \equiv \lambda \overline{y}_{\mathrm{Q}}.\,a(\lambda \overline{z}_{n_i}.\,\mathbf{Y}_i(\overline{y}_{\mathrm{Q}_i})(\overline{z}_{n_i}))_{i \in \mathrm{R}}$$

After the imitation step is performed, equation $\langle A, B \rangle$ is replaced by equation:

$$\langle A[\mathbf{X}{:=}C], B \rangle \tag{5}$$

Simplification may then be applied to (5) provided that

$$f_{A[\mathbf{X}:=C]} = f_B \tag{6}$$

which, by (4), is equivalent to

$$f_C \circ f_A = f_B \tag{7}$$

Note that both f_A and f_B are known, while f_C is uniquely determined by the family of sets $(Q_i)_{i \in \mathrm{R}}$, and conversely, since

$$f_C(i) = j \quad \text{if and only if} \quad y_i \in Q_j$$

Therefore, in order to find an appropriate family of sets $(Q_i)_{i \in \mathrm{R}}$, it suffices to solve (7). Now, it is an elementary theorem of set theory that (7) admits a solution if and only if

$$(\forall i, j \in \mathrm{P}) f_A(i) = f_A(j) \Rightarrow f_B(i) = f_B(j),$$

which gives a condition that may be checked before applying any *imitation*.

Acknowledgement

I would like to thank G. Dowek and M. Rusinowitch for interesting discussions about some of the material in this paper.

References

[1] D. Angluin. Finding patterns common to a set of strings. *Journal of Computer and System Sciences*, 21:46–62, 1980.

[2] I. Cervesato and F. Pfenning. Linear higher-order pre-unification. In *Proceedings of the 12th annual IEEE symposium on logic in computer science*, pages 422–433, 1997.

[3] H. Comon. Completion of rewrite systems with membership constraints. Part I: Deduction rules. *Journal of Symbolic Computation*, 25(4):397–419, 1998.

[4] H. Comon. Completion of rewrite systems with membership constraints. Part II: Constraint solving. *Journal of Symbolic Computation*, 25(4):421–453, 1998.

[5] G. Dowek. Third order matching is decidable. *Annals of Pure and Applied Logic*, 69(2–3):135–155, 1994.

[6] J.-Y. Girard. Linear logic. *Theoretical Computer Science*, 50:1–102, 1987.

[7] W. D. Goldfarb. The undecidability of the second-order unification problem. *Theoretical Computer Science*, 13(2):225–230, 1981.

[8] G. Huet. The undecidability of unification in third order logic. *Information and Control*, 22(3):257–267, 1973.

[9] G. Huet. A unification algorithm for typed λ-calculus. *Theoretical Computer Science*, 1:27–57, 1975.

[10] G. Huet. *Résolution d'équations dans les langages d'ordre* $1, 2, \ldots, \omega$. Thèse de Doctorat d'Etat, Université Paris 7, 1976.

[11] J. Levy. Linear second-order unification. In H. Ganzinger, editor, *Rewriting Techniques and Applications, RTA '96*, volume 1103 of *Lecture Notes in Computer Science*, pages 332–346. Springer-Verlag, 1996.

[12] J. Levy. Decidable and undecidable second-order unification problems. In T. Nipkow, editor, *Rewriting Techniques and Applications, RTA '98*, volume 1379 of *Lecture Notes in Computer Science*, pages 47–60. Springer-Verlag, 1998.

[13] J. M. Merenciano and G. Morrill. Generation as deduction on labelled proof nets. In C. Retoré, editor, *Logical Aspects of Computational Linguistics, LACL '96*, volume 1328 of *Lecture Notes in Artificial Intelligence*, pages 310–328. Springer Verlag, 1997.

[14] D. Miller. A logic programming language with lambda-abstraction, function variables, and simple unification. *Journal of Logic and Computation*, 1(4):497–536, 1991.

[15] M. Moortgat. Categorial type logic. In J. van Benthem and A. ter Meulen, editors, *Handbook of Logic and Language*, chapter 2. Elsevier, 1997.

[16] G. Morrill. *Type Logical Grammar: Categorial Logic of Signs*. Kluwer Academic Publishers, Dordrecht, 1994.

[17] V. Padovani. *Filtrage d'ordre supérieure*. Thèse de Doctorat, Université de Paris 7, 1996.

[18] L. C. Paulson. *Isabelle, a generic theorem prover*, volume 828 of *Lecture Notes in Computer Science*. Springer-Verlag, 1994.

[19] M. Schmidt-Schauss and K. U. Schulz. Solvability of context equations with two context variables is decidable. In H. Ganzinger, editor, *16th International Conference on Automated Deduction*, volume 1632 of *Lecture Notes in Computer Science*, pages 67–81. Springer Verlag, 1999.

[20] W. Snyder and J. Gallier. Higher order unification revisited: Complete sets of transformations. *Journal of Symbolic Computation*, 8(1–2):101–140, 1989.

[21] T. Wierzbicki. Complexity of the higher order matching. In H. Ganzinger, editor, *16th International Conference on Automated Deduction*, volume 1632 of *Lecture Notes in Computer Science*, pages 82–96. Springer Verlag, 1999.

Standardization and Confluence
for a Lambda Calculus
with Generalized Applications

Felix Joachimski[1] and Ralph Matthes[2]

[1] Mathematisches Institut der Ludwig-Maximilians-Universität München
Theresienstraße 39, D-80333 München, Germany
joachski@rz.mathematik.uni-muenchen.de
[2] Institut für Informatik der Ludwig-Maximilians-Universität München
Oettingenstraße 67, D-80538 München, Germany
matthes@informatik.uni-muenchen.de

Abstract. As a minimal environment for the study of permutative reductions an extension ΛJ of the untyped λ-calculus is considered. In this non-terminating system with non-trivial critical pairs, confluence is established by studying triangle properties that allow to treat permutative reductions modularly and could be extended to more complex term systems with permutations. Standardization is shown by means of an inductive definition of standard reduction that closely follows the inductive term structure and captures the intuitive notion of standardness even for permutative reductions.

1 Introduction

Standardization and confluence are phenomena that appear naturally in untyped term rewrite systems such as pure λ-calculus. Permutative reductions usually are studied only in typed systems, e.g., with sum types (see for instance [4]). The calculus ΛJ extends λ-calculus by a generalized application rS that gives rise to permutative reductions of the form $(rS)T \mapsto_\pi r(S\{T\})$ already in the untyped, non-normalizing setting. The resulting term rewrite system is also not orthogonal, rendering confluence and standardization demanding problems. Nevertheless, there is a perspicuous, modular and extensible solution for them, so in fact ΛJ serves as a minimal model for the study of permutation.

The calculus ΛJ has been presented (and its simply-typed version proven strongly normalizing) in [5] as the untyped core of a notation system for von Plato's generalized natural deduction trees [14]. It enjoys a particularly simple characterization of normal forms

$$r, s, t ::= x \mid xR \mid \lambda xr , \qquad R ::= (s, z.t) .$$

With simple types assigned, this grammar represents the cut-free derivations in Gentzen's sequent calculus LJ:

$$\frac{}{\Gamma, x : A \vdash x : A} \qquad \frac{\Gamma \vdash s : A \quad \Gamma, z : B \vdash t : C}{\Gamma, x : A \to B \vdash x(s, z.t) : C}{}^{(L\to)} \qquad \frac{\Gamma, x : A \vdash r : B}{\Gamma \vdash \lambda xr : A \to B}$$

L. Bachmair (Ed.): RTA 2000, LNCS 1833, pp. 141–155, 2000.
© Springer-Verlag Berlin Heidelberg 2000

ΛJ generalizes rule (L$\rightarrow$) by allowing arbitrary terms r instead of the variable x in $x(s, z.t)$. Therefore, ΛJ can be understood as the closure of the computational (i.e., type-free) content of cut-free LJ under substitution—hence the name ΛJ. It also incorporates λ-calculus by setting $rs := r(s, z.z)$.

Standardization. The standardization theorem (see, e.g., [1]) establishes that reduction sequences from one term to another can constructively be transformed into a *standard* reduction sequence. For the pure λ-calculus, henceforward called Λ, the latter concept can be put intuitively as follows: A redex $(\lambda xr)st_1 \ldots t_n$ is never converted after reductions in r, s or the t_i have been performed. Formally, this is usually expressed by requiring that redexes are executed in a strictly left-to-right order as recognized by the position of the λ of a redex $(\lambda xr)s$ in the string representation of terms. This also imposes the restriction that a standard reduction sequence starting with xrs has to perform all reductions on r before those on s, although this is not contained in the intuitive notion given above, which rather implies that the reductions in r and s are independent of each other. There is even more freedom in the presence of permutative reductions: A permutative redex of the form $(\lambda xr)RST$ may permute to $(\lambda xr)(R\{S\})T$ and to $(\lambda xr)R(S\{T\})$ and both possibilities as well as the embedded β-redex should be treated equivalently.

We therefore use an inductive definition $\leadsto$ of standard reduction that closely follows the inductive term structure in a canonical way and captures the intuitive notion of standardness even for permutative reductions.[1] The standardization theorem states $\leadsto = \rightarrow^*$ and thereby yields a new induction principle for $\rightarrow^*$. As a prototypical application we prove a syntax-directed inductive characterization of the weakly normalizing terms of ΛJ.[2]

Confluence. The combination of β-reduction and permutative reductions leads to a system with non-trivial critical pairs,[3] i.e., a non-orthogonal term rewrite relation. Although the critical pairs can be joined and hence the reduction relation is locally confluent by the Critical Pair Lemma [8], this does not suffice for full confluence, as the calculus ΛJ is not (strongly) normalizing and so Newman's lemma is not applicable. Since the critical pairs are not development closed (not even almost, cf. [12]), we also cannot use the extensions of Huet's and Toyama's results in [11].

However, permutations always converge and the interaction between β-reduction and permutation is not too intricate. This allows to prove a commutation property between β-developments and π-reduction. Using sequential composition (instead of the union) of those two relations, we can establish a development for the combined reduction relation and confluence ensues. This proof enjoys a particular modularity in that it derives the triangle property of a combination

---

[1] After our presentation of such a notion for Λ in spring 1998, Tobias Nipkow kindly pointed us to Loader's [6] where this notion had been developed independently.

[2] For Λ such a characterization has been worked out in [13].

[3] Λ with η-reduction has two trivial critical pairs arising from $(\lambda x.rx)s\,(\rightarrow_\eta \cap \mapsto_\beta)\,rs$ and $\lambda x.(\lambda x.rx)x\,(\mapsto_\eta \cap \rightarrow_\beta)\,\lambda x.rx$ for x not free in r.

of reduction relations from the respective property of each part, analogous to the Hindley-Rosen-Lemma for confluence.

As will be shown elsewhere, the method carries over to other systems with permutative reductions as long as they enjoy an introduction/elimination dichotomy under the natural deduction (Curry-Howard) interpretation, e.g., calculi with sum types and even infinitary terms with the ω-rule [9, 7].

Acknowledgements to Henk Barendregt, Herman Jervell, Tobias Nipkow, Vincent van Oostrom, Martin Ruckert and Helmut Schwichtenberg for listening to talks or reading previous versions and giving stimulating advice.

2   A λ-Calculus with Generalized Applications

We briefly recall the definition of the *pure λ-calculus Λ* [1]: Terms are generated from variables x, y, z (of an infinite supply) by the grammar

$$\Lambda \ni r, s ::= x \mid rs \mid \lambda x r \ .$$

x is bound in $\lambda x r$. Terms are identified on the syntax level if they only differ in the names of bound variables. The term rewrite relation is the compatible closure of the β-reduction rule $(\lambda x r)s \mapsto_\beta r[x := s]$, where $r[x := s]$ denotes capture-free substitution of s for x in r.

The term grammar of Λ can be inductively characterized by

$$\Lambda \ni r, s, t ::= x\boldsymbol{s} \mid \lambda x r \mid (\lambda x r)s\boldsymbol{s} \ ,$$

using the vector notation $\boldsymbol{s}$ for a possibly empty list of terms $s_1, \ldots, s_n$ of the grammar. β-normal terms have the grammar

$$r ::= x\boldsymbol{r} \mid \lambda x r \ .$$

The calculus ΛJ is defined by generalizing the application rule of Λ:

$$\Lambda J \ni r, s, t ::= x \mid r(s, z.t) \mid \lambda x r \ .$$

The variable x gets bound in $\lambda x r$ as before, and z gets bound in $z.t$. Again, we consider terms only up to α-equivalence (variable convention). It is clear how to define substitution of terms for variables avoiding the capture of bound variables by choosing their names appropriately. By definition, generalized applications associate to the left like traditional applications in Λ (which do so only by convention).

Embedding. ΛJ can be definitionally embedded into Λ by setting

$$r(s, z.t) := t[z := rs] \ .$$

This motivates the contraction rules of ΛJ, defined below:

- $(\lambda x r)(s, z.t)$ translates to $t[z := (\lambda x r)s]$ which β-reduces to $t[z := r[x := s]]$.

– $r(s, z.t)(s', z'.t')$ and $r(s, z.t(s', z'.t'))$ translate to the same λ-term

$$t'[z' := t[z := rs]s'] = (t'[z' := ts'])[z := rs] \ .$$

Therefore, we want to rewrite $r(s, z.t)(s', z'.t')$ to $r(s, z.t(s', z'.t'))$ in ΛJ. This is clearly a permutative reduction: The outer generalized application is pulled into the inner one.

Hence, the *contractions* of ΛJ are

$$(\lambda x r)(s, z.t) \mapsto_\beta t[z := r[x := s]] \ ,$$
$$r(s, z.t)(s', z'.t') \mapsto_\pi r(s, z.t(s', z'.t')) \ .$$

The respective *one-step reduction relations* $\to_\beta$ and $\to_\pi$ are generated from these contractions by means of the *compatible closure*, defined for $\mapsto \in \{\mapsto_\beta, \mapsto_\pi\}$ by

– If $r \mapsto r'$ then $r \to r'$.
– If $r \to r'$ then $r(s, z.t) \to r'(s, z.t)$, $s(r, z.t) \to s(r', z.t)$, $s(t, z.r) \to s(t, z.r')$ and $\lambda x r \to \lambda x r'$.

Clearly, the embedding of Λ into ΛJ by $rs := r(s, z.z)$ preserves β-reduction.[4] Permutative reductions lead out of the range of this embedding.

The reduction relation $\to := \to_\beta \cup \to_\pi$ coincides with the compatible closure of $\mapsto_\beta \cup \mapsto_\pi$. We use $\mapsto^*$ to denote the reflexive, transitive closure of $\mapsto \in \{\to, \to_\beta, \to_\pi\}$. Since the relations $\mapsto$ are generated by the compatible closure, they are *compatible* in the sense that $s \mapsto s'$ implies $r[x := s] \mapsto^* r[x := s']$ and $r[x := s] \mapsto r[x := s']$ if x occurs exactly once in r. As the underlying contraction rules are closed under substitution the reduction relations are also *substitutive* insofar as $r \mapsto r'$ implies $r[x := s] \mapsto r'[x := s]$. Thus $\mapsto^*$ is *parallel*:

$$r \mapsto^* r' \land s \mapsto^* s' \implies r[x := s] \mapsto^* r'[x := s'] \ .$$

It will be useful to view $r(s, z.t)$ as the application of r to the object $(s, z.t)$ which we call a *generalized argument*,[5] denoted by capital letters R, S, T, U.

Permutative reductions act on generalized arguments: Let $R\{S\}$ be the permutation of S into $R = (s, z.t)$ defined by

$$R\{S\} = (s, z.t)\{S\} := (s, z.tS)$$

(we may assume that z does not occur free in S). Using this notation, permutative contraction reads $rRS \mapsto_\pi rR\{S\}$.

We will also need to consider *multiple generalized arguments* (written as $\boldsymbol{R}, \boldsymbol{S}, \boldsymbol{T}, \boldsymbol{U}$) to denote $n \geq 0$ many generalized arguments. If $\boldsymbol{R}$ represents

---

[4] This should be contrasted with the converse embedding, which neither preserves nor simulates reductions although it respects the equality generated by $\to$.

[5] In the light of the translation of ΛJ into Λ, $(s, z.t)$ is not an argument to r but consists of an argument s and information $z.t$ how to use the result of the application of r to s.

$R_1, \ldots, R_n$ then $r\boldsymbol{R}$ represents the multiple generalized application $rR_1 \ldots R_n$ which can only be parenthesized as $(\ldots(rR_1)\ldots R_n)$.

With this notational machinery we arrive at the following inductive characterization of ΛJ

$$\Lambda\text{J} \ni r, s, t, ::= x\boldsymbol{R} \mid (\lambda xr)\boldsymbol{R} \ .$$

By splitting the two cases into subcases according to the length of the trailing $\boldsymbol{R}$ we can identify the subgrammars $\text{NF}_\beta, \text{NF}_\pi, \text{NF}$ of normal terms w.r.t. $\to_\beta, \to_\pi, \to$, i.e., terms that are not further reducible by the respective reduction relation.

$$
\begin{aligned}
\Lambda\text{J} &\ni r, s, t ::= x \mid x(s, z.t) \mid \lambda xr \mid xRSS \mid (\lambda xr)(s, z.t) \mid (\lambda xr)(s, z.t)SS \\
\text{NF}_\beta &\ni r, s, t ::= x \mid x(s, z.t) \mid \lambda xr \mid xRSS \\
\text{NF}_\pi &\ni r, s, t ::= x \mid x(s, z.t) \mid \lambda xr \qquad\qquad \mid (\lambda xr)(s, z.t) \\
\text{NF} &\ni r, s, t ::= x \mid x(s, z.t) \mid \lambda xr
\end{aligned}
$$

3 Confluence

In contrast to Λ, ΛJ has critical pairs[6]:

$$
\begin{array}{ccc}
(\lambda xr)(s, z.t)R & \overset{\pi}{\longmapsto} & (\lambda xr)(s, z.tR) \\
\big\downarrow{\scriptstyle\beta} & & \big\downarrow{\scriptstyle\beta} \\
t[z := r[x := s]]R & =\!=\!= & (tR)[z := r[x := s]]
\end{array}
$$

$$
\begin{array}{ccc}
rRST & \overset{\pi}{\longmapsto} & rRS\{T\} \\
\big\downarrow{\scriptstyle\pi} & & \big\downarrow{\scriptstyle\pi} \\
rR\{S\}T \overset{\pi}{\longmapsto} rR\{S\}\{T\} & \overset{\pi}{\longrightarrow} & rR\{S\{T\}\}
\end{array}
$$

---

[6] Analogous to the treatment of Λ in Example 3.4 in [8], ΛJ can be modeled in simply-typed lambda calculus with a base type *term* and two constants abs : (*term* $\to$ *term*) $\to$ *term* and app : *term* $\to$ *term* $\to$ (*term* $\to$ *term*) $\to$ *term*. The ΛJ term λxr is represented by $\text{abs}(\lambda x^{term} r^*)$: *term* and $r(s, z.t)$ by $\text{app}(r^*, s^*, \lambda z^{term} t^*)$: *term* with r^*, s^* and t^* the representations of r, s and t. With variables X, Y, Z of type *term* and F, G of type *term* $\to$ *term*, $\mapsto_\beta$ and $\mapsto_\pi$ are represented as

$$\text{app}(\text{abs}(\lambda x^{term}(Fx)), X, \lambda y^{term}(Gy)) \mapsto G(FX) \quad \text{and}$$

$$\text{app}(\text{app}(X, Y, \lambda x^{term}(Fx)), Z, \lambda y^{term}(Gy)) \mapsto \text{app}(X, Y, \lambda x.\text{app}(Fx, Z, \lambda y(Gy))) \ .$$

This is a higher-order left-linear pattern rewrite system. The associated rewrite relation represents that of ΛJ.

As argued in the introduction, we cannot use these joins to infer confluence of $\to$ by standard techniques of higher-order rewriting. Instead, we prove confluence of ΛJ by a modular extension of the proof for Λ in [10] which introduced the beautiful idea of proving confluence by showing a triangle property. First, permutative reductions are shown to be converging by a recursive definition of π-normalization that yields a triangle property for $\to_\pi^*$. Then the Takahashi method for Λ is adapted to prove the triangle property for β-developments $\Rrightarrow_\beta$ in ΛJ. By establishing that $\Rrightarrow_\beta$ and $\to_\pi^*$ commute, we can glue triangles together with the help of Lemma 2 and obtain the triangle property for the sequential composition of $\Rrightarrow_\beta$ and $\to_\pi^*$. Note that the triangle property ("one-sided confluence") gives more information on the reduction relation than mere confluence (see Corollary 4).

Definition 1. *Let $\to$ and $\rightarrowtail$ be binary relations.*

- *We write $\to\!\!\rightarrowtail$ for the composition $\rightarrowtail \circ \to$.*
- *$\to^n$ denotes the n-fold composition of $\to$ with itself.*
- *$\to$ and $\rightarrowtail$ commute (written $\to \square \rightarrowtail$) if $\leftarrowtail\to\,\subseteq\,\to\!\leftarrowtail$.*
- *$\to$ has the triangle property[7] w.r.t. a function f (written $\to \triangle f$), if $a \to a'$ implies $a' \to f(a)$.*
- *$\to$ enjoys the diamond property (written $\to \lozenge$) if $\to \square \to$.*
- *$\to$ is confluent if $\to^* \lozenge$.*
- *$\rightarrowtail$ is a $\to$-development if $\to\,\subseteq\,\rightarrowtail\,\subseteq\,\to^*$ and $\rightarrowtail$ has the triangle property w.r.t. some function f which is then called its complete $\to$-development.[8]*

For a reflexive relation $\to$ with $\to \triangle f$ we have $\forall a \in \mathrm{dom}(\to).a \to f(a)$.

Lemma 1. *If $\to \triangle f$ then*
(i) $\to \lozenge$,
(ii) $a \to a' \implies f(a) \to f(a')$ (Simulation),
(iii) $\to^n \triangle f^n$ (Triangle^n),
(iv) $\to$ is confluent.

Proof. (i). Glue two triangles together.
(ii). $a \to a'$ implies $a' \to f(a)$ by the triangle property, hence $f(a) \to f(a')$ by the triangle property.
(iii). Induction on n. The base case is trivial. For the step case (figure on the right) assume $a \to^n b \to c$. Simulation yields $f(a) \to^n f(b)$, to which we may apply the induction hypothesis in order to get $f(b) \to^n f^n(f(a))$. Since $c \to f(b)$ by the triangle property we get $c \to^{n+1} f^{n+1}(a)$.

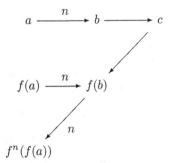

---

[7] The triangle property is the strongest confluence-related property considered in [12].
[8] Notice that this abstract notion of development is not the standard one.

(iv). We show

$$a_1 \;^n\!\!\leftarrow a \rightarrow^m a_2 \implies \exists a'. \; a_1 \rightarrow^m a' \;^n\!\!\leftarrow a_2$$

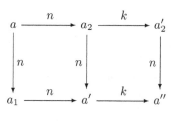

by induction on the maximum of m and n. Without loss of generality $m \geq n$, say $m = n + k$. Assume $a_1 \;^n\!\!\leftarrow a \rightarrow^n a_2 \rightarrow^k a_2'$ (figure on the right). (iii) constructs a square of length n by providing an a' with $a_1 \rightarrow^n a' \;^n\!\!\leftarrow a_2$. If $k \neq 0 \neq n$ then the induction hypothesis gives an a'' such that $a' \rightarrow^k a'' \;^n\!\!\leftarrow a_2'$. □

Corollary 1. *If* $\rightarrowtail$ *is a* $\rightarrow$*-development then* $\rightarrow$ *is confluent.*

Proof. $\rightarrow \subseteq \rightarrowtail \subseteq \rightarrow^*$ entails $\rightarrow^* = \rightarrowtail^*$ and $\rightarrowtail^* \; \diamond$ by (iv). □

Lemma 2 (Triangle Composition).
If $\rightarrow_1 \;\triangle\; f_1$, $\rightarrow_2 \;\triangle\; f_2$ *and* $\rightarrow_1 \;\square\; \rightarrow_2$ *then* $\rightarrow_1\rightarrow_2 \;\triangle\; f_2 \circ f_1$.

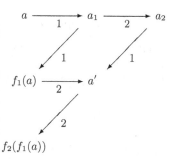

Proof. Assume $a \rightarrow_1 a_1 \rightarrow_2 a_2$. By the triangle property for $\rightarrow_1$ we obtain $a_1 \rightarrow_1 f_1(a)$. Commutation yields an a' such that $f_1(a) \rightarrow_2 a' \;_1\!\!\leftarrow a_2$. By the triangle property for $\rightarrow_2$ we obtain $a' \rightarrow_2 f_2(f_1(a))$. □

3.1   Triangle Property of $\rightarrow^*_\pi$

$\rightarrow_\pi$ is certainly weakly confluent, since its critical pair can be joined. Strong normalization therefore suffices to establish confluence of $\rightarrow_\pi$. This could be achieved as in [3] by assigning a rough upper bound on the length of permutation sequences. Instead, we use a structurally recursive definition of the π-normal form to prove a triangle property for $\rightarrow^*_\pi$ (hence confluence of $\rightarrow_\pi$) and even obtain an exact bound on the height of the π-reduction tree as a byproduct.

Definition 2. *By recursion on terms r define the term $r@R$ as follows:*

$$r@R := \begin{cases} r'@(s, z.t@R) & \textit{if } r = r'(s, z.t), \\ rR & \textit{else.} \end{cases}$$

Lemma 3.
(1) $(r@(s, z.t))@R = r@(s, z.t@R),$
(2) $r, s, t \in NF_\pi \implies r@(s, z.t) \in NF_\pi,$
(3) $rR \rightarrow^*_\pi r@R.$

Proof. Each statement is shown by induction on r. □

Definition 3 (Complete π-Development).

$$
\begin{aligned}
x^\pi &:= x, \\
(\lambda x r)^\pi &:= \lambda x \, r^\pi, \\
(r(s, z.t))^\pi &:= r^\pi @(s^\pi, z.t^\pi).
\end{aligned}
$$

Lemma 4.
  (4) $r^\pi \in NF_\pi$,
  (5) $r \to_\pi^* r^\pi$,
  (6) $r \to_\pi s \Longrightarrow r^\pi = s^\pi$.

Proof. (4) and (5) by induction on r using (2) and (3). (6) by induction on $r \to_\pi s$ using (1) in case of an outer permutation step. □

Corollary 2 (Triangle Property). $\to_\pi^* \, \triangle \, ()^\pi$. *Thus* $\to_\pi$ *is confluent.*

Proof. $r \to_\pi^* s$ implies $r^\pi \overset{(6)}{=} s^\pi \overset{(5)}{{}_\pi{\twoheadleftarrow}} s$. □

A closer look at the previous proofs reveals an exact bound on the height of the π-reduction tree, yielding strong normalization of $\to_\pi$: The constructive content of the proof of (3) yields

$$
@(r, R) := \begin{cases} 1 + @(t, R) + @(r', (s, z.t@R)) & \text{if } r = r'(s, z.t), \\ 0 & \text{else.} \end{cases}
$$

and $rR \to_\pi^{@(r,R)} r@R$. The constructive content of (5) gives

$$
\#x := 0, \quad \#\lambda x r := \#r, \quad \#r(s, z.t) := \#r + \#s + \#t + @(r^\pi, (s^\pi, z.t^\pi))
$$

and $r \to_\pi^{\#r} r^\pi$. (6) can be sharpened to $r \to_\pi s \Longrightarrow \#r > \#s$. This is again shown by induction on $\to_\pi$ where for the case of an outer permutation step we first show by induction on r (using (1)) that

$$
@(r, (s, z.t)) + @(r@(s, z.t), R) > @(t, R) + @(r, (s, z.t@R)) \ .
$$

3.2   Triangle Property of β-Development

It is easy to apply Takahashi's method [10] to $\to_\beta$ in ΛJ. We spell out the (few) details in order to keep the presentation self-contained.

Definition 4 (β-Development). *Inductively define the binary relation* $\Rrightarrow_\beta$ *on terms and simultaneously use the abbreviation*

$$
(s, z.t) \Rrightarrow_\beta (s', z'.t') :\equiv s \Rrightarrow_\beta s' \wedge z = z' \wedge t \Rrightarrow_\beta t' \ .
$$

  (V) $x \Rrightarrow_\beta x$.
  (C) If $r \Rrightarrow_\beta r'$ then $\lambda x r \Rrightarrow_\beta \lambda x r'$.
  (E) If $r \Rrightarrow_\beta r'$ and $R \Rrightarrow_\beta R'$ then $rR \Rrightarrow_\beta r'R'$.

(β) If $u \rightrightarrows_\beta u'$, $s \rightrightarrows_\beta s'$ and $t \rightrightarrows_\beta t'$ then $(\lambda xu)(s, z.t) \rightrightarrows_\beta t'[z := u'[x := s']]$.

Obviously, $\rightrightarrows_\beta$ is reflexive and therefore $\rightarrow_\beta \subseteq \rightrightarrows_\beta$. Also $\rightrightarrows_\beta \subseteq \rightarrow_\beta^*$. By (E) $R \rightrightarrows_\beta R'$ and $S \rightrightarrows_\beta S'$ imply $R\{S\} \rightrightarrows_\beta R'\{S'\}$.

Lemma 5 (Parallelism). *If $r \rightrightarrows_\beta r'$ and $s \rightrightarrows_\beta s'$ then $r[x := s] \rightrightarrows_\beta r'[x := s']$.*
$\square$

Lemma 6 (Inversion). *If $x \rightrightarrows_\beta t$ then $t = x$. If $\lambda xr \rightrightarrows_\beta t$ then $t = \lambda xr'$ with $r \rightrightarrows_\beta r'$.*
$\square$

Definition 5 (Complete β-Development).

$$
\begin{aligned}
x^\beta &:= x, \\
(\lambda xr)^\beta &:= \lambda x\, r^\beta, \\
(r(s, z.t))^\beta &:= \begin{cases} t^\beta[z := u^\beta[x := s^\beta]] & \text{if } r = \lambda xu, \\ r^\beta(s^\beta, z.t^\beta) & \text{otherwise.} \end{cases}
\end{aligned}
$$

Lemma 7 (Triangle Property). $\rightrightarrows_\beta \,\triangle\, ()^\beta$.

Proof. Show $r \rightrightarrows_\beta r' \Rightarrow r' \rightrightarrows_\beta r^\beta$ by induction on $\rightrightarrows_\beta$. The case (β) requires parallelism (Lemma 5). The most interesting case is (E) where $r(s, z.t) \rightrightarrows_\beta r'(s', z.t')$ has been concluded from $r \rightrightarrows_\beta r'$, $s \rightrightarrows_\beta s'$ and $t \rightrightarrows_\beta t'$. By induction hypothesis $r' \rightrightarrows_\beta r^\beta$, $s' \rightrightarrows_\beta s^\beta$ and $t' \rightrightarrows_\beta t^\beta$.

- If $r = \lambda xr_0$ then $r^\beta = \lambda x\, r_0^\beta$ and by inversion r' is of the form $\lambda xr_0'$ with $r_0 \rightrightarrows_\beta r_0'$. $r' = \lambda xr_0' \rightrightarrows_\beta \lambda x\, r_0^\beta = r^\beta$ is derived from $r_0' \rightrightarrows_\beta r_0^\beta$. Using ($\beta$) we obtain $(\lambda xr_0')(s', z.t') \rightrightarrows_\beta t^\beta[z := r_0^\beta[x := s^\beta]]$.
- Otherwise we simply apply (E).
$\square$

Using $r \rightrightarrows_\beta r$ we obtain as a **corollary** $r \rightrightarrows_\beta r^\beta$.

3.3   Confluence of $\rightarrow$

We now establish that $\rightarrowtail := \rightrightarrows_\beta\rightarrow_\pi^*$ is a $\rightarrow$-development. Obviously, $\rightarrow \subseteq \rightarrowtail$ $\subseteq \rightarrow^*$. The triangle property will be shown in the next corollary. Remark that $\rightarrowtail$ is even parallel, since $\rightrightarrows_\beta$ and $\rightarrow_\pi^*$ are.

Lemma 8 (Commutation). $\rightrightarrows_\beta \,\square\, \rightarrow_\pi^*$, *i.e.,* $_\pi^*\!\!\leftarrow\rightrightarrows_\beta \,\subseteq\, \rightrightarrows_\beta{}_\pi^*\!\!\leftarrow$.

Proof. Induction on $\rightrightarrows_\beta$. First note that it suffices to show the claim for one $\rightarrow_\pi$ step in the assumption.

  Case (V). Trivial.
  Case (C). Simple application of the induction hypothesis.
  Case (E). The only interesting subcase is (E) facing an outer permutation

$$rR\{S\} \;_\pi\!\!\leftarrow rRS \rightrightarrows_\beta r'S' \qquad \text{with } rR \rightrightarrows_\beta r' \text{ and } S \rightrightarrows_\beta S' \ .$$

We argue according to the rule used for deriving $rR \rightrightarrows_\beta r'$.

Subsubcase (E). Then r' is of the form $r_0'R'$ and the parallel reduction has been concluded from $r \Rrightarrow_\beta r_0'$ and $R \Rrightarrow_\beta R'$. We get

$$rR\{S\} \Rrightarrow_\beta r_0'R'\{S'\} \, {}_\pi\!\!\leftarrow r_0'R'S'$$

using (E) twice.

Subsubcase (β). Here r is of the form $\lambda x r_0$, R is $(s, z.t)$ and the reductions read

$$(\lambda x r_0)(s, z.tS) \, {}_\pi\!\!\leftarrow (\lambda x r_0)(s, z.t)S \Rrightarrow_\beta t'[z := r_0'[x := s']]S' \ .$$

Using (E) and (β) we obtain

$$(\lambda x r_0)(s, z.tS) \Rrightarrow_\beta t'[z := r_0'[x := s']]S' \ .$$

Case (β). The situation is as follows:

$$(\lambda x u'')(s'', z.t'') \, {}_\pi\!\!\leftarrow (\lambda x u)(s, z.t) \Rrightarrow_\beta t'[z := u'[x := s']]$$

concluded from $u \Rrightarrow_\beta u'$, $s \Rrightarrow_\beta s'$, $t \Rrightarrow_\beta t'$ and with exactly one permutative step in $u \to_\pi^* u''$, $s \to_\pi^* s''$ and $t \to_\pi^* t''$. Choose u''', s''', t''' according to the induction hypothesis. Thus $u'[x := s'] \to_\pi^* u'''[x := s''']$ by parallelism of $\to_\pi^*$ and, similarly,

$$t'[z := u'[x := s']] \to_\pi^* t'''[z := u'''[x := s''']] \, {}_\beta\!\!\Lleftarrow (\lambda x u'')(s'', z.t'') \ . \qquad \square$$

Corollary 3. $\rightarrowtail \triangle \ ()^{\beta\pi}$.

Proof. Apply Lemma 2 to Lemma 7, Corollary 2 and the previous lemma. $\qquad \square$

Corollary 4. *Abbreviate* $r^n := r^{\overbrace{\beta\pi...\beta\pi}^{n \ times}}$.

(i) $\to$ *is confluent,*
(ii) $r \to^n s \Longrightarrow s \to^* r^n$.

Proof. (i). $\rightarrowtail$ is a $\to$-development by the previous corollary. (ii). Use simulation (Lemma 1 (ii)) and $\to \ \subseteq \ \rightarrowtail \ \subseteq \ \to^*$. $\qquad \square$

4   Standardization

We want to show that if $r \to^* r'$ then $r \rightsquigarrow r'$, which expresses that there is some standard reduction sequence from r to r'. For the pure λ-calculus $\rightsquigarrow$ is defined by

(V) *If* $r \rightsquigarrow r'$ *then* $xr \rightsquigarrow xr'$.
(C) *If* $r \rightsquigarrow r'$ *and* $s \rightsquigarrow s'$ *then* $(\lambda x r)s \rightsquigarrow (\lambda x r')s'$.
(β) *If* $r[x := s]s \rightsquigarrow t$ *then* $(\lambda x r)ss \rightsquigarrow t$.

Clearly, we can use a derivation of $r \rightsquigarrow r'$ to devise various reduction sequences from r to r'. Which reduction sequences are captured by this? (β) allows to convert a head β-redex and proceed. Application of rule (C) expresses that such a redex is kept and further standard reductions may only be performed in the constituent terms. Notice that no order is imposed on these subsequent reductions. The same holds for the arguments of a term $x\boldsymbol{r}$ (rule (V)). As an example, a reduction step of the form $(\lambda xx)((\lambda yr)s) \rightarrow (\lambda xx)(r[y := s])$ may occur in a standard reduction sequence and further standard reductions in $r[y := s]$ may follow, but no outer β-reduction of $(\lambda xx)r'$ with r' some descendant of $r[y := s]$ is allowed any more. [9]

4.1 Standard Reductions

Definition 6 (Standard Reduction). *Inductively define the binary relation $\rightsquigarrow$ on terms and simultaneously use the following abbreviations:*

$$(s, z.t) \rightsquigarrow (s', z'.t') \quad :\equiv \quad s \rightsquigarrow s' \wedge z = z' \wedge t \rightsquigarrow t' \ .$$

$$\boldsymbol{R} \rightsquigarrow \boldsymbol{R}' \quad :\equiv \quad \boldsymbol{R} = R_1, \ldots, R_n \wedge \boldsymbol{R}' = R'_1, \ldots, R'_n \wedge \forall i \in \{1, \ldots, n\} : R_i \rightsquigarrow R'_i \ .$$

(V) If $\boldsymbol{R} \rightsquigarrow \boldsymbol{R}'$ then $x\boldsymbol{R} \rightsquigarrow x\boldsymbol{R}'$.
(C) If $r \rightsquigarrow r'$ and $\boldsymbol{S} \rightsquigarrow \boldsymbol{S}'$ then $(\lambda xr)\boldsymbol{S} \rightsquigarrow (\lambda xr')\boldsymbol{S}'$.
(β) If $t[z := r[x := s]]\boldsymbol{S} \rightsquigarrow v$ then $(\lambda xr)(s, z.t)\boldsymbol{S} \rightsquigarrow v$.
(π) If $rR\{S\}\boldsymbol{S} \rightsquigarrow v$ then $rRSS \rightsquigarrow v$.

Given a term $r\boldsymbol{S}$ rule (π) allows standard reduction strategies to start with various permutations of the trailing arguments $\boldsymbol{S}$ until either a possible head β-redex is converted or reduction in the subterms is started by rules (V) and (C). Later use of permutative reductions affecting the outer structure (the term decomposition as displayed in the rules) of the term is forbidden, since the abbreviation $\boldsymbol{R} \rightsquigarrow \boldsymbol{R}'$ is defined by componentwise standard reduction.

By choosing, say, left-to-right order of reductions in (V) and (C) we get

Lemma 9. $\rightsquigarrow \ \subseteq \ \rightarrow^*$. □

4.2 The Standardization Theorem

Our goal is to show that $\rightarrow^* \ \subseteq \ \rightsquigarrow$. The proof follows ideas in [2] (showing standardization for Λ). We profit from our inductive definition of $\rightsquigarrow$ which integrates permutative reductions so smoothly.

Lemma 10.
(1) $\rightsquigarrow$ is reflexive.
(2) If $r \rightsquigarrow r'$ and $\boldsymbol{S} \rightsquigarrow \boldsymbol{S}'$ then $r\boldsymbol{S} \rightsquigarrow r'\boldsymbol{S}'$.
(3) If $\boldsymbol{R} \rightsquigarrow \boldsymbol{R}'$ and $\boldsymbol{S} \rightsquigarrow \boldsymbol{S}'$ then $\boldsymbol{R}\{\boldsymbol{S}\} \rightsquigarrow \boldsymbol{R}'\{\boldsymbol{S}'\}$.

---

[9] Formally, one can inductively characterize the reduction sequences that are captured by a derivation of $r \rightsquigarrow s$ and call them *standard reduction sequences*. An infinite reduction sequence is standard, if all initial finite subsequences are.

Proof. (1). Obvious induction on terms, needing only rules (V) and (C). (2). By induction on $r \rightsquigarrow r'$. Only the cases (β) and (π) need the induction hypothesis. (3). Uses (2). $\qquad\square$

Lemma 11 (Parallelism). *If $r \rightsquigarrow r'$ and $s \rightsquigarrow s'$ then $r[x := s] \rightsquigarrow r'[x := s']$.*[10]

Proof. By induction on $r \rightsquigarrow r'$. The only subcase of interest is (V) with $x\boldsymbol{R} \rightsquigarrow x\boldsymbol{R}'$, so that $s\boldsymbol{R}[x := s] \rightsquigarrow s'\boldsymbol{R}'[x := s']$ has to be shown. This is achieved by the induction hypothesis and repeated applications of (2). $\qquad\square$

Lemma 12. *If $r \rightsquigarrow r' \rightarrow r''$ then $r \rightsquigarrow r''$.*

Proof. Induction on $r \rightsquigarrow r'$.

Case (V). Let $x\boldsymbol{R} \rightsquigarrow x\boldsymbol{R}' \rightarrow r''$ thanks to $\boldsymbol{R} \rightsquigarrow \boldsymbol{R}'$. Either r'' is reached by reducing in one of the $\boldsymbol{R}'$ or by a permutative reduction between the $\boldsymbol{R}'$.
  - In the first case $r'' = x\boldsymbol{R}''$ with $\boldsymbol{R}''$ being derived from $\boldsymbol{R}'$ via reduction of a single term in $\boldsymbol{R}'$. By induction hypothesis $\boldsymbol{R} \rightsquigarrow \boldsymbol{R}''$, hence $x\boldsymbol{R} \rightsquigarrow x\boldsymbol{R}'' = r''$.
  - In the second case $\boldsymbol{R} = \boldsymbol{S}ST\boldsymbol{T}$, $\boldsymbol{R}' = \boldsymbol{S}'S'T'\boldsymbol{T}'$ and $r'' = x\boldsymbol{S}'S'\{T'\}\boldsymbol{T}'$. By Lemma 10 (3) $S\{T\} \rightsquigarrow S'\{T'\}$. Therefore, $x\boldsymbol{S}S\{T\}\boldsymbol{T} \rightsquigarrow x\boldsymbol{S}'S'\{T'\}\boldsymbol{T}'$ and by rule (π) $x\boldsymbol{R} = x\boldsymbol{S}ST\boldsymbol{T} \rightsquigarrow x\boldsymbol{S}'S'\{T'\}\boldsymbol{T}' = r''$.

Case (C). Let $(\lambda xr)\boldsymbol{S} \rightsquigarrow (\lambda xr')\boldsymbol{S}' \rightarrow v$ thanks to $r \rightsquigarrow r'$ and $\boldsymbol{S} \rightsquigarrow \boldsymbol{S}'$. We have to show $(\lambda xr)\boldsymbol{S} \rightsquigarrow v$. Distinguish cases according to the rewrite step $(\lambda xr')\boldsymbol{S}' \rightarrow v$:
  - If it is a reduction inside r' or one of $\boldsymbol{S}'$ then the induction hypothesis applies.
  - If it is an outer β-reduction then $\boldsymbol{S} = (s, z.t)\boldsymbol{T}$, $\boldsymbol{S}' = (s', z.t')\boldsymbol{T}'$ and $v = t'[z := r'[x := s']]\boldsymbol{T}'$ with $s \rightsquigarrow s'$, $t \rightsquigarrow t'$ and $\boldsymbol{T} \rightsquigarrow \boldsymbol{T}'$. By parallelism

    $$r[x := s] \rightsquigarrow r'[x := s'] \quad \text{and} \quad t[z := r[x := s]] \rightsquigarrow t'[z := r'[x := s']] .$$

    Repeated applications of Lemma 10 (2) yield $t[z := r[x := s]]\boldsymbol{T} \rightsquigarrow v$, hence $(\lambda xr)(s, z.t)\boldsymbol{T} \rightsquigarrow v$.
  - The remaining case is a permutative contraction: $\boldsymbol{S} = \boldsymbol{S}_0ST\boldsymbol{T}$, $\boldsymbol{S}' = \boldsymbol{S}'_0S'T'\boldsymbol{T}'$ and $v = (\lambda xr')\boldsymbol{S}'_0S'\{T'\}\boldsymbol{T}'$. This is handled like the second case of (V) above.

Cases (β) and (π). Trivial application of the induction hypothesis. $\qquad\square$

Corollary 5 (Standardization). *If $r \rightarrow^* r'$ then $r \rightsquigarrow r'$.*

Proof. Induction on the number of rewrite steps in $r \rightarrow^* r'$. $\qquad\square$

By Lemma 9 and the previous corollary $\rightarrow^*$ and $\rightsquigarrow$ are (extensionally) the same relation. The virtue of the standardization theorem is that it provides a new inductive characterization of $\rightarrow^*$, namely the inductive definition of $\rightsquigarrow$. This will be used in the next subsection.

---

[10] Note that the lemma is a trivial consequence of $\rightsquigarrow \, = \, \rightarrow^*$, but is necessary to derive it. In contrast to that, parallelism of $\Rightarrow_\beta$ (Lemma 5) is an additional feature for non-trivial developments.

4.3   Inductive Characterization of the Weakly Normalizing Terms

As a typical application of standardization, we characterize the set of weakly normalizing terms

$$wn := \{r \mid \exists s.r \to^* s \in \mathrm{NF}\}$$

of ΛJ by a syntax-directed inductive definition, incorporating a specific standard reduction strategy.

Definition 7. *Inductively define the set WN by*

(V) $x \in WN$, *and if* $s \in WN$ *and* $t \in WN$ *then* $x(s, z.t) \in WN$.
(C)   *If* $r \in WN$ *then* $\lambda x r \in WN$.
(β)   *If* $t[z := r[x := s]]\boldsymbol{S} \in WN$ *then* $(\lambda x r)(s, z.t)\boldsymbol{S} \in WN$.
(π)$^-$ *If* $xR\{S\}\boldsymbol{S} \in WN$ *then* $xRSS \in WN$.

This definition is syntax-directed: A candidate for WN can only enter via the single rule pertaining to its form according to our inductive term characterization.

The reduction strategy underlying WN is: Perform head β-reductions as long as possible. Then reduce below leading λs. If a variable appears in the head position of the term, permute all the generalized arguments into the first one (from left to right), and then continue with the two terms in the remaining generalized argument in parallel. WN defines those terms on which this process succeeds. In this situation parallel and sequential composition of reductions are equivalent.

Our aim is to prove that WN = wn. This is done by help of a restricted notion $\leadsto^-$ of standard reduction which describes the graph of the (partial) normalization function.

Definition 8. *Inductively define the binary relation $\leadsto^-$ on terms:*

(V) $x \leadsto^- x$, *and if* $s \leadsto^- s'$ *and* $t \leadsto^- t'$ *then* $x(s, z.t) \leadsto^- x(s', z.t')$.
(C)   *If* $r \leadsto^- r'$ *then* $\lambda x r \leadsto^- \lambda x r'$.
(β)   *If* $t[z := r[x := s]]\boldsymbol{S} \leadsto^- v$ *then* $(\lambda x r)(s, z.t)\boldsymbol{S} \leadsto^- v$.
(π)$^-$ *If* $xR\{S\}\boldsymbol{S} \leadsto^- v$ *then* $xRSS \leadsto^- v$.

Comparison with $\leadsto$ shows that (V) deals with multiple generalized arguments of length 0 and 1, only. (C) lost its trailing $\boldsymbol{S}$ and permutations (rule $(\pi)^-$) are only allowed between the first two generalized arguments of variables.

Lemma 13. $\leadsto^- \subseteq WN \times NF$.   □

Lemma 14. *If $r \in WN$ then there is a term r' with $r \leadsto^- r'$.*[11]   □

---

[11] As a corollary, $\leadsto^-$ is even a function from WN to NF, since the definition of $\leadsto^-$ is syntax-directed in the left argument.

In order to establish the main lemma of this subsection, we need to show that the restricted version $(\pi)^-$ in the definition of $\leadsto^-$ is sufficient to derive the full rule (π).

Proposition 1. *If* $rR\{S\}S \leadsto^- v$ *then* $rRSS \leadsto^- v$.

Proof. Course-of-generation induction on $\leadsto^-$ (allowing to recourse on the whole derivation). We distinguish cases according to the forms of r in our inductive characterization of terms.

Case x. This is included in $(\pi)^-$.

Case $x(s, z.t)$. S has the form $S_1, \ldots, S_n$. $x(s, z.t)R\{S\}S \leadsto^- v$ has been derived from $x(s, z.tR\{S\})S \leadsto^- v$. This in turn has been derived from $x(s, z.tR\{S\}S_1)S_2 \ldots S_n \leadsto^- v$. Repeating this argument, we find that the relation has been derived from $x(s, z.tR\{S\}S) \leadsto^- v$ which must have been inferred by (V) from $v = x(s', z.v')$ with $s \leadsto^- s'$ and $tR\{S\}S \leadsto^- v'$. By induction hypothesis $tRSS \leadsto^- v'$. Consequently, $x(s, z.tRSS) \leadsto^- v$. Rule $(\pi)^-$ applied $n + 2$ times then yields $x(s, z.t)RSS \leadsto^- v$.

Case $xTUU$. $xTUUR\{S\}S \leadsto^- v$ comes from $xT\{U\}UR\{S\}S \leadsto^- v$. By induction hypothesis $xT\{U\}URSS \leadsto^- v$. Hence, $xTUURSS \leadsto^- v$ by rule $(\pi)^-$.

Case λxr_0. R has the form $(s, z.t)$. $(\lambda xr_0)(s, z.tS)S \leadsto^- v$ has been derived from $(tS)[z := r_0[x := s]]S \leadsto^- v$. We are done by rule (β) since by the variable convention

$$(tS)[z := r_0[x := s]]S = t[z := r_0[x := s]]SS \ .$$

Case $(\lambda xr_0)(s, z.t)R$. Rule (β) has been applied, so the induction hypothesis and rule (β) suffice to prove this case. □

Lemma 15. *If* $r \leadsto r'$ *and* $r' \in NF$ *then* $r \leadsto^- r'$.[12]

Proof. Induction on $\leadsto$.

Case (V). xR' is in NF if R' is empty or a single generalized argument $(s', z.t')$ with $s', t' \in$ NF. So were are done by rule (V) and the induction hypothesis.

Case (C). $(\lambda xr')S'$ is normal if S' is empty and r' is normal. Now apply (C) to the induction hypothesis.

Case (β). By induction hypothesis (the same rule is used in $\leadsto$ and $\leadsto^-$).

Case (π). We have $rRSS \leadsto v \in$ NF thanks to $rR\{S\}S \leadsto v$. By induction hypothesis $rR\{S\}S \leadsto^- v$. The proposition applies. □

Corollary 6. $WN = wn$.

As a consequence, $\leadsto^-$ is indeed the normalization function, defined on wn.

---

[12] In an analogous treatment of the pure λ-calculus any derivation of $r \leadsto r'$ with $r' \in$ NF is already a derivation of $r \leadsto^- r'$. In ΛJ, permutations require remodeling of the derivation.

Proof. WN $\subseteq$ *wn*: Obvious induction on WN. *wn* $\subseteq$ WN: Let $r \to^* r' \in$ NF. By standardization $r \rightsquigarrow r'$. By the preceding lemma $r \rightsquigarrow^- r'$, hence by Lemma 13 $r \in$ WN. $\square$

Notice that the proof of *wn* $\subseteq$ WN essentially uses the inductive characterization of $\to^*$ by the definition of $\rightsquigarrow$ as the main fruit of the standardization theorem. A simple induction on the length of reduction sequences fails to prove *wn* $\subseteq$ WN.

References

[1] Henk P. Barendregt. *The Lambda Calculus: Its Syntax and Semantics*. North–Holland, Amsterdam, second revised edition, 1984.

[2] René David. Une preuve simple de résultats classiques en λ calcul. *C.R. Acad. Sci. Paris*, t. 320, Série I:1401–1406, 1995.

[3] Philippe de Groote. On the strong normalisation of natural deduction with permutation-conversions. In Paliath Narendran and Michaël Rusinowitch, editors, *Rewriting Techniques and Applications, 10th International Conference (RTA '99), Trento, Italy, July 2-4, 1999, Proceedings*, volume 1631 of *Lecture Notes in Computer Science*, pages 45–59. Springer Verlag, 1999.

[4] Jean-Yves Girard, Yves Lafont, and Paul Taylor. *Proofs and Types*, volume 7 of *Cambridge Tracts in Theoretical Computer Science*. Cambridge University Press, 1989.

[5] Felix Joachimski and Ralph Matthes. Short proofs of normalization for the simply-typed lambda-calculus, permutative conversions and Gödel's T. Submitted to the Archive for Mathematical Logic, 1999.

[6] Ralph Loader. Notes on simply typed lambda calculus. Reports of the Laboratory for Foundations of Computer Science ECS-LFCS-98-381, University of Edinburgh, 1998.

[7] Wolfgang Maaß. Church Rosser Theorem für λ-Kalküle mit unendlich langen Termen. In J. Diller and G.H. Müller, editors, *Proof Theory Symposion Kiel 1974*, volume 500 of *Lecture Notes in Mathematics*, pages 257–263. Springer Verlag, 1975.

[8] Richard Mayr and Tobias Nipkow. Higher-order rewrite systems and their confluence. *Theoretical Computer Science*, 192:3–29, 1998.

[9] William W. Tait. Infinitely long terms of transfinite type. In J. Crossley and M. Dummett, editors, *Formal Systems and Recursive Functions*, pages 176–185. North–Holland, Amsterdam, 1965.

[10] Masako Takahashi. Parallel reduction in λ-calculus. *Information and Computation*, 118(1):120–127, 1995.

[11] Vincent van Oostrom. Development closed critical pairs. In G. Dowek, J. Heering, K. Meinke, and B. Möller, editors, *Proceedings of the Second International Workshop on Higher-Order Algebra, Logic and Term Rewriting (HOA '95), Paderborn, Germany*, volume 1074 of *Lecture Notes in Computer Science*, pages 185–200. Springer Verlag, 1996.

[12] Vincent van Oostrom. Developing developments. *Theoretical Computer Science*, 175(1):159–181, 1997.

[13] Femke van Raamsdonk. *Confluence and Normalisation for Higher-Order Rewriting*. Academisch Proefschrift (PhD thesis), Vrije Universiteit te Amsterdam, 1996.

[14] Jan von Plato. Natural deduction with general elimination rules. Submitted, 1998.

Linear Second-Order Unification and Context Unification with Tree-Regular Constraints*

Jordi Levy[1] and Mateu Villaret[2]

[1] IIIA, CSIC
Campus de la UAB, Bellaterra, Barcelona, Spain
http://www.iiia.csic.es/~levy
[2] IMA, UdG
Campus Montilivi, U. de Girona, Girona, Spain
http://www.ima.udg.es/~villaret

Abstract. Linear Second-Order Unification and Context Unification are closely related problems. However, their equivalence was never formally proved. Context unification is a restriction of linear second-order unification. Here we prove that linear second-order unification can be reduced to context unification with tree-regular constraints.
Decidability of context unification is still an open question. We comment on the possibility that linear second-order unification is decidable, if context unification is, and how to get rid of the tree-regular constraints. This is done by reducing rank-bound tree-regular constraints to word-regular constraints.

1 Introduction

Context Unification (CU) [10, 11] is an extension of First-Order Unification where, in addition to the first-order variables, we also have variables that denote contexts. These *context variables* are applied to arguments, thus the term $F(t)$ denotes any term containing t as a subterm, and F denotes the context surrounding such subterm t. *Linear Second-Order Unification* (LSOU) [3, 5] is a restriction of unification in Second-Order Simply Typed λ-Calculus, where only *linear terms* are considered as possible instances of second-order variables. A linear term is a λ-term where the most external λ-bindings bound one and just one occurrence of the variable. CU is a restriction of LSOU, therefore both problems are between the decidable first-order unification problem and the undecidable second-order one. The common assumption is that CU is decidable. This is because various restrictions of this problem [1, 3, 13, 14, 15] make it decidable, while the same restrictions applied to second-order do not [4, 6]. It is also known that, like for the word theory, the context theory is undecidable beyond this existential fragment [9]. The natural question to ask is whether, if CU is decidable, then LSOU will be. This is the main topic of the present paper.

---

* The first author is partially supported by the project MODELOGOS founded by the CICYT.

L. Bachmair (Ed.): RTA 2000, LNCS 1833, pp. 156–171, 2000.
© Springer-Verlag Berlin Heidelberg 2000

CU is a restriction of LSOU where 1) third- or higher-order constants are not allowed, 2) second-order variables are unary, and 3) there are no internal λ-bindings, and external ones are only used to denote the parameter of a second-order variable. The common belief was that third- or higher-order constants do not play an important role w.r.t. the decidability of both problems, neither the use of λ-bindings. The restriction of using unary context variables is not a real restriction because we can replace binary (similarly for n-ary) variables like $F(t_1, t_2)$ by $F_0(f(F_1(t_1), F_2(t_2)))$ introducing a conjectured constant symbol f (see Subsection 3.2 and [12]). However, the equivalence of both problems was never formally proved.

The naive attempt to reduce LSOU to CU by replacing bound variables by new constant symbols does not work. This is because we have to ensure that substitutions avoid *variable capture*. For instance, the following LSOU problem

$$\lambda x.f(x) \doteq_{lsou} \lambda x.f(Y)$$

is not solvable. The substitution $\sigma = [Y \mapsto x]$ gives us:

$$\lambda x.f(x) \doteq_{lsou} \lambda y.f(x)$$

but both terms are not λ-equivalent, because an α-conversion is needed in order to avoid the capture of variable x. However, applying the naive reduction to this problem we get the following solvable CU problem:

$$f(c_x) \doteq_{cu} f(Y)$$

We can try to apply a more sophisticated reduction. Take the original LSOU problem and substitute the bound variables by two distinct constants. However, this method only works for the most external λ-bindings. Applying the reduction to the following solvable LSOU problem with internal λ-bindings:

$$f(g(\lambda x.x), a) \doteq_{lsou} f(Y, Z)$$

we get the following unsolvable CU problem:

$$f(g(c_x), a) \doteq_{cu} f(Y, Z)$$
$$f(g(c'_x), a) \doteq_{cu} f(Y, Z)$$

Bindings can transform free variables into bound variables at different depths. Somehow we have to ensure that if an instance of a (free) variable contains a bound variable, then it also contains its corresponding λ-binding. For instance, given the LSOU problem $F(X) \doteq_{lsou} g(\lambda y.y, a)$ and the following substitutions:

$$\sigma_1 = [\, X \mapsto a, \qquad F \mapsto \lambda z.g(\lambda y.y, z)\,]$$
$$\sigma_2 = [\, X \mapsto y, \qquad F \mapsto \lambda z.g(\lambda y.z, a)\,]$$
$$\sigma_3 = [\, X \mapsto g(\lambda y.y, a), \quad F \mapsto \lambda z.z \qquad\qquad]$$

only σ_1 and σ_3 are unifiers. As we will show, such a restriction can be ensured by means of tree automata [2], but it does not seem easy to be simply encoded in terms of context equations.

On the other hand, context unification and *word unification* are also closely related problems. W ord unification is decidable [8], and can be enriched with regular restrictions without loosing decidability [16]. Tree-regular languages are to terms as regular languages to words. Therefore, if context unification turns out to be decidable, then it seems reasonable to think that context unification could be enriched with tree-regular restrictions without loosing decidability. To support this hypothesis, we would like to prove that membership equations on tree-regular languages can be reduced to mem bership equations on (word) regular languages, by encoding terms as sequences of symbols (the *traversal sequence*). Unfortunately, we can only prove this reduction for a certain subset of tree-regular languages (what we call *rank-bound tree-regular languages*).

There is a proof that, if context most general unifiers are rank-bound, then CU is decidable [7]. If most general context unifiers were proved to be rank-bound, then tree-regular restrictions would also be rank-bound, and we would also have the decidability of LSOU. We commen t on this possibility at the beginning of Section 4.

This paper proceeds as follows. After introducing some basic definitions in Section 2, we reduce LSOU to CU with tree-regular constraints in Section 3. In Section 4 we reduce rank-bound tree-regular restrictions to word-regular restrictions. Finally, in Section 5 we discuss whether this results could be extended to linear third- or higher-order unification.

2   Preliminarie s

Let Σ be a simply typed signature where first-order constants are denoted by $a, b, \ldots$, and higher-order constants by $f, g, h, \ldots$. Let $\mathcal{X}$ be an enumerable set of simply typed variables where first-order variables are denoted by capital letters $X, Y, Z, \ldots$, second-order variables by $F, G, H, \ldots$ and bound variables by lower-case letters $x, y, z, \ldots$. Types and their orders are defined as usual in the simply typed λ-calculus. For simplicity, we can assume that there is only one base type. Other types are built as $\tau_1 \to \tau_2$ where τ_1 and τ_2 are types. A term t is said to have arity n if $t : \tau_1 \to \cdots \tau_n \to \tau_0$ where τ_0 is a base type. Second-order terms are also standard: terms constructed using constants and bound variables of any order, but free variables of order at most two. Normal terms (β-reduced, η-expanded terms) have the form $\lambda \boldsymbol{x}.f(t_1, \ldots, t_n)$ where $\boldsymbol{x}$ is a list of bound variables, f is either a bound, a free variable or a constant, and $t_1, \ldots, t_n$ are normal terms. If $t_1 : \tau_1, \ldots, t_n : \tau_n$ then $f : \tau_1 \to \ldots \to \tau_n \to \tau_0$ where τ_0 is a base type; and, if $x_1 : \tau_1', \ldots, x_m : \tau_m'$ then $\lambda \boldsymbol{x}.f(t_1, \ldots, t_n) : \tau_1' \to \ldots \to \tau_m' \to \tau_0$. Notice that, as we are in second-order, if f is a free second-order variable then $\tau_1, \ldots, \tau_n$ are base types. A term $\lambda \boldsymbol{x}.f(t_1, \ldots, t_n)$ is said to be linear if, written in normal form, any bound variable $x_i \in \boldsymbol{x}$ occurs once and just once in $f(t_1, \ldots, t_n)$. Notice that $t_1, \ldots, t_n$ are not required to be linear.

A linear second-order unification (LSOU) problem is a pair[1] $t =_{lsou} u$ of terms (not necessarily linear). A solution σ of a LSOU problem is a second-

---

[1] Notice that a set of equations is equivalent to just one equation.

order substitution such that $\sigma(t)$ and $\sigma(u)$ are λ-equivalents, and $\sigma(X)$ is a linear term for any free variable X.

A context unification (CU) problem is a pair $t \doteq_{cu} u$ of terms that does not contain λ-bindings neither constants of order higher than two, and where second-order variables are unary. A solution σ of a CU problem $t \doteq_{cu} u$ is a second-order substitution such that $\sigma(t) = \sigma(u)$, and $\sigma(X)$ is a linear term that does not contain n-ary $(n > 1)$ variables, for any free variable X.

A CU problem with tree-regular constraints is a CU problem $t \doteq_{cu} u$ with a tree-regular constraint $v \in \mathcal{A}$.[2] A solution is a ground substitution σ solving the CU problem, and satisfying $\sigma(v) \in L(\mathcal{A})$.

For simplicity we assume that the signature of the problem allows us to ensure the existence of a ground solution whenever a solution exists. Notice that this fact can always be ensured if we extend the signature Σ ensuring that it contains at least a constant $a : \tau_0$ for any base type τ_0 and a binary function $f : \tau_1 \to \tau_2 \to \tau_0$ for any base types τ_0, τ_1 and τ_2.

3   From Linear Second-Order Unification to Context Unification with T ree-Regular Constraints

In this section, we prove that LSOU can be reduced to CU plus tree-regular constraints. This reduction is done in two steps. In subsection 3.1 we reduce the LSOU problem to the n-ary context unification problem by removing λ-bindings and constants with order higher than two. We obtain a context unification problem with n-ary contexts, i.e., second-order variables of arity n, plus tree-regular constraints. In subsection 3.2 we translate n-ary contexts to (1-ary) contexts.

3.1   Reducing Linear Second-Order Unification to n-ary Context Unification plus Tree-Regular Constraints

The translation from LSOU to n-ary CU has to remove λ-bindings from terms. Bound variables will be replaced by new constants. In second-order λ-calculus, λ-bindings of normal terms are alw ays just below higher-order constants or bound variables, or are the most external sym bol. They are never just below free variables. W e can eliminate external λ-bindings by extending the signature Σ with an appropriate new unary constant o (if it does not contain any one) and translating the equation $\lambda x.s \doteq_{lsou} \lambda y.t$ into $o(\lambda x.s) \doteq_{lsou} o(\lambda y.t)$. This new problem does not have external λ-bindings and is equivalent to the original one.

The elimination of in ternal λ-bindings is performed in three steps:

First, we conjecture an α-conversion of bound variables in order to allow unification when they are later translated into constants in the following step. Notice that the second step of this translation procedure depends on the "names" of these bound variables.

---

[2] Notice that a set of tree-regular constraints is equivalent to just one constraint.

Second, let $B \subset \mathcal{X}$ be a finite set of variables and let $A \subset \mathcal{L}istsof(B)$ be a finite set of lists of variables from B. We define a translation function $trans^{A,B}$ that replaces any occurrence of a variable of B by a new first-order constant, and any occurrence of a λ-abstraction, whose list of bound variables is in A, also by a new constant. This set B will be the set of bound variables of the unification problem resulting from step 1, and A will be the set of lists of variables used in the λ-abstractions.

The signature Σ' of the resulting n-ary CU problem also depends on the set B of bound variables conjectured in the previous step, and on the set A of lists of bound variables of the λ-abstractions. It is defined as follows: Σ' contains the same constants as Σ, but every constant h or bound variable z, with order higher than two, is replaced in Σ' by a new second-order constant h' or c_z, respectively. The arity of h' (similarly for c_z) is equal to the arity of h plus its number of non-first-order arguments. Any non-first-order n-ary argument of h with type $\tau_1 \to \cdots \to \tau_n \to \tau_0$ is replaced by two first-order arguments, one with a new special type o, and the other with the base type τ_0. For instance, if $h : \tau_1 \to (\tau_2 \to \tau_3) \to \tau_4$ then $h' : \tau_1 \to o \to \tau_3 \to \tau_4$. The signature Σ' also contains a new constant symbol $b_{[x_1,...,x_n]}$ of type o, for every list $[x_1, ..., x_n] \in A$, and a new constant symbol c_x, for every variable $x \in B$. The set of variables of the resulting problem is $\mathcal{X}' = \mathcal{X} \backslash B$.

Let $t \in T(\Sigma, \mathcal{X})$ be a term, $B \subset \mathcal{X}$ a set of variables, and $A \subset \mathcal{L}istsof(B)$ a set of lists of variables from B. The term $trans^{A,B}(t) \in T(\Sigma', \mathcal{X}')$ is defined by:

$$trans^{A,B}(c) = c$$

$$trans^{A,B}(f(t_1, ..., t_n)) = f(trans^{A,B}(t_1), ..., trans^{A,B}(t_n))$$

$$trans^{A,B}(X) = \begin{cases} X & \text{if } X \notin B \\ c_X & \text{if } X \in B \end{cases}$$

$$trans^{A,B}(F(t_1, ..., t_n)) = \begin{cases} F(trans^{A,B}(t_1), ..., trans^{A,B}(t_n)) & \text{if } F \notin B \\ c_F(trans^{A,B}(t_1), ..., trans^{A,B}(t_n)) & \text{if } F \in B \end{cases}$$

$$trans^{A,B}(h(t_1, ..., t_n, \lambda \boldsymbol{x_1}.u_1, ..., \lambda \boldsymbol{x_m}.u_m)) = ...$$
$$= h'(trans^{A,B}(t_1), ..., trans^{A,B}(t_n), b_{\boldsymbol{x_1}}, trans^{A,B}(u_1), ..., b_{\boldsymbol{x_m}}, trans^{A,B}(u_m))$$

$$trans^{A,B}(z(t_1, ..., t_n, \lambda \boldsymbol{x_1}.u_1, ..., \lambda \boldsymbol{x_m}.u_m)) = ...$$
$$= c_z(trans^{A,B}(t_1), ..., trans^{A,B}(t_n), b_{\boldsymbol{x_1}}, trans^{A,B}(u_1), ..., b_{\boldsymbol{x_m}}, trans^{A,B}(u_m))$$

$$trans^{A,B}(\lambda \boldsymbol{x}.t) = \lambda \boldsymbol{x}.trans^{A',B\backslash \boldsymbol{x}}(t)$$

In the fifth and sixth case, for constants h and variables z with order higher than two, we assume for simplicity that non-first-order parameters are in the last positions. The constant h' is the second-order constant associated to h, c_z is the constant associated to the variable z, and $b_{\boldsymbol{x_i}}$ is the constant associated to the list of variables $\boldsymbol{x_i} \in A$ of the λ-binding $\lambda \boldsymbol{x_i}$. If, for some $i \in [1..m]$, $\boldsymbol{x_i} \notin A$, then the translation is undefined. In the last case, A' is the set of lists A where any list containing variables from $\boldsymbol{x}$ has been removed. Notice that most external λ-bindings are not removed by this translation.

Third, we introduce a set of tree-regular restrictions over the instantiations of variables to prevent them from containing constants associated to bound variables from B without its corresponding λ-bindings.

The tree automata $\mathcal{A}^{A,B} = \langle \Sigma, \mathcal{Q}, \mathcal{Q}_f, \Delta \rangle$ that characterizes the set of terms that do not contain these bound variables from B in *free positions* is defined as follows.

- The signature contains the set of constants Σ'. Remember that Σ' allows us to ensure that, if a certain CU problem S is solvable then, S has a ground solution.
- The set of states is $\mathcal{Q} = \{q_X | X \subseteq B\} \cup \{p_X | X \in A\}$; where B is the set of bound variables and A is the set of λ-bindings.
- There is a single final state $\mathcal{Q}_f = \{q_\emptyset\}$
- The set of transitions Δ is defined as follows:
  - For any first-order constant $a \in \Sigma'$ not associated to a variable from B:

  $$a \to q_\emptyset$$

  - For any first-order constant c_x associated to a bound variable $x \in B$ and any first-order constant b_y associated to a list of bound variables $y \in A$:

  $$c_x \to q_{\{x\}}$$
  $$b_y \to p_y$$

  - For any second-order constant $f \in \Sigma'$ not associated to a variable from B and, for any constant c_x associated to a bound variable $x \in B$, and states $q_{A_1}, \ldots, q_{A_n}$:

  $$f(q_{A_1}, \ldots, q_{A_n}) \to q_{A_1 \cup \ldots \cup A_n}$$
  $$c_x(q_{A_1}, \ldots, q_{A_n}) \to q_{\{x\} \cup A_1 \cup \ldots \cup A_n}$$

  - For any second-order constant $h' \in \Sigma'$ associated to a higher-order constant $h \in \Sigma$ and, for any second-order constant $c_z \in \Sigma'$ associated to a higher-order bound variable $z \in B$:

  $$h'(q_{A_1}, \ldots q_{A_n}, p_{B_1}, q_{C_1}, \ldots, p_{B_m}, q_{C_m}) \to q_D$$
  $$c_z(q_{A_1}, \ldots q_{A_n}, p_{B_1}, q_{C_1}, \ldots, p_{B_m}, q_{C_m}) \to q_E$$

  where

  $$D = \bigcup_{i \in [1..n]} A_i \cup \bigcup_{j \in [1..m]} (C_j \setminus B_j)$$

  $$E = \{z\} \cup \bigcup_{i \in [1..n]} A_i \cup \bigcup_{j \in [1..m]} (C_j \setminus B_j)$$

  Notice that the B_k's are treated in the transitions as sets but they denote lists: $\lambda x, y$ is not the same λ-abstraction as $\lambda y, x$, so they have distinct associated constants but here are treated as the same set.

Then, we introduce a set of tree-regular restrictions over the solution σ of the translated problem.

- For any first-order variable X, the restriction $\sigma(X) \in L(\mathcal{A}^{A,B})$
- For any second-order variable F, the restriction $\sigma(F(a_1, ..., a_n)) \in L(\mathcal{A}^{A,B})$, where $a_i \in \Sigma'$ are first-order constants of the appropriate types.

Example 1. Given the problem $f(X, X) \doteq_{lsou} f(g(\lambda x.F(x)), F(g(\lambda y.y)))$ we can conjecture the following α-equivalent problem (this is the only solvable one) $f(X, X) \doteq_{lsou} f(g(\lambda x.F(x)), F(g(\lambda x.x)))$ and translate it into the following context unification problem with tree-regular constraints

$$f(X, X) \doteq_{cu} f(g'(b_{[x]}, F(c_x)), F(g'(b_{[x]}, c_x)))$$
$$\sigma(X) \in L(\mathcal{A}^{\{[x]\},\{x\}})$$
$$\sigma(F(a)) \in L(\mathcal{A}^{\{[x]\},\{x\}})$$

where the tree automata $\mathcal{A}^{\{[x]\},\{x\}}$ is defined by

$$a \to q_\emptyset \qquad\qquad\qquad b_{[x]} \to p_{\{x\}}$$
$$f(q_A, q_B) \to q_{A \cup B} \qquad\qquad c_x \to q_{\{x\}}$$
$$g'(p_A, q_B) \to q_{B \setminus A}$$

In the following lemmas we assume that all bound variables are in B and bindings in A.

Lemma 1. *For any second-order substitution σ satisfying $\sigma(X)$ does not contain variables of B in free positions, and the domain of σ neither contains variables of B, let $\tau = trans^{A,B}(\sigma)$ be the context substitution defined by $\tau(X) = trans^{A,B}(\sigma(X))$. Then, for any term t we have*

$$trans^{A,B}(\sigma(t)) = \tau(trans^{A,B}(t))$$

Lemma 2. *For any second-order term t the set of free variables of t and B are disjoint if and only if $trans^{A,B}(t) \in L(\mathcal{A}^{A,B})$.*

Theorem 1. *A LSOU problem $s \doteq_{lsou} t$ is unifiable if and only if there exists an α-equivalent unification problem $s' \doteq_{lsou} t'$ such that*

$$trans^{A,B}(s') \doteq_{cu} trans^{A,B}(t')$$

and the corresponding tree-regular constraints are solvable. Here, B is the set of bound variables of s' and t', and A is the set of lists of bound variables corresponding to λ-abstractions of s' and t'.

Corollary 1. *Linear second-order unification is reducible to n-ary context unification plus tree-regular constraints.*

3.2   Reducing n-ary Context Unification to (1-ary) Context Unification

In this subsection we reduce the n-ary CU problem to the (1-ary) CU problem. The same main ideas are also used in other previous papers, like [12]. Given an

n-ary context unification problem S over a signature Σ, if Σ does not contain an n-ary constant with $n \geq 2$ and a first-order constant, we enlarge it with them. We construct a new context unification problem S' by iteratively applying the following rule, until all non-unary context variables of the problem disappear.

For any n-ary context variable F with $n \geq 2$, we guess a p-ary constant symbol g, with $p \geq 2$, from the signature. Then, we guess a partition of $\{1, \ldots, n\}$ into $p \leq n$ many *disjoint* subsets such that $\bigcup_{i \in [1..p]} \{c_1^i, \ldots, c_{q_i}^i\} = \{1, \ldots, n\}$, and at least two of them are non-empty. We instantiate F by the following substitution:

$$[F \mapsto \lambda x_1 \cdots \lambda x_n . F_0(g(F_1(x_{c_1^1}, \ldots, x_{c_{q_1}^1}), \ldots, F_p(x_{c_1^p}, \ldots, x_{c_{q_p}^p})))]$$

where $F_0, \ldots, F_p$ are (maybe non-unary) context or first-order variables.

Example 2. Consider the following n-ary context unification problem:

$$X(Y(a, b)) \doteq_{lsou} Y(X(a), b) \tag{1}$$

where one of its infinitely many minimal solutions is (see Figure 1):

$$\sigma = [\, X \mapsto \lambda x. Z(Z(x, b), b), \\ Y \mapsto \lambda x, y. Z(Z(Z(Z(x, b), y), b), b)] \tag{2}$$

where Z is a fresh context variable. We enlarge our signature to $\Sigma' = \{a, b, g\}$, where g is a new binary constant. Now, we can guess a partition of $\{1, 2\}$ into two disjoint subsets $\{1\}$ and $\{2\}$, where both are non-empty, and instantiate Y by:

$$\tau = [Y \mapsto \lambda x_1, x_2. Y_0(g(Y_1(x_1), Y_2(x_2)))]$$

We obtain a new problem (see Figure 2):

$$X(Y_0(g(Y_1(a), Y_2(b)))) \doteq_{cu} Y_0(g(Y_1(X(a)), Y_2(b))) \tag{3}$$

which is also solvable, and only contains (unary) context variables.

Theorem 2. *n-ary context unification is NP-reducible to (1-ary) context unification.*

4   Translating Tree-Regular Constraints to Regular Constraints over Traversal Sequences

The decidability of context unification with tree-regular constraints, as well as the decidability of context unification, are still open problems. There is a proof that, if most general context unifiers are rank-bound, then CU is decidable [7]. However, it is not known if most general context unifiers are in general rank-bound. In this section we will show that, if this is the case, then the decidability proof of context unification could be extended to context unification with

$$X(Y(a,b))\qquad\qquad Y(X(a),b)$$

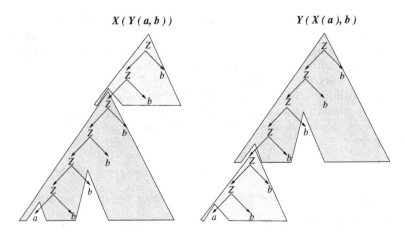

Fig. 1. A solution of the LSOU problem (1)

tree-regular constraints. Therefore, linear second-order unification would also be decidable.

This mentioned proof is based on a reduction of context unification to word unification with regular constraints [16], where terms are translated into sequences of symbols (traversal sequences). In the following we present the main ideas of this reduction.

Definition 1. *Given a signature $\langle \Sigma, \mathcal{X} \rangle$, we define the* extended signature

$$\Sigma^{\Pi} = \{ f^{\rho} \mid f \in \Sigma \cup \mathcal{X} \ \wedge \ \rho \in \Pi_{\mathrm{arity}(f)} \}$$

where Π_n is the group of permutations over n elements.
A sequence $s \in (\Sigma^{\Pi})^*$ is said to be a traversal sequence *of a term t, noted $s \in \mathrm{trav}(t)$, if:*

1. *$s = t$ when $t = c$ is a 0-ary symbol*
2. *$s = f^{\rho} s_{\rho(1)} \cdots s_{\rho(n)}$ when $t = f(t_1, \ldots, t_n)$ being s_i traversal sequences of t_i for any $i \in [1..n]$, and $\rho \in \Pi_n$ a permutation.*

Any traversal sequence of a term characterizes this term. We use an extended signature with permutations in order to allow us the use of distinct traversals, i.e. the traversals of subterms in distinct possible orders.

Definition 2. *The* rank of a term, $\mathrm{rank}(t)$, *is defined by $\mathrm{rank}(a) = 0$, for any constant a, and $\mathrm{rank}(f(t_1, \ldots, t_n)) = c$ where c is the minimum integer satisfying: there exists a permutation τ of indices $1, \ldots, n$ such that, for any $i \in [1..n]$, $\mathrm{rank}(t_{\tau_i}) \leq c - n + i$.*

$X(Y0(g(Y1(a),Y2(b))))$ $Y0(g(Y1(X(a)),Y2(b)))$

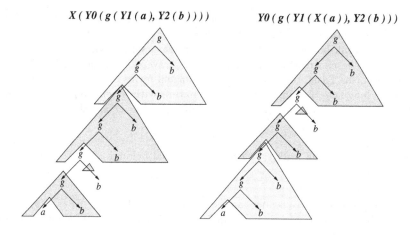

Fig. 2. A reduction of the LSOU problem (1) to context unification

This definition is bizarre, but it can be simplified for binary trees as follows:[3]

$$\text{rank}(a) = 0$$
$$\text{rank}(f(t_1, t_2)) = \begin{cases} \text{rank}(t_1) + 1 & \text{if rank}(t_1) = \text{rank}(t_2) \\ \max\{\text{rank}(t_1), \text{rank}(t_2)\} & \text{otherwise} \end{cases}$$

The rank of a term allows us to define a normal traversal.

Definition 3. *Given a term t, its normal traversal sequence* $\text{NF}(t)$ *is defined recursively as follows:*

1. *If* $t = a$ *then* $\text{NF}(t) = a$.
2. *If* $t = f(t_1, \ldots, t_n)$ *then let be* $\rho \in \Pi_n$ *the permutation satisfying*

$$i < j \Rightarrow \big(\text{rank}(t_{\rho(i)}) < \text{rank}(t_{\rho(j)})$$
$$\vee \text{rank}(t_{\rho(i)}) = \text{rank}(t_{\rho(j)}) \wedge \rho(i) < \rho(j)\big)$$

Then, $\text{NF}(t) = f^\rho \ \text{NF}(t_{\rho(1)}) \ \cdots \ \text{NF}(t_{\rho(n)})$.

Notice that a restriction on the rank of a tree does not imply a restriction on its size. The following conjecture states that the rank of any most general context unifier is bound:

Conjecture 1. For any solvable context unification problem $t \doteq u$ and m.g.u. σ, we have

$$\text{rank}(\sigma(t)) \leq \phi(\text{size}(t \doteq u))$$

where ϕ is a computable function.

---

[3] Alternatively, we can also define the rank of a binary tree t as the depth of the maximum complete tree t' (a tree where all leaves are at the same depth) such that there exist an injective morphism from t' to t.

Traversal sequences of rank-bound terms and regular language are related. For any signature Σ and bound n, there exists a regular language R_Σ^n such that, for any n-rank-bound term t there exists a traversal sequence $w \in \text{trav}(t)$ with $w \in R_\Sigma^n$. We can restrict the choice of the traversal sequence to the traversal normal form. Thus, the set of traversal normal forms of rank-bound terms is a regular language. For instance, for a signature with a constant a and a binary function f, any term satisfying $\text{rank}(t) \leq 1$ has a traversal belonging to:

$$R^1 \equiv ((f^{[1,2]} \mid f^{[2,1]})\, a)^*\, a$$

and those satisfying $\text{rank}(t) \leq 2$ have a traversal belonging to:

$$R^2 \equiv ((f^{[1,2]} \mid f^{[2,1]})\, R^1)^*\, a$$

This allows us to reduce any CU problem, like $X(f(Y(a), Z(b))) \doteq_{cu} Y(f(X(a), Z(b)))$ to some word unification problem plus *traversal equations*, like:

$$X_0\ f^{[1,2]}\ Y_0\ a\ Y_1\ Z_0\ b\ Z_1\ X_1 \doteq_{wu} Y_0'\ f^{[1,2]}\ X_0'\ a\ X_1'\ Z_0'\ b\ Z_1'\ Y_1'$$
$$X_0\ c\ X_1 \equiv X_0'\ c\ X_1'$$
$$Y_0\ c\ Y_1 \equiv Y_0'\ c\ Y_1'$$
$$Z_0\ c\ Z_1 \equiv Z_0'\ c\ Z_1'$$

where the words X_0 and X_1 encode a traversal sequence of the context X, and X_0' and X_1' another traversal of X. The intended meaning of $w_1 \equiv w_2$ is: w_1 and w_2 are *similar* traversal sequences of the same term. By similar we mean that we can bound the number of permutations in which w_1, w_2 and $\text{NF}(t)$ differ, where $w_1, w_2 \in \text{trav}(t)$. If the rank of this term is bound, then we can non-deterministically reduce these traversal equations to word equations plus regular restrictions like

$$\begin{aligned}
X_0\ c\ X_1 &\doteq_{wu} X_0'\ c\ X_1' & X_0\ c\ X_1 &\in R^{\phi(\text{size})} \\
Y_0\ c\ Y_1 &\doteq_{wu} Y_0'\ c\ Y_1' & Y_0\ c\ Y_1 &\in R^{\phi(\text{size})} \\
Z_0\ c\ Z_1 &\doteq_{wu} Z_0'\ c\ Z_1' & Z_0\ c\ Z_1 &\in R^{\phi(\text{size})}
\end{aligned}$$

The restriction $X_0\ c\ X_1 \in R^{\phi(\text{size})}$ ensures that the instances we find for these words are really traversal sequences.

In what follows we show how membership equations on tree-regular languages of rank-bound terms can be reduced to membership equations on (word) regular languages. We start by defining rank-bound tree automata.

Definition 4. *For any tree automata* $\mathcal{A} = \langle \Sigma, \mathcal{Q}, \mathcal{Q}_f, \Delta \rangle$, *and any state* $q_i \in \mathcal{Q}$, *we define*[4]

$$\text{rank}(q_i) = \max\{\text{rank}(t) \mid t \in L(\langle \Sigma, \mathcal{Q}, \{q_i\}, \Delta \rangle)\}$$

For any tree automata $\mathcal{A} = \langle \Sigma, \mathcal{Q}, \mathcal{Q}_f, \Delta \rangle$, *we define*

$$\text{rank}(\mathcal{A}) = \max\{\text{rank}(q_i) \mid q_i \in \mathcal{Q}_f\}$$

A tree automata $\mathcal{A}$ *is said to be bound if* $\text{rank}(\mathcal{A}) < \infty$.

---

[4] Notice that $\langle \Sigma, \mathcal{Q}, \{q_i\}, \Delta \rangle$ is similar to $\mathcal{A}$ but with a unique final state q_i.

Notice that the rank of the states of a tree automata satisfies the following property

- for any state q having only transitions like $c \to q$, where $c \in \Sigma$ is a 0-ary constant, $rank(q) = 0$,
- for any accessible state q_0 having transitions like $f(q_1, q_2) \to q_0$, where $f \in \Sigma$ is a binary function, we have:

$$rank(q_0) \geq \begin{cases} max\{rank(q_1), rank(q_2)\} & \text{if } rank(q_1) \neq rank(q_2) \\ rank(q_1) + 1 & \text{if } rank(q_1) = rank(q_2) \end{cases}$$

We translate tree-regular restrictions to (word) regular restrictions over traversal sequences of terms. For simplicity assume that any symbol of the signature is, at most, binary. The result can easily be extended to any signature.

For any rank-bound tree automata $\mathcal{A} = \langle \Sigma, \mathcal{Q}, \mathcal{Q}_f, \Delta \rangle$, and any state $q \in \mathcal{Q}$, we define a regular language R_q satisfying $R_q \cap \mathrm{trav}(t) \neq \emptyset$, for any term $t \in L(\langle \Sigma, \mathcal{Q}, \{q\}, \Delta \rangle)$, and $R_q \subseteq \bigcup\{\mathrm{trav}(t) \mid t \in L(\langle \Sigma, \mathcal{Q}, \{q\}, \Delta \rangle)\}$.

We will construct the automata that recognizes R_q using the following rules. Assume that $R_{q'}$ is already computed for any state $q' \in \mathcal{Q}$ with $rank(q') < rank(q)$. Let be $n = rank(q)$. The automata R_q has a pair of states p^i and p^f for any state p of the tree automata satisfying $rank(p) = n$, and some additional states that we will specify later. The initial state of R_q is q^i, and there is a single final state and it is q^f. The set of transitions of R_q is defined as follows.

- *Base case*, for any state $p \in \mathcal{Q}$ satisfying $rank(p) = n$, and any transition $a \to p \in \Delta$ we add a transition:

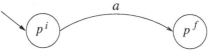

from p^i to p^f labeled with a.
- *Inductive case 1*, for any state p_0 with $rank(p_0) = n$ and any transition $f(p_1, p_2) \to p_0$ satisfying $rank(p_2) < rank(p_1) \leq rank(p_0)$ [5]

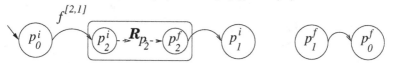

we can assume that R_{p_2} is already computed. We add a copy of the automata R_{p_2}, i.e. a copy of all its states and transitions (these are the unspecified additional states). We also add a transition from p_0^i to the initial state of the copy R_{p_2} labeled with $f^{[2,1]}$, a λ-transition from the final state of R_{p_2} to p_1^i, and another from p_1^f to p_0^f.

For any transition $f(p_1, p_2) \to p_0$ satisfying $rank(p_1) < rank(p_2) \leq rank(p_0)$ we do something similar using the label $f^{[1,2]}$

---

[5] Notice that if $rank(p_1) = rank(p_2) = rank(p_0)$ then $rank(p_0) = \infty$ and the tree automata would be non-rank-bound. Thus the existence of a bound n for the rank of the tree automata is crucial in our translation.

– *Inductive case 2*, for any state p_0 with $\text{rank}(p_0) = n$ and any transition $f(p_1, p_2) \to p_0$ satisfying $\text{rank}(p_1) < \text{rank}(p_0)$ and $\text{rank}(p_2) < \text{rank}(p_0)$

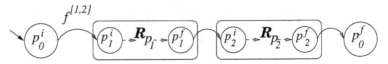

we can assume that R_{p_1} and R_{p_2} have been already computed. We add a copy of each one of these automata, a transition from p_0^i to the initial state of R_{p_1} labeled with $f^{[1,2]}$, a λ-transition from the final state of R_{p_1} to the initial state of R_{p_2}, and another from the final state of R_{p_2} to p_0^f.

Notice that with these cases, transitions like $f(q, q) \to q$ are not considered, because this means that $\text{rank}(q) = \infty$, and q can not lead to a final state. The final automata associated to $\mathcal{A}$ consists of an initial state q_0, a copy of R_q for any final state $q \in \mathcal{Q}_f$, a λ-transitions from q_0 to each one of the initial states of the R_q's. The set of final states is the set of final states of the R_q's.

Example 3. The tree automata defined by the following transitions

$$0 \to q_N, \qquad pair(q_N, q_N) \to q_P, \qquad nil \to q_L,$$
$$s(q_N) \to q_N, \qquad\qquad\qquad cons(q_P, q_L) \to q_L$$

is translated into the following regular automata:

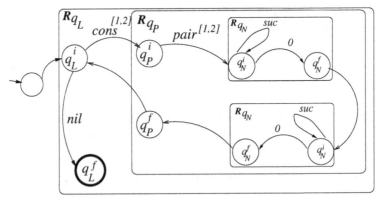

The term $cons(pair(suc(suc(0)), suc(0)), cons(pair(suc(0), 0), nil))$ recognized by the tree automata, has a traversal sequence

$$cons^{[1,2]} \; pair^{[1,2]} \; suc \; suc \; 0 \; suc \; 0 \; cons^{[1,2]} \; pair^{[1,2]} \; suc \; 0 \; 0 \; nil$$

recognized by the regular automata.

Theorem 3. *For any tree-regular language $L(\mathcal{A})$ of a rank-bound tree automata $\mathcal{A}$, let $L(B)$ be the regular language recognized by the automata B resulting from applying the previous translation. The following properties hold.*

1. *If $t \in L(\mathcal{A})$ then there exist a sequence $l \in L(B)$ such that $l \in \text{trans}(t)$.*
2. *If $l \in L(B)$ then there exist a term $t \in L(\mathcal{A})$ such that $l \in \text{trans}(t)$.*

The set of terms satisfying $\text{rank}(t) \leq n$ defines a tree-regular language. Moreover, this language can be recognized by a rank-bound tree automata $\mathcal{A}_n$. For instance, if $\Sigma = \{a, b, h(), f(,)\}$, then

$$
\mathcal{A}_n = \begin{cases}
\Sigma = \{a, b, h(), f(,)\}, \quad \mathcal{Q} = \{q_0, ..., q_n, q_{n+1}\}, \quad \mathcal{Q}_f = \{q_0, ..., q_n\}, \\
\Delta = \begin{cases}
a \to q_0, \quad b \to q_0, \\
h(q_i) \to q_i & \text{for any } i \in [0..n+1] \\
f(q_i, q_j) \to q_{max\{i,j\}} & \text{for any } i, j \in [0..n+1] \text{ with } i \neq j \\
f(q_i, q_i) \to q_{i+1} & \text{for any } i \in [0..n] \\
f(q_{n+1}, q_{n+1}) \to q_{n+1}
\end{cases}
\end{cases}
$$

Definition 5. *A tree-regular language L is said to be n rank-bound if for any term $t \in L$ we have $\text{rank}(t) \leq n$.*

Theorem 4. *Any rank-bound regular language is recognized by a rank-bound tree automata. The language recognized by a rank-bound tree automata is a rank-bound tree-regular language.*

5 Extending the Results to Higher-Order Unification

In Section 3 we have shown how linear second-order unification can be reduced to context unification with tree-regular constraints. In this section we discuss whether this result could be extended to linear higher-order unification.

Higher-order unification can be defined as the problem of finding a substitution σ making the normal form of t wo instances of terms $\sigma(s)$ and $\sigma(t)$ equal. When w e try to find such a substitution we have to take into account how this terms will β-reduce after being instantiated. The problem is simple in linear second-order. We know that any instance of $F(t_1, ..., t_n)$, after β-reduction, will contain $\sigma(t_i)$ as subterms, and representing $\sigma(F(t_1, ..., t_n))$ as a tree, all nodes corresponding to $\sigma(F)$ will be connected, forming a context. In third-order the situation is more complicate. First, we have to require instances of variables to be linear in all λ-bindings, i.e. not only in the most external λ-bindings. If we apply the substitution $F \mapsto \lambda y.\lambda z.g_1(y(g_2(z)))$ to $F(\lambda x.f(x), a)$ we get $g_1(f(g_2(a)))$. The nodes corresponding to F are no longer connected: $\sigma(F)$ is broken into pieces, and some of the argumen ts can also disappear. Each one of such pieces forms a kind of context. For instance, if $F : (o \to o) \to o \to o$, any instance of $F(t_1, t_2)$ has one of the following forms:

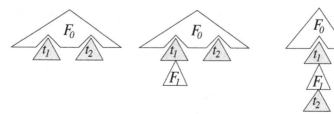

Each one of these situations is captured respectively by:

$$F \mapsto \lambda x.\lambda y.F_0(\lambda z.x(z), y)$$
$$F \mapsto \lambda x.\lambda y.F_0(\lambda z.x(F_1(z)), y)$$
$$F \mapsto \lambda x.\lambda y.F_0(\lambda z.x(F_1(y, z)))$$

In this example, F_0 is still a third-order typed variable. Moreover, the first instantiation is $F \mapsto F_0$ in normal form, so it subsumes the other two. The second one is equal to the first one, if z contains a single variable and we instantiate $F_1 \mapsto \lambda z.z$. In fact, this classification only makes sense if we translate F_0 into a context variable using the method described in Section 3 for higher-order constants. We would get:

$$F \mapsto \lambda x.\lambda y.F_0'(d_z, x(c_z), y)$$
$$F \mapsto \lambda x.\lambda y.F_0'(X_z, x(F_1(c_z)), y)$$
$$F \mapsto \lambda x.\lambda y.F_0'(X_z, x(F_1(y, c_z)))$$

The variable X_z encodes the binding λz. If we were able to know a priori how long and, with which types, can be these λ-bindings, then the translation would not seem much more complicate than in the second-order case.

Acknowledgments

We thank M. Bonet and all the anonymous referees for their helpful comments on the paper.

References

[1] H. Comon. Completion of rewrite systems with membership constraints. *J. of Symbolic Computation*, 25(4):397–453, 1998.

[2] H. Comon, M. Dauchet, R. Gilleron, F. Jacquemard, D. Lugiez, S. Tison, and M. Tommasi. Tree automata techniques and applications. Available on: http://www.grappa.univ-lille3.fr/tata, 1997.

[3] J. Levy. Linear second-order unification. In *7th Int. Conf. on Rewriting Techniques and Applications, RTA'96*, volume 1103 of *LNCS*, pages 332–346, New Jersey, USA, 1996.

[4] J. Levy. Decidable and undecidable second-order unification problems. In *9th Int. Conf. on Rewriting Techniques and Applications, RTA'98*, volume 1379 of *LNCS*, pages 47–60, Tsukuba, Japan, 1998.

[5] J. Levy and J. Agustí. Bi-rewrite systems. *J. of Symbolic Computation*, 22:279–314, 1996.

[6] J. Levy and M. Veanes. On the undecidability of second-order unification. *Information and Computation*, 2000. (to appear).

[7] J. Levy and M. Villaret. On the decidability of context unification. Technical report, IIIA, CSIC, 2000.

[8] G. S. Makanin. The problem of solvability of equations in a free semigroup. *Math. USSR Sbornik*, 32(2):129–198, 1977.

[9] J. Niehren, M. Pinkal, and P. Ruhrberg. On equality up-to constraints over finite trees, context unification, and one-step rewriting. In *14th International Conference on Automated Deduction, CADE-14*, volume 1249 of *LNCS*, pages 34–48, Townsville, North Queensland, Australia, 1997.

[10] M. Schmidt-Schauß. An algorithm for distributive unification. In *7th Int. Conf. on Rewriting Techniques and Applications, RTA '96*, volume 1103 of *LNCS*, pages 287–301, New Jersey, USA, 1996.

[11] M. Schmidt-Schauß. A decision algorithm for distributive unification. *Theoretical Computer Science*, 208:111–148, 1998.

[12] M. Schmidt-Schauß. Decidability of bounded second-order unification. Technical Report Frank-report-11, FB Informatik, J.W. Goethe Universität Frankfurt, 1999.

[13] M. Schmidt-Schauß. A decision algorithm for stratified context unification. Technical Report Frank-report-12, FB Informatik, J.W. Goethe Universität Frankfurt, 1999.

[14] M. Schmidt-Schauß and K.U. Schulz. On the exponent of periodicity of minimal solutions of context equations. In *9th Int. Conf. on Rewriting Techniques and Applications, RTA '98*, volume 1379 of *LNCS*, pages 61–75, Tsukuba, Japan, 1998.

[15] M. Schmidt-Schauß and K.U. Schulz. Solvability of context equations with two context variables is decidable. In *Conference on Automated Deduction, CADE'99*, LNCS, pages 67–81, 1999.

[16] K. U. Schulz. Makanin's algorithm, two improvements and a generalization. Technical Report CIS-Bericht-91-39, Centrum für Informations und Sprachverarbeitung, Universität München, 1991.

Word Problems and Confluence Problems for Restricted Semi-Thue Systems

Markus Lohrey

Universität Stuttgart, Institut für Informatik
Breitwiesenstr. 20–22, 70565 Stuttgart, Germany
lohreyms@informatik.uni-stuttgart.de
Phone: +49-711-7816408,   Fax: +49-711-7816310

Abstract. We investigate word problems and confluence problems for the following four classes of terminating semi-Thue systems: length-reducing systems, weight-reducing systems, length-lexicographic systems, and weight-lexicographic systems. For each of these four classes we determine the complexity of several variants of the word problem and confluence problem. Finally we show that the variable membership problem for quasi context-sensitive grammars is EXPSPACE-complete.

1 Introduction

The main purpose of semi-Thue systems is to solve word problems for finitely presented monoids. But since there exists a fixed semi-Thue system $\mathcal{R}$ such that the word problem for the monoid presented by the set of equations that corresponds to $\mathcal{R}$ (in the following we will briefly speak of the word problem for $\mathcal{R}$) is undecidable [Mar47, Pos47], also semi-Thue systems cannot help for the effective solution of arbitrary word problems. This motivates the investigation of restricted classes of semi-Thue systems which give rise to decidable word problems. One of the most prominent class of semi-Thue systems with a decidable word problem is the class of all terminating and confluent semi-Thue systems. But if we want to have efficient algorithms for the solution of word problems also this class might be too large: It is known that for every $n \geq 3$ there exists a terminating and confluent semi-Thue system $\mathcal{R}$ such that the (characteristic function of the) word problem for $\mathcal{R}$ is contained in the nth Grzegorczyk class but not in the $(n-1)$th Grzegorczyk class [BO84]. Thus the complexity of the word problem for a terminating and confluent semi-Thue system can be extremely high. One way to reduce the complexity of the word problem is to force bounds on the length of derivation sequences. Such a bound can be forced by restricting to certain subclasses of terminating systems. For instance it is known that for a length-reducing and confluent semi-Thue system the word problem can be solved in linear time [Boo82]. On the other hand in [Loh99] it was shown that a uniform variant of the word problem for length-reducing and confluent semi-Thue systems (where the semi-Thue system is also part of the input) is P-complete.

In this paper we will continue the investigation of the word problem for restricted classes of terminating and confluent semi-Thue systems. We will study

L. Bachmair (Ed.): RTA 2000, LNCS 1833, pp. 172–186, 2000.
© Springer-Verlag Berlin Heidelberg 2000

the following four classes of semi-Thue systems, see e.g. also [BO93], pp 41–42: length-reducing systems, weight-reducing systems, length-lexicographic systems, and weight-lexicographic systems. Let $\mathcal{C}$ be one of these four classes. We will study the following five decision problems for $\mathcal{C}$: (i) the word problem for a fixed confluent $\mathcal{R} \in \mathcal{C}$, where the input consists of two words, (ii) the uniform word problem for a fixed alphabet Σ, where the input consists of a confluent semi-Thue system $\mathcal{R} \in \mathcal{C}$ over the alphabet Σ and two words, (iii) the uniform word problem, where the input consists of a confluent $\mathcal{R} \in \mathcal{C}$ and two words, (iv) the confluence problem for a fixed alphabet Σ, where the input consists of a semi-Thue system $\mathcal{R} \in \mathcal{C}$ over the alphabet Σ, and finally (v) the confluence problem, where the input consists of an arbitrary $\mathcal{R} \in \mathcal{C}$. For each of the resulting 20 decision problems we will determine its complexity, see Table 1, p 182, and Table 2, p 183. Finally we consider a problem from [BL94], the variable membership problem for quasi context-sensitive grammars. This problem was shown to be in EXPSPACE but NEXPTIME-hard in [BL94]. In this paper we will prove that this problem is EXPSPACE-complete. We assume that the reader is familiar with the basic notions of complexity theory, in particular with the complexity classes P, PSPACE, EXPTIME, and EXPSPACE, see e.g. [Pap94].

2   Preliminaries

In the following let Σ be a finite alphabet. The empty word will be denoted by ϵ. A *weight-function* is a homomorphism $f : \Sigma^* \to \mathbb{N}$ from the free monoid Σ^* with concatenation to the natural numbers with addition such that $f(s) = 0$ if and only if $s = \epsilon$. The weight-function f with $f(a) = 1$ for all $a \in \Sigma$ is called the *length-function*. In this case for a word $s \in \Sigma^*$ we abbreviate $f(s)$ by $|s|$ and call it the *length* of s. Furthermore for every $a \in \Sigma$ we denote by $|s|_a$ the number of different occurrences of the symbol a in s. For a binary relation $\to$ on some set, we denote by $\xrightarrow{+} (\xrightarrow{*})$ the transitive (reflexive and transitive) closure of $\to$.

In this paper, a *deterministic Turing-machine* is a tuple $\mathcal{M} = (Q, \Sigma, \delta, q_0, q_f)$, where Q is the finite set of states, Σ is the tape alphabet with $\Sigma \cap Q = \emptyset$, $\delta : (Q \backslash \{q_f\}) \times \Sigma \to Q \times \Sigma \times \{-1, +1\}$ is the transition function, where -1 $(+1)$ means that the read-write head moves to the left (right), $q_0 \in Q$ is the initial state, and $q_f \in Q$ is the unique final state. The tape alphabet Σ always contains a blank symbol $\square$. We assume that $\mathcal{M}$ has a one-sided infinite tape, whose cells can be identified with the natural numbers. Note that $\mathcal{M}$ cannot perform any transition out of the final state q_f. These assumptions do not restrict the computational power of Turing-machines and will always be assumed in this paper. An input for $\mathcal{M}$ is a word $w \in (\Sigma \backslash \{\square\})^*$. A word of the form uqv, where $u, v \in \Sigma^*$ and $q \in Q$, codes the configuration, where the machine is in state q, the cells 0 to $|uv| - 1$ contain the word uv, all cells k with $k \geq |uv|$ contain the blank symbol $\square$, and the read-write head is scanning cell $|u|$. We write $sqt \Rightarrow_{\mathcal{M}} upv$ if $\mathcal{M}$ can move in one step from the configuration sqt to the configuration upv, where $q, p \in Q$ and $st, uv \in \Sigma^*$. The language that is accepted by $\mathcal{M}$ is defined by $L(\mathcal{M}) = \{w \in (\Sigma \backslash \{\square\})^* \mid \exists u, v \in \Sigma^* : q_0 w \xrightarrow{+}_{\mathcal{M}} uq_f v\}$. Note that $w \in L(\mathcal{M})$

if and only if $\mathcal{M}$ terminates on the input w. A *deterministic linear bounded automaton* is a deterministic Turing-machine that operates in space $n+1$ on an input of length n.

A *semi-Thue system* $\mathcal{R}$ over Σ, briefly STS, is a finite set $\mathcal{R} \subseteq \Sigma^* \times \Sigma^*$, whose elements are called rules. See [BO93] for a good introduction to the theory of semi-Thue systems. A rule (s,t) will also be written as $s \to t$. The sets $\mathrm{dom}(\mathcal{R})$ of all left-hand sides and $\mathrm{ran}(\mathcal{R})$ of all right-hand sides are defined by $\mathrm{dom}(\mathcal{R}) = \{s \mid \exists t : (s,t) \in \mathcal{R}\}$ and $\mathrm{ran}(\mathcal{R}) = \{t \mid \exists s : (s,t) \in \mathcal{R}\}$. We define the two binary relations $\to_{\mathcal{R}}$ and $\leftrightarrow_{\mathcal{R}}$ as follows, where $x, y \in \Sigma^*$:

- $x \to_{\mathcal{R}} y$ if there exist $u, v \in \Sigma^*$ and $(s,t) \in \mathcal{R}$ with $x = usv$ and $y = utv$.
- $x \leftrightarrow_{\mathcal{R}} y$ if $(x \to_{\mathcal{R}} y$ or $y \to_{\mathcal{R}} x)$.

The relation $\overset{*}{\leftrightarrow}_{\mathcal{R}}$ is a congruence relation with respect to the concatenation of words, it is called the *Thue-congruence* associated with $\mathcal{R}$. Hence we can define the quotient monoid $\Sigma^*/ \overset{*}{\leftrightarrow}_{\mathcal{R}}$, which is briefly denoted by $\Sigma^*/\mathcal{R}$. We say that $\mathcal{R}$ is *terminating* if there does not exist an infinite sequence of words $s_i \in \Sigma^*$ ($i \in \mathbb{N}$) with $s_0 \to_{\mathcal{R}} s_1 \to_{\mathcal{R}} s_2 \to_{\mathcal{R}} \cdots$. The set of *irreducible words* with respect to $\mathcal{R}$ is $\mathrm{IRR}(\mathcal{R}) = \Sigma^*\backslash\{stu \in \Sigma^* \mid s, u \in \Sigma^*, t \in \mathrm{dom}(\mathcal{R})\}$. A word t is a *normal form* of s if $s \overset{*}{\to}_{\mathcal{R}} t \in \mathrm{IRR}(\mathcal{R})$. We say that $\mathcal{R}$ is *confluent* if for all $s, t, u \in \Sigma^*$ with $s \overset{*}{\to}_{\mathcal{R}} t$ and $s \overset{*}{\to}_{\mathcal{R}} u$ there exists $w \in \Sigma^*$ with $t \overset{*}{\to}_{\mathcal{R}} w$ and $u \overset{*}{\to}_{\mathcal{R}} w$. We say that $\mathcal{R}$ is *locally confluent* if for all $s, t, u \in \Sigma^*$ with $s \to_{\mathcal{R}} t$ and $s \to_{\mathcal{R}} u$ there exists $w \in \Sigma^*$ with $t \overset{*}{\to}_{\mathcal{R}} w$ and $u \overset{*}{\to}_{\mathcal{R}} w$. If $\mathcal{R}$ is terminating then by Newman's lemma [New43] $\mathcal{R}$ is confluent if and only if $\mathcal{R}$ is locally confluent.

Two decision problems that are of fundamental importance in the theory of semi-Thue systems are the (uniform) word problem and the confluence problem. Let $\mathcal{C}$ be a class of STSs. The *uniform word problem*, briefly UWP, for the class $\mathcal{C}$ is the following decision problem:

INPUT: An STS $\mathcal{R} \in \mathcal{C}$ (over some alphabet Σ) and two words $u, v \in \Sigma^*$.

QUESTION: Does $u \overset{*}{\leftrightarrow}_{\mathcal{R}} v$ hold?

The *confluence problem*, briefly CP, for the class $\mathcal{C}$ is the following decision problem:

INPUT: An STS $\mathcal{R} \in \mathcal{C}$.

QUESTION: Is $\mathcal{R}$ confluent?

The UWP for a singleton class $\{\mathcal{R}\}$ is called the *word problem*, briefly WP, for the STS $\mathcal{R}$.

For the class of all terminating STSs the CP is known to be decidable [NB72]. This classical result is based on the so called critical pairs of an STS, which result from overlapping left-hand sides. A pair $(s_1, s_2) \in \Sigma^* \times \Sigma^*$ is a critical pair of $\mathcal{R}$ if there exist rules $(t_1, u_1), (t_2, u_2) \in \mathcal{R}$ such that one of the following two cases holds:

- $t_1 = vt_2w$, $s_1 = u_1$, and $s_2 = vu_2w$ for some $v, w \in \Sigma^*$ (here the word $t_1 = vt_2w$ is an overlapping of t_1 and t_2).
- $t_1 = vt$, $t_2 = tw$, $s_1 = u_1w$, and $s_2 = vu_2$ for some $t, v, w \in \Sigma^*$ with $t \neq \epsilon$ (here the word vtw is an overlapping of t_1 and t_2).

Note that there are only finitely many critical pairs of $\mathcal{R}$. In order to check whether a terminating STS $\mathcal{R}$ is confluent it suffices to calculate for every critical pair (s, t) of $\mathcal{R}$ arbitrary normal forms of s and t. If for some critical pair these normal forms are not identical then $\mathcal{R}$ is not confluent, otherwise $\mathcal{R}$ is confluent.

Similarly for the class of all terminating and confluent STSs the UWP is decidable [KB67]: In order to check whether $s \overset{*}{\leftrightarrow}_{\mathcal{R}} t$ holds for given words $s, t \in \Sigma^*$ we compute arbitrary normal forms of s and t. Then $s \overset{*}{\leftrightarrow}_{\mathcal{R}} t$ if and only if these normal forms are the same.

In this paper we consider the following classes of terminating systems. An STS $\mathcal{R}$ is *length-reducing* if $|s| > |t|$ for all $(s, t) \in \mathcal{R}$. An STS $\mathcal{R}$ is *weight-reducing* if there exists a weight-function f such that $f(s) > f(t)$ for all $(s, t) \in \mathcal{R}$. An STS $\mathcal{R}$ is *length-lexicographic* if there exists a linear order $\succ$ on the alphabet Σ such that for all $(s, t) \in \mathcal{R}$ it holds $|s| > |t|$ or ($|s| = |t|$ and there exist $u, v, w \in \Sigma^*$ and $a, b \in \Sigma$ with $s = uav$, $t = ubw$, and $a \succ b$). An STS $\mathcal{R}$ is *weight-lexicographic* if there exist a linear order $\succ$ on the alphabet Σ and a weight-function f such that for all $(s, t) \in \mathcal{R}$ it holds $f(s) > f(t)$ or ($f(s) = f(t)$ and there exist $u, v, w \in \Sigma^*$ and $a, b \in \Sigma$ with $s = uav$, $t = ubw$, and $a \succ b$).

For all these classes, restricted to confluent STSs, the UWP is decidable. Since we want to determine the complexity of the UWP for these classes, we have to define the *length of an STS*. For an STS $\mathcal{R}$ it is natural to define its length $\|\mathcal{R}\|$ by $\|\mathcal{R}\| = \sum_{(s,t) \in \mathcal{R}} |st|$.

3   Length-Reducing Semi-Thue Systems

In [Loh99] it was shown that the UWP for the class of all length-reducing and confluent STSs over $\{a, b\}$ is P-complete. In this section we prove that for a fixed STS the complexity decreases to LOGCFL. Recall that LOGCFL is the class of all problems that are log space reducible to the membership problem for a context-free language [Sud78]. It is strongly conjectured that LOGCFL is a proper subset of P.

Theorem 1. *Let $\mathcal{R}$ be a fixed length-reducing and confluent STS. Then the WP for $\mathcal{R}$ is in LOGCFL.*

Proof. Let $\mathcal{R}$ be a fixed length-reducing STS over Σ and let $s \in \Sigma^*$. From the results of [DW86] [1] it follows immediately that the following problem is in LOGCFL:
INPUT: A word $t \in \Sigma^*$.
QUESTION: Does $t \overset{*}{\rightarrow}_{\mathcal{R}} s$ hold?
Now let $\mathcal{R}$ be a fixed length-reducing and confluent STS over Σ and let $u, v \in \Sigma^*$. Let $\overline{\Sigma} = \{\overline{a} \mid a \in \Sigma\}$ be a disjoint copy of Σ. For a word $s \in \Sigma^*$ define the word

---

[1] The main result of [DW86] is that the membership problem for a fixed growing context-sensitive grammar is in LOGCFL. Note that the uniform variant of this problem is NP-complete [KN87, CH90, BL92].

$\overline{s}^{\mathrm{rev}} \in \overline{\Sigma}^*$ inductively by $\overline{\epsilon}^{\mathrm{rev}} = \epsilon$ and $\overline{at}^{\mathrm{rev}} = \overline{t}^{\mathrm{rev}}\overline{a}$ for $a \in \Sigma$ and $t \in \Sigma^*$. Define the length-reducing STS $\mathcal{P}$ by

$$\mathcal{P} = \mathcal{R} \cup \{\overline{s}^{\mathrm{rev}} \to \overline{t}^{\mathrm{rev}} \mid (s,t) \in \mathcal{R}\} \cup \{a\overline{a} \to \epsilon \mid a \in \Sigma\}.$$

Since $\mathcal{R}$ is confluent, it holds $u \overset{*}{\leftrightarrow}_{\mathcal{R}} v$ if and only if $u\overline{v}^{\mathrm{rev}} \overset{*}{\to}_{\mathcal{P}} \epsilon$. The later property can be checked in LOGCFL. Clearly $u\overline{v}^{\mathrm{rev}}$ can be constructed in log space from u and v. $\qquad\square$

4   Weight-Reducing Semi-Thue Systems

Weight-reducing STSs were investigated for instance in [Die87, Jan88, NO88] and [BL92] as a grammatical formalism. The WP for a fixed weight-reducing and confluent STS can be easily reduced to the WP for a fixed length-reducing and confluent STS. Thus the WP for every fixed weight-reducing and confluent STS can also be solved in LOGCFL:

Theorem 2. *Let $\mathcal{R}$ be a fixed weight-reducing and confluent STS. Then the WP for $\mathcal{R}$ is in LOGCFL.*

Proof. Let $\mathcal{R}$ be a weight-reducing and confluent STS over Σ and let $u, v \in \Sigma^*$. Let f be a weight-function such that $f(s) > f(t)$ for all $(s,t) \in \mathcal{R}$. Let $\$ \notin \Sigma$ and define the morphism $\varphi : \Sigma^* \to (\Sigma \cup \{\$\})^*$ by $\varphi(a) = \$^{f(a)}a^{f(a)}$ for all $a \in \Sigma$. Note that non-trivial overlappings between two words $\varphi(a)$ and $\varphi(b)$ are not possible. It follows that the STS $\varphi(\mathcal{R}) = \{\varphi(s) \to \varphi(t) \mid (s,t) \in \mathcal{R}\}$ is length-reducing and confluent, and we see that $u \overset{*}{\leftrightarrow}_{\mathcal{R}} v$ if and only if $\varphi(u) \overset{*}{\leftrightarrow}_{\varphi(\mathcal{R})} \varphi(v)$. Since $\varphi(u)$ and $\varphi(v)$ can be constructed in log space, the theorem follows from Theorem 1. $\qquad\square$

Next we will consider the UWP for weight-reducing and confluent STSs over a fixed alphabet Σ. In order to get an upper bound for this problem we need the following lemma, which we state in a slightly more general form for later applications.

Lemma 1. *Let Σ be a finite alphabet with $|\Sigma| = n$ and let $\mathcal{R}$ be an STS over Σ with $\alpha = \max\{|s|_a \mid s \in dom(\mathcal{R}) \cup ran(\mathcal{R}), a \in \Sigma\}$. Let g be a weight-function with $g(s) \geq g(t)$ for all $(s,t) \in \mathcal{R}$. Then there exists a weight-function f such that for all $(s,t) \in \mathcal{R}$ the following holds:*

- *If $g(s) > g(t)$ then $f(s) > f(t)$, and if $g(s) = g(t)$ then $f(s) = f(t)$.*
- *$f(a) \leq (n+1)(\alpha n)^n$ for all $a \in \Sigma$.*

Proof. We use the following result about solutions of integer (in)equalities from [vZGS78]: Let A, B, C, D be $(m \times n)$-, $(m \times 1)$-, $(p \times n)$-, $(p \times 1)$-matrices, respectively, with integer entries. Let $r = \mathrm{rank}(A)$, $s = \mathrm{rank}\begin{pmatrix} A \\ C \end{pmatrix}$. Let M be an upper bound on the absolute values of all $(s-1) \times (s-1)$- or $(s \times s)$-subdeterminants of the $(m+p) \times (n+1)$-matrix $\begin{pmatrix} A & B \\ C & D \end{pmatrix}$, which are formed with at least r rows from the matrix $(A\ B)$. Then the system $Ax = B$, $Cx \geq D$ has an integer solution if and only if it has an integer solution x such that the absolute value of every entry of x is bounded by $(n+1)M$.

Now let Σ, n, $\mathcal{R}$, α, and g be as specified in the lemma. Let $\Sigma = \{a_1, \ldots, a_n\}$ and $\mathcal{R} = \{(s_i, t_i) \mid 1 \leq i \leq k\} \cup \{(u_i, v_i) \mid 1 \leq i \leq \ell\}$, where $g(s_i) = g(t_i)$ for $1 \leq i \leq k$ and $g(u_i) > g(v_i)$ for $1 \leq i \leq \ell$. Define the $(k \times n)$-matrix A by $A_{i,j} = |s_i|_{a_j} - |t_i|_{a_j}$ and define the $(\ell \times n)$-matrix C' by $C'_{i,j} = |u_i|_{a_j} - |v_i|_{a_j}$. Let $C = \begin{pmatrix} C' \\ \mathrm{Id}_n \end{pmatrix}$, where Id_n is the $(n \times n)$-identity matrix. Finally let $(j)_i$ be the i-dimensional column vector with all entries equal to j. Then the n-dimensional column vector x with $x_j = g(a_j)$ is a solution of the following system:

$$Ax = (0)_k \qquad Cx \geq (1)_{\ell+n} \tag{1}$$

Note that $r = \mathrm{rank}(A) \leq n$ and $s = \mathrm{rank}\begin{pmatrix} A \\ C \end{pmatrix} \leq n$. Furthermore every entry of the matrix $E = \begin{pmatrix} A & (0)_k \\ C & (1)_{\ell+n} \end{pmatrix}$ is bounded by α. Thus the absolute value of every $(s-1) \times (s-1)$- or $(s \times s)$-subdeterminant of E is bounded by $s! \cdot \alpha^s \leq (\alpha n)^n$. By the result of [vZGS78] the system (1) has a solution y with $y_j \leq (n+1)(\alpha n)^n$ for all $1 \leq j \leq n$. If we define the weight-function f by $f(a_j) = y_j$ then f has the properties stated in the lemma. □

Theorem 3. *Let Σ be a fixed alphabet with $|\Sigma| \geq 2$. Then the UWP for the class of all weight-reducing and confluent STSs over Σ is P-complete.*

Proof. Let $|\Sigma| = n \geq 2$. Let $\mathcal{R}$ be a weight-reducing and confluent STS over Σ and let $u, v \in \Sigma^*$. By Lemma 1 there exists a weight-function f such that $f(s) > f(t)$ for all $(s, t) \in \mathcal{R}$ and $f(a) \leq (n+1)(\alpha n)^n$ for all $a \in \Sigma$. Thus every derivation that starts from the word u has a length bounded by $|u| \cdot (n+1) \cdot (\alpha n)^n$, which is polynomial in the input length $\|\mathcal{R}\| + |uv|$. Thus a normal form of u can be calculated in polynomial time and similarly for v. This proves the upper bound. P-hardness follows from the fact that the UWP for the class of all length-reducing and confluent STSs over $\{a, b\}$ is P-complete [Loh99]. □

Finally for the class of all weight-reducing and confluent STSs the complexity of the UWP increases to EXPTIME:

Theorem 4. *The UWP for the class of all weight-reducing and confluent STSs is EXPTIME-complete.*

Proof. The EXPTIME-upper bound can be shown by using the arguments from the previous proof. Just note that this time the upper bound of $(n+1)(\alpha n)^n$ for a weight-function is exponential in the length of the input. For the lower bound let $\mathcal{M} = (Q, \Sigma, \delta, q_0, q_f)$ be a deterministic Turing-machine such that for some polynomial p it holds: If $w \in L(\mathcal{M})$ then $\mathcal{M}$, started on w, reaches the final state q_f after at most $2^{p(|w|)}$ many steps. Let $w \in (\Sigma \backslash \{\Box\})^*$ be an arbitrary input for $\mathcal{M}$. Let $m = p(|w|)$ and let

$$\Gamma = Q \cup \Sigma \cup \bigcup_{i=0}^{m} (\Sigma_i \cup \{\triangleright_i, 2^{(i)}, A_i, B_i\}) \cup \{\#, \triangleright\}.$$

Here $2^{(i)}$ is a single symbol and $\Sigma_i = \{a_i \mid a \in \Sigma\}$ is a disjoint copy of Σ for $0 \le i \le m$. Let $\Sigma_\triangleright = \Sigma \cup \{\triangleright\}$ and let $\mathcal{R}$ be the STS over Γ that consists of the following rules:

(1)	$2^{(i)}ab \rightarrow a_m 2^{(i-1)} \cdots 2^{(1)}2^{(0)}b$	for $0 \le i \le m$, $a \in \Sigma_\triangleright$, $b \in \Sigma$
(2)	$2^{(i)}a_k \rightarrow a_{k-1}2^{(i)}$	for $0 \le i \le m$, $1 \le k \le m$, $a \in \Sigma_\triangleright$
(3)	$\#a_k \rightarrow a\#$	for $0 \le k \le m$, $a \in \Sigma_\triangleright$
(4)	$\#x \rightarrow x$	for $x \in \Sigma \cup Q \backslash \{q_f\}$
(5)	$2^{(i)}q \rightarrow q$	for $q \in Q \backslash \{q_f\}$
(6)	$2^{(i)}cqa \rightarrow cbp$	for $0 \le i \le m$, $c \in \Sigma_\triangleright$, $\delta(q,a) = (p,b,+1)$
(7)	$2^{(i)}cqa \rightarrow pcb$	for $0 \le i \le m$, $c \in \Sigma_\triangleright$, $\delta(q,a) = (p,b,-1)$
(8)	$A_i \rightarrow A_{i+1}A_{i+1}$	for $0 \le i < m$
(9)	$B_i \rightarrow B_{i+1}B_{i+1}$	for $0 \le i < m$
(10)	$A_m \rightarrow \#2^{(m)}$	
(11)	$B_m \rightarrow \square$	
(12)	$xq_f \rightarrow q_f$	for $x \in \Gamma$
(13)	$q_f x \rightarrow q_f$	for $x \in \Gamma$

We claim that $\mathcal{R}$ is weight-reducing. For this we define the weight-function f as follows: [2]

$$f(A_i) = 2 \cdot f(A_{i+1}) + 1 \text{ for } 0 \le i < m \qquad f(A_m) = 2^m + 2$$
$$f(B_i) = 2 \cdot f(B_{i+1}) + 1 \text{ for } 0 \le i < m \qquad f(B_m) = 2$$
$$f(x) = 1 \text{ for } x \in Q \cup \Sigma_\triangleright \cup \{\#\} \qquad f(2^{(i)}) = 2^i \text{ for } 0 \le i \le m$$
$$f(a_i) = 1 + \frac{i+1}{m+2} \text{ for } 0 \le i \le m, a \in \Sigma_\triangleright$$

Then it is easy to check that $f(s) > f(t)$ for all $(s,t) \in \mathcal{R}$. All non-trivial critical pairs of $\mathcal{R}$ are of the form (sq_f, tq_f) (where $(sx,t) \in \mathcal{R}$, $x \in \Sigma$), $(q_f s, q_f t)$ (where $(xs,t) \in \mathcal{R}$, $x \in \Sigma$), or $(xq_f, q_f y)$ (where $x, y \in \Sigma$). By the rules in (12) and (13) both components of these critical pairs can be reduced to q_f. Thus $\mathcal{R}$ is confluent. Finally we claim that $A_0 \triangleright q_0 w B_0 \overset{*}{\leftrightarrow}_{\mathcal{R}} q_f$ if and only if $w \in L(\mathcal{M})$.

Before we prove this claim let us first explain the effect of the rules from $\mathcal{R}$. For $0 \le i \le 2^m$ let $\text{sum}(i) = 2^{(i_1)} \cdots 2^{(i_k)} \in \Gamma^*$ if $i_1 > \cdots > i_k$ and $i = 2^{i_1} + \cdots + 2^{i_k}$ (note that $\text{sum}(0) = \epsilon$). Let us call a word of the form $\#\text{sum}(i) \in \Gamma^*$ a counter with value i. The effect of the rules in (1), (2), and (3) is to move counters to the right in words from $\triangleright \Sigma^*$. Here the symbol $\triangleright$ is a left-end marker. If a whole counter moves one step to the right, its value is decreased by one. More generally for all $u \in \Sigma^*$, $b \in \Sigma$, and all $|u| < i \le 2^m$ we have $\#\text{sum}(i) \triangleright ub \overset{*}{\rightarrow}_{\mathcal{R}} \triangleright u \#\text{sum}(i - |u| - 1)b$. If a counter has reached the value 0, i.e, it consists only of the symbol $\#$ then the counter is deleted with a rule in (4). Also if a counter collides with a state symbol from Q at its right end, then the counter is deleted with the rules in (4) and (5). Note that such a collision may occur after an application of a rule in (7). The rules in (6) and (7) simulate the machine $\mathcal{M}$. In order to be weight-reducing, these rules consume the right-most

---

[2] Here we use rational weights, but of course they can be replaced by integer weights.

symbol of the right-most counter. The rules in (8) and (10) produce 2^m many counters of the form $\#2^{(m)}$. Each of these counters can move at most 2^m cells to the right. But since $\mathcal{M}$ terminates after at most 2^m many steps, the distance between the left end of the tape and the read-write head is also at most 2^m. This implies that with each of the 2^m many counters that are produced from A_0, at least one step of $\mathcal{M}$ can be simulated. The rules in (9) and (11) produce 2^m many blank symbols, which is enough in order to simulate 2^m many steps of $\mathcal{M}$. Finally the rules in (12) and (13) make the final state q_f absorbing.

Now if $w \in L(\mathcal{M})$ then $A_0 \triangleright q_0 w B_0 \xrightarrow{*}_{\mathcal{R}} (\#2^{(m)})^{2^m} \triangleright q_0 w \square^{2^m} \xrightarrow{*}_{\mathcal{R}} u q_f v \xrightarrow{*}_{\mathcal{R}} q_f$ for some $u, v \in \Gamma^*$. On the other hand if $w \notin L(\mathcal{M})$ then $\mathcal{M}$ does not terminate on w. By simulating $\mathcal{M}$ long enough, and thereby consuming all 2^m many initial counters, we obtain $A_0 \triangleright q_0 w B_0 \xrightarrow{*}_{\mathcal{R}} (\#2^{(m)})^{2^m} \triangleright q_0 w \square^{2^m} \xrightarrow{*}_{\mathcal{R}} \triangleright u q v \in \mathrm{IRR}(\mathcal{R})$ for some $q \in Q \backslash \{q_f\}$, $u, v \in \Sigma^*$. Since also $q_f \in \mathrm{IRR}(\mathcal{R})$ and $\mathcal{R}$ is confluent, $A_0 \triangleright q_0 w B_0 \overset{*}{\leftrightarrow}_{\mathcal{R}} q_f$ cannot hold.   □

Since P is a proper subclass of EXPTIME, it follows from Theorem 3 and Theorem 4 that in general it is not possible to encode the alphabet of a weight-reducing and confluent STS into a fixed alphabet with a polynomial blow-up such that the resulting STS is still weight-reducing and confluent. For length-reducing systems this is always possible, see [Loh99] and the coding function from the proof of Theorem 5.

5   Length-Lexicographic Semi-Thue Systems

In this section we consider length-lexicographic semi-Thue systems, see for instance [KN85]. The complexity bounds that we will achieve in this section are the same that are known for preperfect systems. An STS $\mathcal{R}$ is *preperfect* if for all $s, t \in \Sigma^*$ it holds $s \overset{*}{\leftrightarrow}_{\mathcal{R}} t$ if and only if there exists $u \in \Sigma^*$ with $s \xrightarrow{*}_{\mathcal{R}} u$ and $t \xrightarrow{*}_{\mathcal{R}} u$, where the relation $\mapsto_{\mathcal{R}}$ is defined by $v \mapsto_{\mathcal{R}} w$ if $v \leftrightarrow_{\mathcal{R}} w$ and $|v| \geq |w|$. Since every length-preserving STS is preperfect and every linear bounded automaton can easily be simulated by a length-preserving STS, there exists a fixed preperfect STS $\mathcal{R}$ such that the WP for $\mathcal{R}$ is PSPACE-complete [BJM+81]. The following theorem may be seen as a stronger version of this well-known fact in the sense that a deterministic linear bounded automaton can even be simulated by a length-lexicographic, length-preserving, and confluent STS.

Theorem 5. *The WP for a length-lexicographic and confluent STS is contained in PSPACE. Furthermore there exists a fixed length-lexicographic and confluent STS $\mathcal{R}$ over $\{a, b\}$ such that the WP for $\mathcal{R}$ is PSPACE-complete.*

Proof. The first statement of the theorem is obvious. For the second statement let $\mathcal{M} = (Q, \Sigma, \delta, q_0, q_f)$ be a deterministic linear bounded automaton such that the question whether $w \in L(\mathcal{M})$ is PSPACE-complete. Such a linear bounded automaton exists, see e.g. [BO84]. We may assume that $\mathcal{M}$ operates in phases, where a single phase consists of a sequence of $2 \cdot n$ transitions of the form $q_1 w_1 \overset{*}{\Rightarrow}_{\mathcal{M}} w_2 q_2 \overset{*}{\Rightarrow}_{\mathcal{M}} q_3 w_3$, where $w_1, w_2, w_3 \in \Sigma^*$ and $q_1, q_2, q_3 \in Q$. During the sequence $q_1 w_1 \overset{*}{\Rightarrow}_{\mathcal{M}} w_2 q_2$ only right-moves are made, and during the sequence

$w_2 q_2 \overset{*}{\Rightarrow}_{\mathcal{M}} q_3 w_3$ only left-moves are made. A similar trick is used for instance also in [CH90]. Let $c > 0$ be constant such that if $w \in L(\mathcal{M})$ then $\mathcal{M}$, started on w, reaches the final state q_f after at most $2^{c \cdot n}$ phases. Let $w \in (\Sigma \backslash \{\square\})^*$ be an input for $\mathcal{M}$ with $|w| = n$. As usual let $\overline{\Sigma}$ be a disjoint copy of Σ and similarly for $\overline{Q}$. Let $\Gamma = Q \cup \overline{Q} \cup \Sigma \cup \overline{\Sigma} \cup \{\triangleleft, 0, 1, \overline{1}\}$ and let $\mathcal{R}$ be the STS over Γ that consists of the following rules: [3]

$$
\begin{array}{ll}
0\overline{q} \to \overline{q}\overline{1} & \text{for all } q \in Q \\
1\overline{q} \to 0q & \text{for all } q \in Q \\
q\overline{1} \to 1q & \text{for all } q \in Q \\
qa \to \overline{b}p & \text{if } \delta(q,a) = (p, b, +1) \\
q\triangleleft \to \overline{q}\triangleleft & \text{for all } q \in Q \backslash \{q_f\} \\
\overline{a}\,\overline{q} \to \overline{p}b & \text{if } \delta(q,a) = (p, b, -1) \\
xq_f \to q_f & \text{for all } x \in \Gamma \\
q_f x \to q_f & \text{for all } x \in \Gamma
\end{array}
$$

First we claim that $\mathcal{R}$ is length-lexicographic. For this choose a linear order $\succ$ on the alphabet Γ that satisfies $Q \succ 1 \succ 0 \succ \overline{\Sigma} \succ \overline{Q}$ (here for instance $Q \succ 1$ means that $q \succ 1$ for every $q \in Q$). Furthermore $\mathcal{R}$ is confluent. Finally we claim that $10^{c \cdot n} q_0 w \triangleleft \overset{*}{\leftrightarrow}_{\mathcal{R}} q_f$ if and only if $w \in L(\mathcal{M})$. For $v = b_k \cdots b_0 \in \{0,1\}^*$ ($b_i \in \{0,1\}$) let $\text{val}(v) = \sum_{i=0}^{k} b_i \cdot 2^i$. Note that for every $q \in Q$ and $s, t \in \{0,1\}^+$ with $s \neq 0^{|s|}$ it holds $s\overline{q} \overset{*}{\to}_{\mathcal{R}} tq$ if and only if $|s| = |t|$ and $\text{val}(t) = \text{val}(s) - 1$. First assume that $w \in L(\mathcal{M})$. Then $10^{c \cdot n} q_0 w \triangleleft \overset{*}{\to}_{\mathcal{R}} vq_f u \triangleleft \overset{*}{\to}_{\mathcal{R}} q_f$ for some $u \in \Sigma^*$ and $v \in \{0,1\}^+$. Now assume that $w \notin L(\mathcal{M})$. Then $\mathcal{M}$ does not terminate on w and we obtain $10^{c \cdot n} q_0 w \triangleleft \overset{*}{\to}_{\mathcal{R}} 0^{c \cdot n + 1} \overline{q} u \triangleleft \overset{*}{\to}_{\mathcal{R}} \overline{q} \overline{1}^{c \cdot n + 1} u \triangleleft \in \text{IRR}(\mathcal{R})$, where $u \in \Sigma^*$ and $q \in Q \backslash \{q_f\}$. Since also $q_f \in \text{IRR}(\mathcal{R})$ and $\mathcal{R}$ is confluent, $10^{c \cdot n} q_0 w \triangleleft \overset{*}{\leftrightarrow}_{\mathcal{R}} q_f$ cannot hold.

Finally, we have to encode the alphabet Γ into the alphabet $\{a, b\}$. For this let $\Gamma = \{a_1, \ldots, a_k\}$ and let $a_1 \succ a_2 \succ \cdots \succ a_k$ be the chosen linear order on Γ. Define a morphism $\varphi : \Gamma^* \longrightarrow \{a, b\}^*$ by $\varphi(a_i) = ab^i ab^{2k+1-i}$ and let $a \succ b$. Then the STS $\varphi(\mathcal{R})$ is also length-lexicographic and confluent and for all $u, v \in \Gamma^*$ it holds $u \overset{*}{\leftrightarrow}_{\mathcal{R}} v$ if and only if $\varphi(u) \overset{*}{\leftrightarrow}_{\varphi(\mathcal{R})} \varphi(v)$, see [BO93], p 60. □

6 Weight-Lexicographic Semi-Thue Systems

The widest class of STSs that we study in this paper are weight-lexicographic STSs. Let $\mathcal{R}$ be a weight-lexicographic STS over an alphabet Σ with $|\Sigma| = n$ and let $u \in \Sigma^*$. Thus there exists a weight-function f with $f(s) \geq f(t)$ for all $(s, t) \in \mathcal{R}$. If $u = u_0 \to_{\mathcal{R}} u_1 \to_{\mathcal{R}} \cdots \to_{\mathcal{R}} u_n$ is some derivation then for all $0 \leq i \leq n$ it holds $|u_i| \leq f(u_i) \leq f(u)$. By Lemma 1 we may assume that $f(a) \leq (n+1)(\alpha n)^n$ for all $a \in \Sigma$ and thus $|u_i| \leq |u| \cdot (n+1)(\alpha n)^n$. Together with Theorem 5 it follows that the UWP for weight-lexicographic and confluent STSs over a fixed alphabet is PSPACE-complete and furthermore that there

---

[3] It will be always clear from the context whether e.g. 1 denotes the symbol $1 \in \Gamma$ or the natural number 1.

exists a fixed weight-lexicographic and confluent STS whose WP is PSPACE-complete. For arbitrary weight-lexicographic and confluent STSs we have the following result.

Theorem 6. *The UWP for the class of all weight-lexicographic and confluent STSs is EXPSPACE-complete.*

Proof. The EXPSPACE-upper bound can be shown by using the arguments above. For the lower bound let $\mathcal{M} = (Q, \Sigma, \delta, q_0, q_f)$ be a deterministic Turing-machine which uses for every input w at most space $2^{p(|w|)}$, where p is some polynomial. Similarly to the proof of Theorem 5 we may assume that $\mathcal{M}$ operates in phases. There exists a polynomial q such that if $w \in L(\mathcal{M})$ then $\mathcal{M}$, started on w, reaches q_f after at most $2^{2^{q(|w|)}}$ many phases. Let $w \in (\Sigma \backslash \{\square\})^*$ be an arbitrary input for $\mathcal{M}$. Let $m = p(|w|)$, $n = q(|w|)$, and

$$\Gamma = Q \cup \overline{Q} \cup \Sigma \cup \overline{\Sigma} \cup \{\triangleleft, 0, 1, \overline{1}\} \cup \{A_i \mid 0 \leq i \leq n\} \cup \{B_i \mid 0 \leq i \leq m\}.$$

Let $\mathcal{R}$ be the STS over Γ that consists of the following rules:

$0\overline{q} \rightarrow \overline{q}\overline{1}$	for all $q \in Q$
$1\overline{q} \rightarrow 0q$	for all $q \in Q$
$q\overline{1} \rightarrow 1q$	for all $q \in Q$
$qa \rightarrow \overline{b}p$	if $\delta(q, a) = (p, b, +1)$
$q\triangleleft \rightarrow \overline{q}\triangleleft$	for all $q \in Q \backslash \{q_f\}$
$\overline{a}\,\overline{q} \rightarrow \overline{p}b$	if $\delta(q, a) = (p, b, -1)$
$A_i \rightarrow A_{i+1}A_{i+1}$ for $0 \leq i < n$	
$B_i \rightarrow B_{i+1}B_{i+1}$ for $0 \leq i < m$	
$A_n \rightarrow 0$	
$B_m \rightarrow \square$	
$xq_f \rightarrow q_f$	for all $x \in \Gamma$
$q_f x \rightarrow q_f$	for all $x \in \Gamma$

Note that the first six rules are exactly the same rules that we used for the simulation of a linear bounded automaton in the proof of Theorem 5. We claim that $\mathcal{R}$ is weight-lexicographic. For this define the weight-function f by $f(x) = 1$ for all $x \in Q \cup \overline{Q} \cup \Sigma \cup \overline{\Sigma} \cup \{\triangleleft, 0, 1, \overline{1}, A_n, B_m\}$ and $f(A_i) = 2 \cdot f(A_{i+1})$, $f(B_j) = 2 \cdot f(B_{j+1})$ for $0 \leq i < n$, $0 \leq j < m$. Then the last two rules are weight-reducing and all other rules are weight-preserving. Now choose a linear order $\succ$ on Γ that satisfies $Q \succ 1 \succ 0 \succ \overline{\Sigma} \succ \overline{Q}$, $A_0 \succ A_1 \succ \cdots \succ A_n \succ 0$, and $B_0 \succ B_1 \succ \cdots \succ B_m \succ \square$. It is easy to see that $\mathcal{R}$ is confluent. Finally we have $w \in L(\mathcal{M})$ if and only if $1A_0q_0wB_0\triangleleft \overset{*}{\leftrightarrow}_{\mathcal{R}} q_f$. This can be shown by using the arguments from the proof of Theorem 5. Just note that this time from the word $1A_0$ we can generate the word 10^{2^n} which allows the simulation of 2^{2^n} many phases. Analogously to the proof of Theorem 4 the symbol B_0 generates enough blank symbols in order to satisfy the space requirements of $\mathcal{M}$. $\square$

7   Confluence Problems

The CP for the class of all STSs is undecidable [BO84]. On the other hand, the CP for the class of all terminating STSs is decidable [NB72]. For length-reducing

STSs the CP is in P [BO81], the best known algorithm is the $O(\|\mathcal{R}\|^3)$-algorithm from [KKMN85]. Furthermore in [Loh99] it was shown that the CP for the class of all length-reducing STSs is P-complete. This was shown by using the following log space reduction from the UWP for length-reducing and confluent STSs to the CP for length-reducing STSs, see also [VRL98], Theorem 24: Let $\mathcal{R}$ be a length-reducing and confluent STS over Σ. Furthermore let A and B be new symbols. Then for all $s, t \in \Sigma^*$ the length-reducing STS $\mathcal{R} \cup \{A^{|st|}B \to s, A^{|st|}B \to t\}$ is confluent if and only if $s \overset{*}{\leftrightarrow}_\mathcal{R} t$ holds. Finally the alphabet $\Sigma \cup \{A, B\}$ can be reduced to the alphabet $\{a, b\}$ by using the coding function from the end of the proof of Theorem 5. The same reduction can be also used for weight-reducing, length-lexicographic, and weight-lexicographic STSs. Thus a lower bound for the UWP for one of the classes considered in the preceding sections carrys over to the CP for this class. Furthermore also the given upper bounds hold for the CP for the corresponding class: Our upper bound algorithms for UWPs are all based on the calculation of normal forms. But since every STS has only polynomially many critical pairs, any upper bound for the calculation of normal forms also gives an upper bound for the CP. The resulting complexity results are summarized in Table 1.

Table 1. Complexity results for confluence problems

	length-reducing STSs	weight-reducing STSs	length-lexicographic STSs	weight-lexicographic STSs
CP for a fixed alphabet	P-complete	P-complete	PSPACE-complete	PSPACE-complete
CP	P-complete	EXPTIME-complete	PSPACE-complete	EXPSPACE-complete

8   Quasi Context-Sensitive Grammars

A *quasi context-sensitive grammar*, briefly QCSG, is a (type-0) grammar $G = (N, T, S, P)$ (here N is the set of non-terminals, T is the set of terminals, $S \in N$ is the start non-terminal, and $P \subseteq (N \cup T)^* N (N \cup T)^* \times (N \cup T)^*$ is a finite set of productions) such that for some weight-function $f : (N \cup T)^* \to \mathbb{N}$ we have $f(u) \leq f(v)$ for all $(u, v) \in P$, see [BL94]. The *variable membership problem for QCSGs* it the following problem:

INPUT: A QCSG G with terminal alphabet T and a terminal word $v \in T^*$.
QUESTION: Does $v \in L(G)$ hold?

In [BL94] it was shown that this problem is in EXPSPACE and furthermore that it is NEXPTIME-hard. Using some ideas from Section 4 we can prove that this problem is in fact EXPSPACE-hard.

Theorem 7. *The variable membership problem for QCSGs is EXPSPACE-complete.*

Proof. It remains to show that the problem is EXPSPACE-hard. For this let $\mathcal{M} = (Q, \Sigma, \delta, q_0, q_f)$ be a Turing-machine, which uses for every input w at most space $2^{p(|w|)} - 2$ for some polynomial p. Let $w \in (\Sigma \backslash \{\Box\})^*$ be an input for $\mathcal{M}$ and let $m = p(|w|)$. We will construct a QCSG $G = (N, T, S, P)$ and a word $v \in T^*$ such that $w \in L(\mathcal{M})$ if and only if $v \in L(G)$. The non-terminal and terminal alphabet of G are $N = \{S, B\} \cup \{A_i \mid 0 \leq i < m\} \cup Q \cup \Sigma$ and $T = \{A_m\}$. The set P consists of the following productions:

$$
\begin{array}{ll}
S \rightarrow q_0 w B & \\
B \rightarrow \Box B & \\
qa \rightarrow bp & \text{if } \delta(q, a) = (p, b, +1) \\
cqa \rightarrow pcb & \text{if } \delta(q, a) = (p, b, -1), c \in \Sigma \\
q_f \rightarrow A_0 & \\
xA_0 \rightarrow A_0 A_0 \text{ for } x \in \Sigma & \\
A_0 x \rightarrow A_0 A_0 \text{ for } x \in \Sigma \cup \{B\} & \\
A_i A_i \rightarrow A_{i+1} \text{ for } 0 \leq i < m &
\end{array}
$$

In order to show that G is quasi context-sensitive we define the weight-function $f : (N \cup T)^* \rightarrow \mathbb{N}$ by $f(x) = 1$ for all $x \in \{S, B\} \cup \Sigma \cup Q$ and $f(A_i) = 2^i$ for all $0 \leq i \leq m$. Then $f(s) \leq f(t)$ for all $(s, t) \in P$. If $w \in L(\mathcal{M})$ then $A_m \in L(G)$ by the following derivation:

$$
S \rightarrow q_0 w B \xrightarrow{*}_P q_0 w \Box^{2^m - |w| - 2} B \xrightarrow{*}_P s q_f t B \rightarrow_P s A_0 t B \xrightarrow{*}_P A_0^{2^m} \xrightarrow{*}_P A_m,
$$

where $s, t \in \Sigma^*$ and $|st| = 2^m - 2$. On the other hand, if $A_m \in L(G)$ then a sentential form $u q_f v$ with $uv \in \Gamma^*$ and $|uv| = 2^m - 1$ must be reachable from S, i.e, reachable from $q_0 w B$. This is only possible if $w \in L(\mathcal{M})$. Furthermore G and v can be calculated from $\mathcal{M}$ and w in log space. This concludes the proof.   □

Note that from the previous proof it follows immediately that the following problem is also EXPSPACE-complete:

INPUT: A context-sensitive grammar G with terminal alphabet $\{a\}$ and a number $n \in \mathbb{N}$ coded in binary.

QUESTION: Does $a^n \in L(G)$ hold?

The same problem for context-free grammars is NP-complete [Huy84].

9   Summary and Open Problems

The complexity results for WPs are summarized in Table 2. Here the statement in the first row that e.g. the WP for length-lexicographic and confluent STSs is PSPACE-complete means that for every length-lexicographic and confluent STS the WP is in PSPACE and furthermore there exists a fixed length-lexicographic and confluent STS whose WP is PSPACE-complete. Furthermore the completeness results in the second row already hold for the alphabet $\{a, b\}$.

Table 2. Complexity results for word problems

	length-reducing & confluent STSs	weight-reducing & confluent STSs	length-lexicographic & confluent STSs	weight-lexicographic & confluent STSs
WP	LOGCFL	LOGCFL	PSPACE-complete	PSPACE-complete
UWP for a fixed alphabet	P-complete	P-complete	PSPACE-complete	PSPACE-complete
UWP	P-complete	EXPTIME-complete	PSPACE-complete	EXPSPACE-complete

One open question that remains concerns the WP for a fixed length-reducing (weight-reducing) and confluent STS. Does there exist such a system whose WP is LOGCFL-complete or are these WPs always contained for instance in the subclass LOGDCFL, the class of all languages that are log space reducible to a deterministic context-free language? Since there exits a fixed deterministic context-free language whose membership problem is LOGDCFL-complete [Sud78], Theorem 2.2 of [MNO88] implies that there exists a fixed length-reducing and confluent STS whose WP is LOGDCFL-hard.

Another interesting open problem is the descriptive power of the STSs considered in this paper. Let $\mathcal{M}_{\ell r}$ ($\mathcal{M}_{wr}$, $\mathcal{M}_{\ell\ell}$, $\mathcal{M}_{w\ell}$) be the class of all monoids (modulo isomorphism) of the form $\Sigma^*/\mathcal{R}$, where $\mathcal{R}$ is a length-reducing (weight-reducing, length-lexicographic, weight-lexicographic) and confluent STS over Σ. In [Die87] it was shown that the monoid $\{a, b, c\}^*/\{ab \to c^2\}$ is not contained in $\mathcal{M}_{\ell r}$. Since the STS $\{ab \to c^2\}$ is of course confluent, weight-reducing, and length-lexicographic, it follows that $\mathcal{M}_{\ell r}$ is strictly contained in $\mathcal{M}_{wr}$ and $\mathcal{M}_{\ell\ell}$. Furthermore the monoid $\{a, b\}^*/\{ab \to ba\}$ is contained in $\mathcal{M}_{\ell\ell}\setminus\mathcal{M}_{wr}$ [Die90], p 90. If there exists a monoid in $\mathcal{M}_{wr}\setminus\mathcal{M}_{\ell\ell}$ then $\mathcal{M}_{wr}$ and $\mathcal{M}_{\ell\ell}$ are incomparable and both are proper subclasses of $\mathcal{M}_{w\ell}$. But we do not know whether this holds.

Finally, another interesting class of rewriting systems, for which (uniform) word problems and confluence problems were studied, is the class of commutative semi-Thue systems, see for instance [Car75, Huy85, Huy86, Loh99, MM82, VRL98] for several decidability and complexity results. But there are still many interesting open questions, see for instance the remarks in [Huy85, Huy86, Loh99].

Acknowledgments

I would like to thank Gerhard Buntrock, Volker Diekert, and Yuji Kobayashi for valuable comments.

References

[BJM+81] R. Book, M. Jantzen, B. Monien, C. P. O'Dunlaing, and C. Wrathall. On the complexity of word problems in certain Thue systems. In J. Gruska and M. Chytil, editors, *Proceedings of the 10rd Mathematical Foundations of Computer Science (MFCS'81), Štrbské Pleso (Czechoslovakia)*, number 118 in Lecture Notes in Computer Science, pages 216–223. Springer, 1981.

[BL92]   G. Buntrock and K. Loryś. On growing context-sensitive languages. In W. Kuich, editor, *Proceedings of the19th International Colloquium on Automata, Languages and Programming (ICALP 92), Vienna (Austria)*, number 623 in Lecture Notes in Computer Science, pages 77–88. Springer, 1992.

[BL94]   G. Buntrock and K. Loryś. The variable membership problem: Succinctness versus complexity. In P. Enjalbert, E. W. Mayr, and K. W. Wagner, editors, *Proceedings of the11th Annual Symposium on Theoretical Aspects of Computer Science (STACS 94), Caen (France)*, number 775 in Lecture Notes in Computer Science, pages 595–606. Springer, 1994.

[BO81]   R.V. Book and C. P. O'Dunlaing. Testing for the Church–Rosser property (note). *Theoretical Computer Science*, 16:223–229, 1981.

[BO84]   G. Bauer and F. Otto. Finite complete rewriting systems and the complexity of the word problem. *Acta Informatica*, 21:521–540, 1984.

[BO93]   R.V. Book and F. Otto. *String–Rewriting Systems*. Springer, 1993.

[Boo82]   R.V. Book. Confluent and other types of Thue systems. *Journal of the Association for Computing Machinery*, 29(1):171–182, 1982.

[Car75]   E. W. Cardoza. Computational complexity of the word problem for commutative semigroups. Technical Report MAC Technical Memorandum 67, MIT, 1975.

[CH90]   S. Cho and D. T. Huynh. The complexity of membership for deterministic growing context-sensitive grammars. *International Journal of Computer Mathematics*, 37:185–188, 1990.

[Die87]   V. Diekert. Some properties of weight-reducing presentations. Report TUM-I8710, Technical University Munich, 1987.

[Die90]   V. Diekert. *Combinatorics on Traces*. Number 454 in Lecture Notes in Computer Science. Springer, 1990.

[DW86]   E. Dahlhaus and M. K. Warmuth. Membership for growing context-sensitive grammars is polynomial. *Journal of Computer and System Sciences*, 33:456–472, 1986.

[Huy84]   D. T. Huynh. Deciding the inequivalence of context-free grammars with 1-letter terminal alphabet is Σ_2^p-complete. *Theoretical Computer Science*, 33:305–326, 1984.

[Huy85]   D. T. Huynh. Complexity of the word problem for commutative semigroups of fixed dimension. *Acta Informatica*, 22:421–432, 1985.

[Huy86]   D. T. Huynh. The complexity of the membership problem for two subclasses of polynomial ideals. *SIAM Journal on Computing*, 15(2):581–594, 1986.

[Jan88]   M. Jantzen. Confluent string rewriting. In *EATCS Monographs on theoretical computer science*, volume 14. Springer, 1988.

[KB67]   D. E. Knuth and P. B. Bendix. Simple Word Problems in Universal Algebras. In J. Leech, editor, *Computational Problems in Abstact Algebras*, pages 263–297. Pergamon Press, Elmsford N.Y., 1967.

[KKMN85] D. Kapur, M. S. Krishnamoorthy, R. McNaughton, and P. Narendran. An $O(|T|^3)$ algorithm for testing the Church-Rosser property of Thue systems. *Theoretical Computer Science*, 35(1):109–114, 1985.

[KN85]   D. Kapur and P. Narendran. The Knuth-Bendix completion procedure and Thue systems. *SIAM Journal on Computing*, 14(3):1052–1072, 1985.

[KN87]   K. Krithivasan and P. Narendran. On the membership problem for some grammars. Technical Report CS-TR-1787, University of Maryland, 1987.

[Loh99]   M. Lohrey. Complexity results for confluence problems. In M. Kutylowski and L. Pacholski, editors, *Proceedings of the 24th International Symposium on Mathematical Foundations of Computer Science (MFCS'99), Szklarska Poreba (Poland)*, number 1672 in Lecture Notes in Computer Science, pages 114–124. Springer, 1999.

[Mar47]   A. Markov. On the impossibility of certain algorithms in the theory of associative systems. *Doklady Akademii Nauk SSSR*, 55, 58:587–590, 353–356, 1947.

[MM82]   E. W. Mayr and A. R. Meyer. The complexity of the word problems for commutative semigroups and polynomial ideals. *Advances in Mathematics*, 46:305–329, 1982.

[MNO88]   R. McNaughton, P. Narendran, and F. Otto. Church-Rosser Thue systems and formal languages. *Journal of the Association for Computing Machinery*, 35(2):324–344, 1988.

[NB72]   M. Nivat and M. Benois. Congruences parfaites et quasi–parfaites. *Seminaire Dubreil*, 25(7–01–09), 1971–1972.

[New43]   M. H. A. Newman. On theories with a combinatorial definition of "equivalence". *Annals Mathematics*, 43:223–243, 1943.

[NO88]   P. Narendran and F. Otto. Elements of finite order for finite weigthreducing and confluent Thue systems. *Acta Informatica*, 25:573–591, 1988.

[Pap94]   C. H. Papadimitriou. *Computational Complexity*. Addison Wesley, 1994.

[Pos47]   E. Post. Recursive unsolvability of a problem of Thue. *Journal of Symbolic Logic*, 12(1):1–11, 1947.

[Sud78]   I. H. Sudborough. On the tape complexity of deterministic context–free languages. *Journal of the Association for Computing Machinery*, 25(3):405–414, 1978.

[VRL98]   R. M. Verma, M. Rusinowitch, and D. Lugiez. Algorithms and reductions for rewriting problems. In *Proceedings of the 9th Conference on Rewriting Techniques and Applications (RTA-98), Tsukuba (Japan)*, number 1379 in Lecture Notes in Computer Science, pages 166–180. Springer, 1998.

[vZGS78] J. von Zur Gathen and M. Sieveking. A bound on solutions of linear integer equalities and inequalities. *Proceedings of the American Mathematical Society*, 72(1):155–158, 1978.

The Explicit Representability
of Implicit Generalizations

Reinhard Pichler

Technische Universität Wien
reini@logic.at

Abstract. In [9], implicit generalizations over some Herbrand universe H were introduced as constructs of the form $I = t/t_1 \vee \ldots \vee t_m$ with the intended meaning that I represents all H-ground instances of t that are not instances of any term t_i on the right-hand side. More generally, we can also consider disjunctions $\mathcal{I} = I_1 \vee \ldots \vee I_n$ of implicit generalizations, where $\mathcal{I}$ contains all ground terms from H that are contained in at least one of the implicit generalizations I_j. Implicit generalizations have applications to many areas of Computer Science. For the actual work, the so-called *finite explicit representability problem* plays an important role, i.e.: Given a disjunction of implicit generalizations $\mathcal{I} = I_1 \vee \ldots \vee I_n$, do there exist terms $r_1, \ldots, r_l$, s.t. the ground terms represented by $\mathcal{I}$ coincide with the union of the H-ground instances of the terms r_j? In this paper, we prove the coNP-completeness of this decision problem.

1 Introduction

In [9], *implicit generalizations* over a Herbrand universe H were introduced as constructs of the form $I = t/t_1 \vee \ldots \vee t_m$ with the intended meaning that I represents all H-ground instances of t that are not instances of any term t_i on the right-hand side. The usefulness of implicit generalizations mainly comes from their increased expressive power w.r.t. *explicit generalizations*, which are defined as disjunctions of terms $E = r_1 \vee \ldots \vee r_l$, s.t. E represents all ground terms $s \in H$ that are instances of some r_j. In particular, implicit generalizations allow us to finitely represent certain sets of ground terms, which have no finite representation via explicit generalizations. For the actual work with implicit generalizations, two decision problems have to be solved, namely: The *emptiness problem* (i.e.: Does a given implicit generalization $I = t/t_1 \vee \ldots \vee t_m$ contain no ground term $s \in H$?) and the *finite explicit representability problem* (i.e.: Given an implicit generalization $I = t/t_1 \vee \ldots \vee t_m$, does there exist an explicit generalization $E = r_1 \vee \ldots \vee r_l$, s.t. I and E represent the same set of ground terms in H?). More generally, we can also consider disjunctions $\mathcal{I} = I_1 \vee \ldots \vee I_n$ of implicit generalizations, where $\mathcal{I}$ contains all ground terms from H that are contained in at least one of the implicit generalizations I_j.

Implicit generalizations have applications to many areas of Computer Science like machine learning, unification, specification of abstract data types, logic programming, functional programming, etc. A good overview is given in [8].

L. Bachmair (Ed.): RTA 2000, LNCS 1833, pp. 187–202, 2000.

Both the emptiness problem and the finite explicit representability problem have been shown to be coNP-complete in case of a single implicit generalization (cf. [6], [7], [5] and [11], respectively). Extending the coNP-completeness (and, in particular, the coNP-membership) of the emptiness problem to disjunctions of implicit generalizations is trivial, i.e.: $\mathcal{I} = I_1 \vee \ldots \vee I_n$ is empty, iff all the disjuncts I_j are empty. On the other hand, the finite explicit representability problem for disjunctions of implicit generalizations is a bit more tricky. In particular, $\mathcal{I} = I_1 \vee \ldots \vee I_n$ may well have a finite explicit representation, even though some of the disjuncts I_j possibly do not, e.g.: The implicit generalization $I_1 = f(x,y)/f(x,x)$ over the Herbrand universe H with signature $\Sigma = \{a, f\}$ has no finite explicit representation, while the disjunction $\mathcal{I} = I_1 \vee I_2$ with $I_2 = f(x,x)$ is clearly equivalent to $E = f(x,y)$. In this paper, we present a new algorithm for the finite explicit representability problem, which will allow us to prove the coNP-completeness also in case of disjunctions of implicit generalizations.

This paper is structured as follows: In Sect. 2, we review some basic notions. An alternative algorithm for the finite explicit representability problem of a single implicit generalization is presented in Sect. 3, which will be extended to disjunctions of implicit generalizations in Sect. 4. In Sect. 5, this algorithm will be slightly modified in order to prove the coNP-completeness of the finite explicit representability problem. A comparison with related works will be provided in Sect. 6. Finally, in Sect. 7, we give a short conclusion.

2   Preliminaries

Throughout this paper, we only consider the case of an infinite Herbrand universe (i.e.: a universe with a finite signature that contains at least one proper function symbol), since otherwise the finite explicit representability problem is trivial. The property of terms which ultimately decides whether an implicit generalization actually has an equivalent explicit representation is the so-called "linearity", i.e.: We say that a term t (or a tuple $\boldsymbol{t}$ of terms) is *linear*, iff every variable in t (or in $\boldsymbol{t}$, respectively) occurs at most once. Otherwise t is called *non-linear*. Let $\boldsymbol{x}$ denote the vector of variables that occur in some term t. Then we call an instance $t\vartheta$ of t a *linear instance of t* (or *linear w.r.t. t*), iff $\boldsymbol{x}\vartheta$ is linear. Otherwise $t\vartheta$ is called *non-linear w.r.t. t*. In general, the *range of a substitution* denotes a set of terms. However, in this paper, we usually have to deal with substitutions in the context of instances $t\vartheta$ of another term t. It is therefore more convenient to consider the range of a substitution as a *vector of terms*, namely: Let $\boldsymbol{x}$ denote the vector of variables in some term t. Then we refer to the vector $\boldsymbol{x}\vartheta$ as the *range* of ϑ. In particular, we can then say that an instance $t\vartheta$ of t is non-linear, iff there exists a multiply occurring variable in the range of ϑ.

W.l.o.g. we can assume that all terms t_i on the right-hand side of an implicit generalization $I = t/t_1 \vee \ldots \vee t_m$, are instances of t, since otherwise we would replace t_i by the most general instance $\mathrm{mgi}(t, t_i)$. If all the terms t_i are linear w.r.t. t, then an equivalent explicit representation can be immediately obtained via the complement representation of linear instances from

[9], i.e.: Let $P_i = \{p_{i1}, \ldots, p_{iM_i}\}$ be a set of pairwise variable disjoint terms, s.t. every H-ground instance of t that is not an instance of $t\vartheta_i$ is contained in some term $p_{i\alpha}$ and vice versa. Then I is equivalent to the explicit generalization $E = \bigvee_{\alpha_1=1}^{M_1} \cdots \bigvee_{\alpha_m=1}^{M_m} mgi(p_{1\alpha_1}, \ldots, p_{m\alpha_m})$ (cf. [12], Corollary 3.5).

The implicit generalization $I = t/t\vartheta$ can be considered as the complement of $t\vartheta$ w.r.t. t, i.e.: I contains all H-ground instances of t that are not instances of $t\vartheta$. In [9], an algorithm is presented which computes an explicit representation of the complement of $t\vartheta$ w.r.t. t, if $t\vartheta$ is a linear instance of t. However, for our purposes, we need an appropriate representation of the complement $t\vartheta$ w.r.t. t also in case of a non-linear instance $t\vartheta$. In [4], such a representation is given via terms with equational constraints. Recall that a constrained term over some Herbrand universe H is a pair $[t : X]$ consisting of a term t and an equational formula X, s.t. an H-ground instance $t\sigma$ of t is also an instance of $[t : X]$, iff σ is a solution of X. Moreover, a term t can be considered as a constrained term by adding the trivially true formula $\top$ as a constraint, i.e.: t and $[t : \top]$ are equivalent. Then the complement of $t\vartheta$ w.r.t. t can be constructed in the following way: Consider the tree representation of ϑ, "deviate" from this representation at some node and close all other branches of σ as early as possible with new, pairwise distinct variables. Depending on the label of a node, this deviation can be done in two different ways: If a node is labelled by a constant or function symbol, then this node has to be labelled by a different constant or function symbol. If a node is labelled by a variable which also occurs at some other position, then the two occurrences of this variable have to be replaced by two fresh variables x, y and the constraint $x \neq y$ has to be added. However, if a node is labelled by a variable which occurs nowhere else, then no deviation at all is possible at this node. We only need the following properties of this construction (for a proof of these properties, see [4]):

Theorem 2.1. (Complement of a Term) *Let t be a term over the Herbrand universe H and let $t\vartheta$ be an instance of t. Then there exists a set of constrained terms $P = \{[p_1 : X_1], \ldots, [p_n : X_n]\}$ with the following properties:*

1. *$t/t\vartheta = [p_1 : X_1] \vee \ldots \vee [p_n : X_n]$, i.e.: Every H-ground instance of the complement of $t\vartheta$ w.r.t. t is contained in some $[p_i : X_i]$ and vice versa.*
2. *For every $i \in \{1, \ldots, n\}$, p_i is a linear instance of t and X_i is either the trivially true formula $\top$ or a (quantifier-free) disequation.*
3. *The size of every constrained term $[p_i : X_i] \in P$ and the number n of such terms are polynomially bounded by the number of positions in $t\vartheta$.*

Note that the equational constraints are only needed in order to finitely express the complement of a *non-linear* instance. Hence, the two approaches from [9] and [4] are quite similar, when only linear instances are considered. The representation of the complement of a term from Theorem 2.1 can be easily extended to implicit generalizations, i.e.: Let $I = t/t_1 \vee \ldots \vee t_m$ be an implicit generalization and let $P = \{[p_1 : X_1], \ldots, [p_n : X_n]\}$ be the complement of t w.r.t. some variable x, then $P \cup \{t_1, \ldots, t_m\}$ is a representation of the complement of I. By writing the terms t_i as constrained terms $[t_i : \top]$ with trivially true constraint, we end up again with a set of constrained terms.

A *conjunction of terms* $t_1 \wedge \ldots \wedge t_n$ over H represents the set of all terms $s \in H$ that are instances of every term t_i. Suppose that the terms t_i are pairwise variable disjoint. Then the set of ground terms contained in such a conjunction corresponds to the H-ground instances of the most general instance $mgi(t_1, \ldots, t_n)$. The most general unifier of terms $t_1, \ldots, t_n$ is denoted by $mgu(t_1, \ldots, t_n)$. When computing the intersection of the complement of various terms $t\vartheta_1, \ldots, t\vartheta_m$, we shall have to deal with *conjunctions of constrained terms* of the form $[p_1 : X_1] \wedge \ldots \wedge [p_m : X_m]$, where the p_i's are pairwise variable disjoint. Analogously to the case of terms without constraints, the set of ground terms contained in such a conjunction is equivalent to a single constrained term, which can be obtained via unification, namely: $[p_1\mu : Z\mu]$ with $\mu = mgu(p_1, \ldots, p_m)$ and $Z \equiv X_1 \wedge \ldots \wedge X_m$. Hence, for testing whether the intersection of constrained terms is empty, we have to check, whether $\mu = mgu(p_1, \ldots, p_m)$ exists and whether $Z\mu$ has at least one solution. Since the constraint part $Z\mu$ will always be a conjunction of disequations, the latter condition holds, iff $Z\mu$ contains no trivial disequation of the form $t \neq t$. (For a proof see [1], Lemma 2).

3   The Basic Algorithm

In [9], it is shown that an implicit generalization $I = t/t\vartheta_1 \vee \ldots \vee t\vartheta_m$ has an equivalent explicit representation, iff every non-linear instance $t\vartheta_i$ of t can be replaced by a finite number of linear instances of t (cf. [9], Proposition 4.7). A non-linear instance $t\vartheta_i$ of t which cannot be replaced by a finite number of linear ones will be referred to as *"essentially non-linear"* in this paper, e.g.:

Example 3.1. (Essential Non-linearity) Let $I = f(x,y)/f(x,x)$ be an implicit generalization over the Herbrand universe H with signature $\Sigma = \{f, a\}$. It is shown in [9], that I has no equivalent explicit representation. In particular, the non-linear term $f(x,x)$ cannot be replaced by finitely many linear ones. In other words, $f(x,x)$ is *essentially non-linear*. On the other hand, consider the implicit generalization $J = f(y_1, y_2)/[f(x,x) \vee f(f(x_1, x_2), x_3)]$. Then $J' = f(y_1, y_2)/[f(a,a) \vee f(f(x_1, x_2), x_3)]$ is equivalent to J, i.e.: the term $f(x,x)$ is *inessentially non-linear*, since it can be replaced by the linear term $f(a, a)$.

In this section, we provide a new decision procedure for the finite explicit representability problem of a single implicit generalization by formalizing the notion of "essential non-linearity". To this end, we identify in Theorem 3.1 a disjunction of terms $s_1 \vee \ldots \vee s_n$, by which a non-linear term $t\vartheta_i$ on the right-hand side of an implicit generalization $I = t/t\vartheta_1 \vee \ldots \vee t\vartheta_m$ may be replaced. In Theorem 3.2, we shall then show that I has no finite explicit representation, if at least one of these new terms s_j is non-linear w.r.t. t. The idea of the replacement step in Theorem 3.1 is that we may restrict the term $t\vartheta_i$ on the right-hand side of I to those instances, which are in the complement of the remaining terms $t\vartheta_j$ with $j \neq i$. The correctness of this step corresponds to the equality $A - [B \cup C] = A - [(B - C) \cup C]$, which holds for arbitrary sets A, B and C.

Theorem 3.1. (Replacing Non-linear Terms) *Let $I = t/t\vartheta_1 \vee \ldots \vee t\vartheta_m$ be an implicit generalization and suppose that the term $t\vartheta_1$ is non-linear w.r.t. t. Moreover, let $P_i = \{[p_{i1} : X_{i1}], \ldots, [p_{iM_i} : X_{iM_i}]\}$ represent the complement of $t\vartheta_i$ w.r.t. t according to Theorem 2.1 and let all terms in $\{t\vartheta_1\} \cup \{p_{i\alpha_i} \mid 2 \leq i \leq m$ and $1 \leq \alpha_i \leq M_i\}$ be pairwise variable disjoint. Then I is equivalent to the implicit generalization $I' = t/t\vartheta_2 \vee \ldots \vee t\vartheta_m \vee \bigvee_{\mu \in M} t\vartheta_1\mu$ with*

$$M = \{\, \mu \mid \exists \alpha_2 \ldots \exists \alpha_m, \text{ s.t. } \mu = \mathrm{mgu}(t\vartheta_1, p_{2\alpha_2}, \ldots, p_{m\alpha_m}) \text{ and}$$
$$(X_{2\alpha_2} \wedge \ldots \wedge X_{m\alpha_m})\mu \text{ contains no trivial disequation}\,\}.$$

Proof. All terms $t\vartheta_1\mu$ with $\mu \in M$ are instances of $t\vartheta_1$. Hence, $I \subseteq I'$ trivially holds. We only have to prove the opposite subset relation: Let $u \in I'$, i.e.: u is an instance of t but not of $t\vartheta_2 \vee \ldots \vee t\vartheta_m$ nor of any term $t\vartheta_1\mu$. We have to show that then u is not an instance of $t\vartheta_1$. Suppose on the contrary that u is an instance of $t\vartheta_1$. Moreover, by assumption, u is not an instance of any term $t\vartheta_i$ with $i \geq 2$. But for every i, $P_i = \{[p_{i1} : X_{i1}], \ldots, [p_{iM_i} : X_{iM_i}]\}$ completely covers the complement of $t\vartheta_i$ w.r.t. t. Hence, there exist indices $\alpha_2, \ldots, \alpha_m$, s.t. $u \in t\vartheta_1 \wedge [p_{2\alpha_2} : X_{2\alpha_2}] \wedge \ldots \wedge [p_{m\alpha_m} : X_{m\alpha_m}]$. By the considerations from Sect. 2, this conjunction is equivalent to $[t\vartheta_1\mu : Z\mu]$ with $\mu = \mathrm{mgu}(t\vartheta_1, p_{2\alpha_2}, \ldots, p_{m\alpha_m})$ and $Z \equiv X_{2\alpha_2} \wedge \ldots \wedge X_{m\alpha_m}$. Moreover, $[t\vartheta_1\mu : Z\mu] \subseteq t\vartheta_1\mu$ holds. But then u is not an instance of I', which contradicts our original assumption. $\quad\square$

Theorem 3.2. (Essential Non-linearity) *Let $I = t/t\vartheta_1 \vee \ldots \vee t\vartheta_m$ be an implicit generalization. Moreover, let $P_i = \{[p_{i1} : X_{i1}], \ldots, [p_{iM_i} : X_{iM_i}]\}$ represent the complement of $t\vartheta_i$ w.r.t. t according to Theorem 2.1 and let all terms in $\{t\vartheta_1\} \cup \{p_{i\alpha_i} \mid 2 \leq i \leq m$ and $1 \leq \alpha_i \leq M_i\}$ be pairwise variable disjoint. Finally, suppose that the term $t\vartheta_1$ is non-linear w.r.t. t and that there exist indices $\alpha_2, \ldots, \alpha_m$ with $1 \leq \alpha_i \leq M_i$ for all i, s.t. the following conditions hold:*

1. *$\mu = \mathrm{mgu}(t\vartheta_1, p_{2\alpha_2}, \ldots, p_{m\alpha_m})$ exists.*
2. *$Z\mu$ with $Z \equiv X_{2\alpha_2} \wedge \ldots \wedge X_{m\alpha_m}$ contains no trivial disequation.*
3. *There exists a multiply occurring variable y in the range of ϑ_1, s.t. $y\mu$ is a non-ground term.*

Then I has no finite explicit representation.

Proof. (indirect) For our proof, we have to make use of certain terms $C_0(D)$, $C_1(D), \ldots$ from [9], which are defined as follows: Let f be a function symbol in the signature of H with arity q and let a be a constant. $F_D(l_1, \ldots, l_{q^D})$ denotes the term whose tree representation has the label f at all nodes down to depth $D - 1$ and whose leaf nodes at depth D are labelled with l_1 through l_{q^D}. G_D corresponds to the special case where all leaves l_i are labelled with the constant a, i.e.: $G_D = F_D(a, \ldots, a)$, e.g.: For a binary function symbol f, $G_2 = f(f(a, a), f(a, a))$ holds. Then, for every $i \geq 0$, the term $C_i(D)$ is defined as $C_i(D) = F_D(G_{D \times (q^D \times i+1)}, G_{D \times (q^D \times i+2)}, \ldots, G_{D \times q^D \times (i+1)})$. These terms have the following property: *If there exists an index i, s.t. $C_i(D)$ is an instance of a term t of depth smaller than D, then t contains no multiple variable occurrences and $C_j(D)$ is also an instance of t for every index j.* (For any details, see [9]).

Now let $y\mu = f[z]$, where $f[z]$ is some term containing the variable z. We modify ϑ_1 to ϑ'_1 by replacing one occurrence of y in the range of ϑ_1 by a fresh variable y' and extend μ to μ', s.t. $y'\mu' = f[z']$ for another fresh variable z'. Now suppose that I is equivalent to the explicit generalization $E = r_1 \vee \ldots \vee r_M$. Then we derive a contradiction in the following way:

1. $t\vartheta'_1\mu'$ is an instance of $p_{2\alpha_2}\pi$ with $\pi = \mathrm{mgu}(p_{2\alpha_2}, \ldots, p_{m\alpha_m})$: By the definition of the complement from Theorem 2.1, all terms $p_{i\alpha_i}$ are linear w.r.t. t and, therefore, also $p_{2\alpha_2}\pi$ is linear w.r.t. t. Hence, $p_{2\alpha_2}\pi = t\varphi$ for some substitution φ which has no multiple variable occurrences in its range. Moreover, $t\vartheta_1\mu$ is an instance of $[p_{2\alpha_2}\pi : Z\pi]$ with $Z \equiv X_{2\alpha_2} \wedge \ldots \wedge X_{m\alpha_m}$. Thus, there exists a substitution λ, s.t. $\vartheta_1\mu = \varphi\lambda$. Now consider the occurrence of $f[z]$ in $\vartheta_1\mu$ which is replaced by $f[z']$ in $\vartheta'_1\mu'$. There must be some variable x in the range of φ which is instantiated by λ to some term containing this occurrence of the variable z. But then, since all variables in the range of φ occur only once, λ can be modified to λ' s.t. this occurrence of z in $x\lambda$ is replaced by z' and λ' coincides with λ everywhere else. Thus $\varphi\lambda' = \vartheta'_1\mu'$ with $\lambda = \lambda' \circ \{z' \leftarrow z\}$ holds. Hence, in particular, $t\vartheta'_1\mu'$ is an instance of $p_{2\alpha_2}\pi$ with $p_{2\alpha_2}\pi\lambda' = t\vartheta'_1\mu'$ and $p_{2\alpha_2}\pi\lambda' \circ \{z' \leftarrow z\} = t\vartheta_1\mu$. Moreover, since the equational problems $X_{i\alpha_i}$ only contain variables from $p_{i\alpha_i}$, the equivalence $Z\mu \equiv Z\pi\lambda' \circ \{z' \leftarrow z\}$ also holds.

2. Construction of two ground terms s' and s, s.t. s' is in I and s is outside: By assumption, $Z\mu \equiv Z\pi\lambda' \circ \{z' \leftarrow z\}$ contains no trivial disequation. Hence, $Z\pi\lambda'$ contains no trivial disequation either and, therefore there exists a solution σ' of $Z\pi\lambda'$, i.e.: $Z\pi\lambda'\sigma'$ is yet another conjunction of disequations with no trivial disequation. Now let D be an integer, s.t. D is greater than the depth of any term occurring in $Z\pi\lambda'\sigma'$ and greater than the depth of $\mathrm{mgi}(t\vartheta'_1\mu', r_\gamma)$ for all terms r_γ from the explicit representation E of I. Then we can modify σ' to τ', s.t. both substitutions coincide on all variables except for z' and z, where we define $z\tau' = C_0(D)$ and $z'\tau' = C_1(D)$. By the definition of $C_0(D)$ and $C_1(D)$, no trivial disequation can be introduced into $Z\pi\lambda'\tau'$ by this transformation from $Z\pi\lambda'\sigma'$. Let the term s' be defined as $s' = p_{2\alpha_2}\pi\lambda'\tau'$. Then, on the one hand, s' is an instance of $[p_{2\alpha_2}\pi : Z\pi]$ and, on the other hand, s' is not an instance of $t\vartheta_1\mu = p_{2\alpha_2}\pi\lambda' \circ \{z' \leftarrow z\}$, since τ' assigns different values to z and z'. However, if we define the substitution τ in such a way, that it instantiates both variables z and z' to $C_0(D)$ and τ coincides with σ' everywhere else, then $s = p_{2\alpha_2}\pi\lambda'\tau$ is indeed an instance of $t\vartheta_1\mu = p_{2\alpha_2}\pi\lambda' \circ \{z' \leftarrow z\}$ and, in particular, of $t\vartheta_1$. Thus, s' is an instance of $I = t/\vartheta_1 \vee \ldots \vee t\vartheta_m$ whereas s is not.

3. If s' is an instance of r_γ, then s is also an instance of r_γ: By assumption, I is equivalent to $E = r_1 \vee \ldots \vee r_M$. Hence, there exists a term r_γ, s.t. s' is an instance of r_γ and, therefore, also of $\mathrm{mgi}(t\vartheta'_1\mu', r_\gamma)$. Thus, there exist substitutions ρ' and η', s.t. $t\vartheta'_1\mu'\rho' = \mathrm{mgi}(t\vartheta'_1\mu', r_\gamma)$ and $t\vartheta'_1\mu'\rho'\eta' = s'$. By construction, $C_1(D)$ is an instance of $z'\rho'$ and the term depth of $z'\rho'$ is smaller than D. Moreover, all subterms of $C_1(D)$ with root at depth smaller than D occur nowhere else in the range of $\vartheta'_1\mu'\rho'\eta'$. Hence, the variables in $z'\rho'$ occur nowhere else in the range of ρ' and, by the properties of the terms $C_i(D)$ recalled above, $C_0(D)$ is also an instance of $z'\rho'$. We can thus modify η' to η, s.t. $z'\rho'\eta = C_0(D)$ holds and η coincides with η' on all variables not occurring in $z'\rho'$. Thus, in particular,

$t\vartheta'_1\mu'\rho'\eta = s$ holds. But then, s is an instance of $t\vartheta'_1\mu'\rho' = mgi(t\vartheta'_1\mu', r_\gamma)$ and, therefore, also of r_γ, which contradicts the assumption that r_γ is a disjunct from the explicit representation E of I.   □

By combining the two theorems above, we immediately get a new algorithm for deciding the finite explicit representability problem of a single implicit generalization, namely: Let $I = t/t\vartheta_1\vee\ldots\vee t\vartheta_m$ be an implicit generalization. If *all terms* $t\vartheta_i$ are *linear* w.r.t. t, then I clearly has an equivalent explicit representation, which can be easily obtained via the complement representation of linear terms from [9] (cf. Sect. 2). On the other hand, if one of the terms $t\vartheta_i$ is a non-linear instance of t, then we may assume w.l.o.g. that $i = 1$ holds. Hence, we can either apply Theorem 3.1 or Theorem 3.2. In the former case, the non-linear instance $t\vartheta_i$ of t is replaced by a disjunction of linear instances. In the latter case, it has been shown that no equivalent explicit representation exists. The termination of this algorithm is obvious, since the number of non-linear instances of t on the right-hand side of I is strictly decreased, whenever we apply Theorem 3.1. This algorithm is put to work in the following example:

Example 3.2. Let $I = t/t\vartheta_1 \vee \ldots \vee t\vartheta_4$ be an implicit generalization over the Herbrand universe H with signature $\Sigma = \{a, f, g\}$, s.t. $t = f(f(x_1, x_2), x_3))$ and

$\vartheta_1 = \{x_1 \leftarrow f(y_{11}, y_{12}), x_2 \leftarrow y_{13}, x_3 \leftarrow y_{13}\}$
$\vartheta_2 = \{x_1 \leftarrow y_{21}, x_2 \leftarrow y_{21}, x_3 \leftarrow g(y_{22})\}$
$\vartheta_3 = \{x_1 \leftarrow y_{31}, x_2 \leftarrow f(y_{32}, y_{33}), x_3 \leftarrow y_{34}\}$
$\vartheta_4 = \{x_1 \leftarrow y_{41}, x_2 \leftarrow g(y_{42}), x_3 \leftarrow y_{43}\}$

Note that every instance of t is of the form $t\sigma$ with $\sigma = \{x_1 \leftarrow s_1, x_2 \leftarrow s_2, x_3 \leftarrow s_3\}$. In order to keep the notation simple, we denote such an instance by $t(s_1, s_2, s_3)$. Then the complement of the terms $t\vartheta_2$, $t\vartheta_3$ and $t\vartheta_4$, respectively, is represented by the following sets:

$P_2 = \{[t(z_{21}, z_{22}, a) : \top], [t(z_{21}, z_{22}, f(z_{23}, z_{24})) : \top], [t(z_{21}, z_{22}, z_{23}) : z_{21} \neq z_{22}]\}$
$P_3 = \{[t(z_{31}, a, z_{32}) : \top], [t(z_{31}, g(z_{32}), z_{33}) : \top]\}$
$P_4 = \{[t(z_{41}, a, z_{42}) : \top], [t(z_{41}, f(z_{42}, z_{43}), z_{44}) : \top]\}$

in order to apply Theorem 3.1, we have to compute the set M of certain unifiers. We shall write $\mu_{(\alpha_2, \alpha_3, \alpha_4)}$ to denote the *mgu* of $t\vartheta_1$ with the terms $p_{2\alpha_2}$, $p_{3\alpha_3}$ and $p_{4\alpha_4}$ from the sets P_2, P_3 and P_4. Then M according to Theorem 3.1 consists of a single element, namely $\mu_{(1,1,1)} = \mu_{(3,1,1)} = \{y_{13} \leftarrow a\}$. Hence, I may be transformed into $I' = t/t\vartheta_2 \vee t\vartheta_3 \vee t\vartheta_4 \vee t(f(y_{11}, y_{12}), a, a)$.

We already know the representations $P'_2 = P_3$ and $P'_3 = P_4$ of the complement of $t\vartheta_3$ and $t\vartheta_4$. Moreover, we need the complement P'_4 of $t(f(y_{11}, y_{12}), a, a)$, namely:

$P'_4 = \{[t(a, z_{51}, z_{52}) : \top], [t(g(z_{51}), z_{52}, z_{53}) : \top], [t(z_{51}, f(z_{52}, z_{53}), z_{54}) : \top],$
$\quad [t(z_{51}, g(z_{52}), z_{53}) : \top], [t(z_{51}, z_{52}, f(z_{53}, z_{54})) : \top], [t(z_{51}, z_{52}, g(z_{53})) : \top]\}$

Then M (for I') from Theorem 3.1 again consists of a single element only, namely: $\mu_{(1,1,1)} = \mu_{(1,1,6)} = \{y_{21} \leftarrow a\}$. The non-linear instance $t\vartheta_2$ in I' may therefore be replaced by the linear instance $t\vartheta_2\mu_{(1,1,1)} = t(a, a, g(y_{22}))$. We have thus transformed I into $I'' = t/t(y_{31}, f(y_{32}, y_{33}), y_{34})\vee t(y_{41}, g(y_{42}), y_{43})\vee t(f(y_{11}, y_{12}), a, a)$ $\vee t(a, a, g(y_{22}))$, which has only linear instances of t on the right-hand side. Hence,

I has a finite explicit representation. In order to actually compute an equivalent explicit generalization, we need the representation of the complement of all terms on the right-hand side. $P_1'' = P_3$, $P_2'' = P_4$ and $P_3'' = P_4'$ have already been computed. The complement of $t(a, a, g(y_{22}))$ can be represented by

$$P_4'' = \{t(g(z_{51}), z_{52}, z_{53}), t(f(z_{51}, z_{52}), z_{53}), t(z_{51}, g(z_{52}), z_{53}),$$
$$t(z_{51}, f(z_{52}, z_{53}), z_{54}), t(z_{51}, z_{52}, a), t(z_{51}, z_{52}, f(z_{53}, z_{54}))\}$$

Hence, the explicit representation E of I is of the form

$$E = \left(\bigvee_{p_1 \in P_1''} p_1\right) \wedge \left(\bigvee_{p_2 \in P_2''} p_2\right) \wedge \left(\bigvee_{p_3 \in P_3''} p_3\right) \wedge \left(\bigvee_{p_4 \in P_4''} p_4\right) =$$
$$\bigvee_{p_1 \in P_1''} \bigvee_{p_2 \in P_2''} \bigvee_{p_3 \in P_3''} \bigvee_{p_4 \in P_4''} (p_1 \wedge p_2 \wedge p_3 \wedge p_4)$$

By computing all possible conjunctions $p_1 \wedge p_2 \wedge p_3 \wedge p_4$ and deleting those terms which are a proper instance of another conjunction, we get

$$E = t(a, a, a) \vee t(f(y_1, y_2), a, g(y_3)) \vee t(y_1, a, f(y_2, y_3)) \vee t(g(y_1), a, y_2) = f(f(a, a), a)) \vee f(f(f(y_1, y_2), a), g(y_3))) \vee f(f(y_1, a), f(y_2, y_3)) \vee f(f(g(y_1), a), y_2))$$

4   Disjunctions of Implicit Generalizations

We shall now extend the notion of essential non-linearity to disjunctions of implicit generalizations $I = I_1 \vee \ldots \vee I_n$ with $I_i = t_i/t_{i1} \vee \ldots \vee t_{im_i}$ for $i \in \{1, \ldots, n\}$. To this end, we again provide a disjunction of terms $s_1 \vee \ldots \vee s_l$, by which a non-linear term $t\vartheta_{ij}$ on the right-hand side of an implicit generalization I_i may be replaced. Moreover, we claim that I has no finite explicit representation, if at least one of these new terms s_k is non-linear w.r.t. t_i. The idea of the replacement step in Theorem 4.1 is twofold: On the one hand, we may restrict the term $t_i\vartheta_{ij}$ to those instances, which are in the complement of the remaining terms $t_i\vartheta_{ik}$ on the right-hand side of I_i with $k \neq j$. On the other hand, we may further restrict the term $t_i\vartheta_{ij}$ to those instances, which are in the complement of the other I_k's with $k \neq i$. The correctness of this replacement step is due to the equalities $A - [B \cup C] = A - [(B - C) \cup C]$ and $[A - B] \cup C = [A - (B - C)] \cup C$, respectively, which hold for arbitrary sets A, B and C.

Theorem 4.1. (Replacing Non-linear Terms) Let $I = I_1 \vee \ldots \vee I_n$ with $I_j = t_j/t_j\vartheta_{j1} \vee \ldots \vee t_j\vartheta_{jm_j}$ for $j \in \{1, \ldots, n\}$ be a disjunction of implicit generalizations. Suppose that the term $t_1\vartheta_{11}$ is non-linear w.r.t. t_1 and let y denote the vector of variables that occur more than once in the range of ϑ_{11}. Moreover, let $P_i = \{[p_{i1} : X_{i1}], \ldots, [p_{iM_i} : X_{iM_i}]\}$ represent the complement of $t_1\vartheta_{11}$ w.r.t. t_1 according to Theorem 2.1 and let $Q_j = \{[q_{j1} : Y_{j1}], \ldots, [q_{jN_j} : Y_{jN_j}]\}$ represent the complement of I_j. Finally, let all terms in $\{t_1\vartheta_{11}\} \cup \{p_{i\alpha_i} \mid 2 \leq i \leq m_1$ and $1 \leq \alpha_i \leq M_i\} \cup \{q_{j\beta_j} \mid 2 \leq j \leq n$ and $1 \leq \beta_j \leq N_j\}$ be pairwise variable disjoint. Then I is equivalent to $I' = I_1' \vee I_2 \vee \ldots \vee I_n$, where I_1' is defined as

$$I_1' = t_1/t_1\vartheta_{12} \vee \ldots \vee t\vartheta_{1m_1} \vee \bigvee_{\mu \in M} t_1\vartheta_{11}\mu \quad \text{with}$$

$$M = \{\, \mu \mid \exists \alpha_2 \ldots \exists \alpha_{m_1} \exists \beta_2 \ldots \exists \beta_n, \text{ s.t. } \nu = \text{mgu}(t_1\vartheta_{11}, p_{2\alpha_2}, \ldots, p_{m\alpha_{m_1}}, q_{2\beta_2}, \ldots$$
$$\ldots, q_{n\beta_n}), (X_{2\alpha_2} \wedge \ldots \wedge X_{m_1\alpha_{m_1}} \wedge Y_{2\beta_2} \wedge \ldots \wedge Y_{n\beta_n})\nu \text{ contains no}$$
$$\text{trivial disequation and } \mu = \nu|_y \,\}.$$

Proof. All terms $t_1\vartheta_{11}\mu$ with $\mu \in M$ are instances of $t_1\vartheta_{11}$. Hence, $I_1 \subseteq I_1'$ and, therefore, also $\mathcal{I} \subseteq \mathcal{I}'$ trivially holds. So we only have to prove the opposite subset relation: Let $u \in \mathcal{I}'$. If $u \in I_2 \vee \ldots \vee I_n$, then u is of course also contained in $\mathcal{I}$. Thus, the only interesting case to consider is that $u \in I_1'$ and $u \notin I_2 \vee \ldots \vee I_n$. Hence, in particular, u is in the complement of every $t_1\vartheta_{1i}$ with $2 \leq i \leq m_1$ and in the complement of every I_j with $2 \leq j \leq n$. But for every i, $P_i = \{[p_{i1} : X_{i1}], \ldots, [p_{iM_i} : X_{iM_i}]\}$ completely covers the complement of $t_1\vartheta_{1i}$ w.r.t. t_1. Likewise, for every j, $Q_j = \{[q_{j1} : Y_{j1}], \ldots, [q_{jN_j} : Y_{jN_j}]\}$ completely covers the complement of I_j. Hence, there exist indices $\alpha_2, \ldots, \alpha_{m_1}$ and $\beta_2, \ldots, \beta_n$, s.t. $u \in \bigwedge_{i=2}^{m_1}[p_{i\alpha_i} : X_{i\alpha_i}] \wedge \bigwedge_{j=2}^{n}[q_{j\beta_j} : Y_{j\beta_j}]$. Then, analogously to the proof of Theorem 3.1, it can be shown that if u is not an instance of any term $t_1\vartheta_{11}\mu$ on the right-hand side of I_1', then u is not an instance of $t_1\vartheta_{11}$ either. In other words, if u is an instance of I_1' but not of any I_i with $i \geq 2$, then u is actually an instance of I_1.   □

Theorem 4.2. (Essential Non-linearity) *Let $\mathcal{I} = I_1 \vee \ldots \vee I_n$ with $I_j = t_j/t_j\vartheta_{j1} \vee \ldots \vee t_j\vartheta_{jm_j}$ for $j \in \{1, \ldots, n\}$ be a disjunction of implicit generalizations. Moreover, let $P_i = \{[p_{i1} : X_{i1}], \ldots, [p_{iM_i} : X_{iM_i}]\}$ represent the complement of $t_1\vartheta_{1i}$ w.r.t. t_1 and let $Q_j = \{[q_{j1} : Y_{j1}], \ldots, [q_{jN_j} : Y_{jN_j}]\}$ represent the complement of I_j. Let all terms in $\{t_1\vartheta_{11}\} \cup \{p_{i\alpha_i} \mid 2 \leq i \leq m_1 \text{ and } 1 \leq \alpha_i \leq M_i\}$ $\cup \{q_{j\beta_j} \mid 2 \leq j \leq n \text{ and } 1 \leq \beta_j \leq N_j\}$ be pairwise variable disjoint. Now suppose that the term $t_1\vartheta_{11}$ is non-linear w.r.t. t_1 and let $\mathbf{y}$ denote the vector of variables that occur more than once in the range of ϑ_{11}. Finally suppose that there exist indices $\alpha_2, \ldots, \alpha_{m_1}$ with $1 \leq \alpha_i \leq M_i$ for all i, and $\beta_2, \ldots, \beta_n$ with $1 \leq \beta_j \leq N_j$ for all j, s.t. the following conditions hold:*

1. $\nu = \text{mgu}(t_1\vartheta_{11}, p_{2\alpha_2}, \ldots, p_{m\alpha_{m_1}}, q_{2\beta_2}, \ldots, q_{n\beta_n})$ exists.
2. $Z\nu$ with $Z \equiv X_{2\alpha_2} \wedge \ldots \wedge X_{m_1\alpha_{m_1}} \wedge Y_{2\beta_2} \wedge \ldots \wedge Y_{n\beta_n}$ contains no trivial disequation.
3. There exists a variable y in $\mathbf{y}$ s.t. $y\nu$ is a non-ground term.

Then $\mathcal{I}$ has no finite explicit representation.

Proof. (Rough Sketch): Similarly to the proof of Theorem 3.2, we can use the terms $C_0(D)$ and $C_1(D)$ in order to construct ground terms s and s', s.t. s' is inside $\mathcal{I}$ and s is outside. Now suppose that $\mathcal{I}$ has a finite explicit representation $r_1 \vee \ldots \vee r_l$. Then, in particular, s' is an instance of some r_j. Analogously to the proof of Theorem 3.2, we can derive a contradiction by showing that then also s is an instance of r_j. For any details, see [14].   □

5   coNP-Completeness

In [11], the coNP-completeness of the finite explicit representability problem of a single implicit generalization was proven. In this section we show the coNP-membership (and, hence, the coNP-completeness) also in case of disjunctions of implicit generalizations. In Theorem 4.2 we have provided a criterion for checking that a given disjunction of implicit generalizations has no finite explicit representation. It would be tempting to prove the coNP-membership via a non-deterministic algorithm based on this criterion, i.e.: Guess a non-linear term $t_i\vartheta_{ij}$ on the right-hand side of an implicit generalization I_i and check that the conditions from Theorem 4.2 are fulfilled. Unfortunately, our criterion from Theorem 4.2 is sufficient but not necessary as the following example illustrates:

Example 5.1. Let $I_1 = f(f(x_1,x_2), f(x_3,x_4))/[f(f(y_1,y_1), f(y_2, a)) \vee f(f(y_1, y_1), f(a,y_2)) \vee f(y_1, f(f(y_2,y_3),y_4))]$ and $I_2 = f(x_1, f(x_2, f(x_3,x_4)))$ be implicit generalizations over H with signature $\Sigma = \{a, f\}$ and let $\mathcal{I} = I_1 \vee I_2$. If we apply Theorem 4.1 to either $f(f(y_1,y_1), f(y_2,a))$ or $f(f(y_1,y_1), f(a,y_2))$ on the right-hand side of I_1, then this term may be deleted. However, after the deletion of one non-linear term, Theorem 4.1 only allows us to restrict the other term to the non-linear term $f(f(y_1,y_1), f(a,a))$. But then it is actually possible to apply Theorem 4.2 thus detecting that $\mathcal{I}$ has no finite explicit representation.

Rather than looking for a single term on the right-hand side of some implicit generalization I_i which fulfills the criterion from Theorem 4.2, we have to find a subset of the terms on the right-hand side of I_i for which the conditions from Theorem 4.2 hold simultaneously. We thus get the following criterion, which will then be used in Theorem 5.2 to prove the main result of this work.

Theorem 5.1. *Let* $\mathcal{I} = I_1 \vee \ldots \vee I_n$ *with* $I_i = t_i/t_i\vartheta_{i1} \vee \ldots \vee t_i\vartheta_{im_i}$ *be a disjunction of implicit generalizations, s.t.* $P_{ij} = \{[p_{(ij),1} : X_{(ij),1}], \ldots, [p_{(ij),M_{ij}} : X_{(ij),M_{ij}}]\}$ *represents the complement of* $t_i\vartheta_{ij}$ *w.r.t.* t_i *and* $Q_k = \{[q_{k1} : Y_{k1}], \ldots, [q_{kN_k} : Y_{kN_k}]\}$ *represents the complement of* I_k. *Moreover, we assume that all terms in* $[\bigcup_{i=1}^{n} \bigcup_{j=1}^{m_i} \{t_i\vartheta_{ij}\}] \cup [\bigcup_{i=1}^{n} \bigcup_{j=1}^{m_i} \bigcup_{\alpha=1}^{M_{ij}} \{p_{(ij),\alpha}\}] \cup [\bigcup_{k=1}^{n} \bigcup_{\beta=1}^{N_k} \{q_{k\beta}\}]$ *are pairwise variable disjoint. Then* $\mathcal{I}$ *has no finite explicit representation, iff:*

1. $\exists i \in \{1, \ldots, n\}$ *and* $\exists K \subseteq \{1, \ldots, m_i\}$ *with* $K \neq \emptyset$
2. *For every* $j \in (\{1, \ldots, m_i\} - K)$, *there exists an index* $\alpha_j \in \{1, \ldots, M_{ij}\}$
3. *For every* $j \in \{1, \ldots, i-1, i+1, \ldots, n\}$, *there exists an index* $\beta_j \in \{1, \ldots, N_j\}$

s.t. the following conditions hold for all $k \in K$:

1. *The terms in* $\{t_i\vartheta_{ik}\} \cup \{p_{(ij),\alpha_j} \mid j \in (\{1, \ldots, m_i\} - K)\} \cup \{q_{j\beta_j} \mid j \in \{1, \ldots, i-1, i+1, \ldots, n\}\}$ *are unifiable, with mgu* ν_k *say.*
2. $Z\nu_k$ *with* $Z \equiv [\bigwedge_{j \in (\{1, \ldots, m_i\} - K)} X_{(ij),\alpha_j}] \wedge [\bigwedge_{j \in \{1, \ldots, i-1, i+1, \ldots, n\}} Y_{j\beta_j}]$ *contains no trivial disequation.*
3. *There exists a multiply occurring variable* y_k *in the range of* ϑ_{ik}, *s.t.* $y_k\nu_k$ *is a non-ground term.*

Proof. (Sketch): For the **"if"-direction**, we may assume w.l.o.g. that $i = 1$ and $K = \{1, \ldots, \kappa\}$ holds for some $\kappa \in \{1, \ldots, m_1\}$. Furthermore, suppose that $\mathcal{I}$ has a finite explicit representation $r_1 \vee \ldots \vee r_l$. Similarly to the proof of Theorem 3.2, we can construct terms s_1 and s_1', s.t. these terms are instances of t_1 but they are neither contained in $I_2 \vee \ldots \vee I_n$ nor in $t_1\vartheta_{1(\kappa+1)} \vee \ldots \vee t_1\vartheta_{1m_1}$. Moreover, s_1 is an instance of $t_1\vartheta_{11}$, while s_1' is not. If s_1' is not an instance of any term $t_1\vartheta_{1k}$ with $1 \leq k \leq \kappa$, then s_1' is in fact an instance of I_1 and therefore of $\mathcal{I}$, while s_1 is not. Of course, there is no guarantee that s_1' is not an instance of the terms $t_1\vartheta_{12}, \ldots, t_1\vartheta_{1\kappa}$. But then, similarly to the proof of Proposition 4.6 in [9], we can iteratively construct terms s_k and s'_k for $k \in \{1, \dots, \kappa\}$, s.t. s'_k is not an instance of the terms $t_1\vartheta_{11}, \dots, t_1\vartheta_{1k}$, while s_k is an instance of $t_1\vartheta_{1k}$. Hence, eventually we shall indeed get terms s_k and s'_k, s.t. s'_k is an instance of I_1 and therefore of $\mathcal{I}$, while s_k is not. In particular, s'_k is an instance of some r_γ. But then, analogously to the proof of Theorem 3.2, we can derive a contradiction by showing that also s_k is an instance of r_γ. The details are worked out in [14].

In order to prove also the **"only if"-direction**, we provide an algorithm which, for every $i \in \{1, \dots, n\}$, inspects all possible triples (K, A, B) with the following properties: $K \subseteq \{1, \dots, m_i\}$ with $K \neq \emptyset$, A contains an index $\alpha_j \in \{1, \dots, M_{ij}\}$ for every $j \in (\{1, \dots, m_i\} - K)$ and B contains an index $\beta_j \in \{1, \dots, N_j\}$ for every $j \in \{1, \dots, i-1, i+1, \dots, n\}$. Moreover, this algorithm computes sets $\mathcal{R}_i$ of instances of the terms $t_i\vartheta_{ij}$, s.t. all terms $r \in \mathcal{R}_i$ are linear w.r.t. t_i:

Let $i \in \{1, \dots, n\}$. Moreover, suppose that there exists $\kappa \in \{1, \dots, m_i\}$, s.t. the terms $t_i\vartheta_{ij}$ on the right-hand side of I_i are arranged in such a way that the terms $t_i\vartheta_{i1}, \dots, t_i\vartheta_{i\kappa}$ are non-linear instances of t_i and the terms $t_i\vartheta_{i(\kappa+1)}, \dots, t_i\vartheta_{im_i}$ are linear w.r.t. t_i. Then we start off with $(\mathcal{T}_i, \mathcal{R}_i)$, where the set $\mathcal{R}_i$ of linear instances of t_i consists of the terms $\{t_i\vartheta_{i(\kappa+1)}, \dots, t_i\vartheta_{im_i}\}$ and $\mathcal{T}_i$ denotes the set of all possible triples (K, A, B) with $K = \{1, \dots, \kappa\}$, $A = \{\alpha_{\kappa+1}, \dots, \alpha_{m_i}\}$ and $B = \{\beta_1, \dots, \beta_{i-1}, \beta_{i+1}, \dots, \beta_n\}$.

end condition: If we encounter a triple (K, A, B) in $\mathcal{T}_i$, s.t. for all $k \in K$, the above three conditions of this theorem hold or if there is no triple at all left in $\mathcal{T}_i$, then we may stop.

shrinking K: Suppose that there is a triple (K, A, B) in $\mathcal{T}_i$ with $K \neq \emptyset$ and that there is at least one element $k \in K$, s.t. one of the conditions of this theorem does not hold for (K, A, B). Let k denote the minimum in K, for which at least one condition of the theorem is violated. Then we delete (K, A, B) from $\mathcal{T}_i$ and add to $\mathcal{T}_i$ all triples of the form (K', A', B) with $K' = K - \{k\}$ and $A' = A \cup \{\alpha_k\}$ for all possible values of $\alpha_k \in \{1, \dots, M_{ik}\}$. Furthermore, if ν_k exists and $Z\nu_k$ contains no trivial disequation (i.e.: the first two conditions of this theorem hold for k, while the third one does not), then we add $t_i\vartheta_{ik}\mu_k$ to $\mathcal{R}_i$, where μ_k is the restriction of ν_k to the variables occurring in the range of ϑ_{ik}.

base case: All triples (K, A, B) in $\mathcal{T}_i$ with $K = \emptyset$ may be deleted.

The termination of this algorithm is clear since, in every step, a triple (K, A, B) in $\mathcal{T}_i$ is either deleted or replaced by finitely many triples of the form (K', A', B), where K' is a proper subset of K. Now suppose that this algorithm is applied to every $i \in \{1, \dots, n\}$. In Proposition 5.1 below, we show that either this algorithm eventually detects an index $i \in \{1, \dots, n\}$ together with a triple (K, A, B), s.t. for all $k \in K$, the three conditions of this theorem hold or, for every $i \in \{1, \dots, n\}$, the terms $t_i\vartheta_{i1} \vee \dots \vee t_i\vartheta_{im_i}$ on the right-hand side of I_i may be replaced by the terms in $\mathcal{R}_i$. Note that all the terms in $\mathcal{R}_i$ are linear instances of t_i, since our algorithm only adds those terms $t_i\vartheta_{ik}\mu_k$ to $\mathcal{R}_i$, for which the third condition of this theorem is violated. However, the replacement of the terms on the right-hand side of every I_i by linear instances of t_i is impossible by assumption. But then we must eventually encounter an index $i \in \{1, \dots, n\}$ together with a triple (K, A, B) with the desired properties. $\square$

Proposition 5.1. *The algorithm given in the proof of Theorem 5.1 is correct, i.e.: Suppose that for every index $i \in \{1, \ldots, n\}$, this algorithm stops without finding a triple (K, A, B), s.t. for all $k \in K$ the three conditions of Theorem 5.1 hold. Then, for every $i \in \{1, \ldots, n\}$, the terms $t_i \vartheta_{i1} \vee \ldots \vee t_i \vartheta_{im_i}$ on the right-hand side of I_i may in fact be replaced by the terms in $\mathcal{R}_i$, i.e.: Let $\mathcal{I} = I_1 \vee \ldots \vee I_n$ and $\mathcal{I}' = I_1' \vee \ldots \vee I_n'$ with $I_i' = t_i / \bigvee_{r \in \mathcal{R}_i} r$. Then $\mathcal{I}$ and $\mathcal{I}'$ are equivalent.*

Proof. By the construction of $\mathcal{R}_i$, every term $r \in \mathcal{R}_i$ is an instance of some term $t_i \vartheta_{ij}$. Hence, $I_i \subseteq I_i'$ and, therefore, also $\mathcal{I} \subseteq \mathcal{I}'$ clearly holds. In order to prove also the opposite subset relation, we choose an arbitrary $s \in \mathcal{I}'$ and show that $s \in \mathcal{I}$ also holds. W.l.o.g. we may assume that s is an instance of I_1'. Then, in particular, s is an instance of t_1. If s is an instance of $I_2 \vee \ldots \vee I_n$ then, of course, $s \in \mathcal{I}$ holds and we are done. So suppose that s is in the complement of $I_2, \ldots, I_n$, i.e.: There exist indices $\beta_2, \ldots, \beta_n$, s.t. $s \in \bigwedge_{j=2}^{n} [q_{j\beta_j} : Y_{j\beta_j}]$. Note that by the construction of $\mathcal{R}_1$, all linear instances $t_1 \vartheta_{1(\kappa+1)}, \ldots, t_1 \vartheta_{1m_1}$ of t_1 from the right-hand side of I_1 also appear on the right-hand side of I_1'. By the condition $s \in I_1'$ we thus know that s is in the complement of the terms $t_1 \vartheta_{1(\kappa+1)}, \ldots, t_1 \vartheta_{1m_1}$, i.e.: There exist indices $\alpha_{\kappa+1}, \ldots, \alpha_{m_1}$, s.t. $s \in \bigwedge_{j=\kappa+1}^{m_1} [p_{(1j),\alpha_j} : X_{(1j),\alpha_j}]$. Now consider the triple (K, A, B) from the initial set $\mathcal{T}_i$, where A and B consist exactly of those indices $\alpha_{1(\kappa+1)}, \ldots, \alpha_{1m_1}$ and $\beta_2, \ldots, \beta_n$, respectively, s.t. s is an instance of $\bigwedge_{j=\kappa+1}^{m_1} [p_{(1j),\alpha_j} : X_{(1j),\alpha_j}] \wedge \bigwedge_{j=2}^{n} [q_{j\beta_j} : Y_{j\beta_j}]$. If $K = \emptyset$, then $\kappa = 0$ holds and all terms on the right-hand side of I_1 are linear instances of t_1. But then, s is indeed an instance of I_1 since we only consider the case where s is an instance of t_1 and s is in the complement of $t_1 \vartheta_{1(\kappa+1)}, \ldots, t_1 \vartheta_{1m_1}$. So let $K \neq \emptyset$ and let k denote the minimum in K, for which at least one condition of Theorem 5.1 is violated. We claim that then s is not an instance of $t_1 \vartheta_{1k}$. For suppose on the contrary that s is an instance of $t_1 \vartheta_{1k}$. Then s is also contained in $t_1 \vartheta_{1k} \wedge \bigwedge_{j=\kappa+1}^{m_1} [p_{(1j),\alpha_j} : X_{(1j),\alpha_j}] \wedge \bigwedge_{j=2}^{n} [q_{j\beta_j} : Y_{j\beta_j}]$. Hence, s is an instance of $t_1 \vartheta_{1k} \mu_k$, where μ_k is the restriction of the unifier $\nu_k = mgu(\{t_1 \vartheta_{1k}\} \cup \{p_{(1j),\alpha_j} \mid \kappa + 1 \leq j \leq m_1\} \cup \{q_{j\beta_j} \mid 2 \leq j \leq n\}$ to the variables occurring in the range of ϑ_{1k}. However, $t_1 \vartheta_{1k} \mu_k$ is added to $\mathcal{R}_1$ by our algorithm and, therefore, $t_1 \vartheta_{1k} \mu_k$ occurs on the right-hand side of I_1'. But then s is not an instance of I_1', which is a contradiction.

Thus s is in the complement of $t_1 \vartheta_{1k}$ and, therefore, there exists an index $\alpha_k \in \{1, \ldots, M_{1k}\}$, s.t. $s \in [p_{(1k),\alpha_k} : X_{(1k),\alpha_k}]$. Note that our algorithm actually adds the triple (K', A', B) to $\mathcal{T}_1$, where $K' = K - \{k\}$ and $A' = A \cup \{\alpha_k\}$ holds. By assumption, our algorithm never finds a triple for which all the three conditions from Theorem 5.1 hold. Hence, eventually we shall have to select this new triple (K', A', B). But then we can repeat the same argument as above also for the triple (K', A', B): Let k' denote the minimum in K', for which at least one condition of Theorem 5.1 is violated. We claim that then s is not an instance of $t_1 \vartheta_{1k'}$. For suppose on the contrary that s is an instance of $t_1 \vartheta_{1k'}$ Then s is also in $t_1 \vartheta_{1k'} \wedge [p_{(1k),\alpha_k} : X_{(1k),\alpha_k}] \wedge \bigwedge_{j=\kappa+1}^{m_1} [p_{(1j),\alpha_j} : X_{(1j),\alpha_j}] \wedge \bigwedge_{j=2}^{n} [q_{j\beta_j} : Y_{j\beta_j}]$. Hence, s is an instance of $t_1 \vartheta_{1k'} \mu_{k'}$, where $\mu_{k'}$ is defined as the restriction of $\nu_{k'} = mgu(\{t_1 \vartheta_{1k'}\} \cup \{p_{(1k),\alpha_k}\} \cup \{p_{(1j),\alpha_j} \mid \kappa + 1 \leq j \leq m_1\} \cup \{q_{j\beta_j} \mid 2 \leq j \leq n\})$ to the variables occurring in the range of $\vartheta_{1k'}$. This is again impossible, since $t_1 \vartheta_{1k'} \mu_{k'}$ is added to $\mathcal{R}_1$ by our algorithm and, therefore, $t_1 \vartheta_{1k'} \mu_{k'}$ occurs on the right-hand side of I_1'. But then s is not an instance of I_1', which is a contradiction.

Note that by iterating this argument κ-times, we can show that s is not an instance of any term $t_1\vartheta_{1j}$ with $j \in \{1,\ldots,\kappa\}$. Moreover, recall that we consider the case where s is an instance of t_1 and s is in the complement of $t_1\vartheta_{1(\kappa+1)},\ldots,t_1\vartheta_{1m_1}$. Hence, s is an instance of I_1 and, therefore, also of $\mathcal{I}$. $\square$

Theorem 5.2. (coNP-Completeness) *The finite explicit representability problem for disjunctions of implicit generalizations is coNP-complete.*

Proof. By the coNP-hardness result from [12], we only have to prove the coNP-membership. To this end, we consider the following algorithm for checking that $\mathcal{I} = I_1\vee\ldots\vee I_n$ with $I_i = t_i/t_i\vartheta_{i1}\vee\ldots\vee t_i\vartheta_{im_i}$ has no finite explicit representation:

1. Guess i, K, A and B.
2. Check for all $k \in K$, that the conditions from Theorem 5.1 hold.

This algorithm clearly works in non-deterministically polynomial time, provided that an efficient unification algorithm is used (cf. [13]). Moreover, the correctness of this algorithm follows immediately from Theorem 5.1. $\square$

By the coNP-hardness of the finite explicit representability problem, we can hardly expect to find a decision procedure with a significantly better worst case complexity than a deterministic version of the NP-algorithm from Theorem 5.2 above, i.e.: Test for all possible values of i, K, $A = \{\alpha_{ij}\,|\,j \in (\{1,\ldots,m_i\} - K)\}$ and $B = \{\beta_1,\ldots,\beta_{i-1},\beta_{i+1},\ldots,\beta_n\}$, whether the conditions from Theorem 5.1 hold. However, for practical purposes, one will clearly prefer the algorithm from the proof of Theorem 5.1, which has a major advantage, namely: If the conditions from Theorem 5.1 do not hold for any value of i, K, A and B, then this algorithm actually computes for every $i \in \{1,\ldots,n\}$ a set $\mathcal{R}_i$ of linear instances of the term t_i, s.t. the terms on the right-hand side of I_i may be replaced by the terms in $\mathcal{R}_i$. By the considerations on the complement of linear terms from Sect. 2, it is then easy to convert the resulting implicit generalizations I_i' into explicit ones.

6   Related Works

A decision procedure for the finite explicit representability problem of a single implicit generalization $I = t/t_1 \vee \ldots \vee t_m$ was first presented in [9]. Its basic idea is a division into subproblems via the complement of a *linear* instance t_i of t on the right-hand side of I, i.e.: Let $P = \{p_1,\ldots,p_M\}$ represent the complement of t_i w.r.t. t and suppose that all terms t_i, p_j and t are pairwise variable disjoint. Then I is equivalent to the disjunction $\bigvee_{j=1}^{M} I_j$ with $I_j = p_j/mgi(t_1,p_j) \vee \ldots \vee mgi(t_{i-1},p_j)\vee mgi(t_{i+1},p_j)\vee\ldots\vee mgi(t_n,p_j)$. The subproblems I_j thus produced have strictly fewer terms on the right-hand side. Hence, by applying this splitting step recursively to each subproblem, we either manage to remove all terms from the right-hand side of all subproblems or we eventually encounter a subproblem with only non-linear instances on the right-hand side. For the latter case, it has been shown in [9], that I has no finite explicit representation.

In Sect. 3 we have provided a new algorithm for deciding the finite explicit representability problem of a single implicit generalization. Note that this algorithm starts from the "opposite direction", i.e.: Rather than removing the linear terms from the right-hand side until finally only non-linear ones are left, our algorithm tries to replace the non-linear terms by linear ones until it finally detects a non-linear term which cannot be replaced by linear ones. If an implicit generalization has only few non-linear terms on the right-hand side, then our algorithm from Sect. 3 may possibly be advantageous. However, in general, the algorithm from [9] will by far outperform our algorithm from Sect. 3 for the following reason: Suppose that there are several non-linear terms on the right-hand side of an implicit generalization I and that we may replace one such non-linear term t_1 by linear ones $u_1, \ldots, u_M$ via Theorem 3.1. This replacement step, in general, yields *exponentially many* terms $u_1, \ldots, u_M$ (w.r.t. the number m of terms and also w.r.t. the size of these terms). Suppose that we next apply Theorem 3.1 to another non-linear term t_2 on the right-hand side of I and that we may replace t_2 by the linear terms $v_1, \ldots, v_N$. Then N is actually exponential w.r.t. M. But then N is doubly exponential w.r.t. the size of the original implicit generalization. Of course, having to restrict yet another non-linear term t_3 to the complement of the terms $v_1, \ldots, v_N$ makes the situation even worse, etc.

In [15], the algorithm from [9] is extended to disjunctions of implicit generalizations. This algorithm consists of two rewrite rules: One is exactly the splitting rule from [9]. The other one basically allows us to restrict the instances of a non-linear term t_{ij} on the right-hand side of an implicit generalization I_i to the complement of another implicit generalization $I_k = t_k/t_{k1} \lor \ldots \lor t_{km_k}$, provided that $mgi(t_{ij}, t_k)$ is a linear instance of t_{ij}. Hence, when a disjunction of implicit generalizations contains several disjuncts with non-linear terms on the right-hand side, then the algorithm of [15] also suffers from the above mentioned exponential blow-up whenever a term is restricted to the complement of terms which are themselves the result of such a restriction step. Of course, our algorithm from Sect. 4 has this problem as well. However, the algorithm from Sect. 5 is clearly better than this, i.e.: The terms that we add to the sets $\mathcal{R}_i$ in the proof of Theorem 5.1 result from restricting a non-linear term t_{ij} on the right-hand side of I_i to the complement of some other terms $t_{ij'}$ on the right-hand side of I_i and to the complement of the other implicit generalizations $I_{i'}$. However, we never restrict a term w.r.t. terms which are themselves the result of a previous restriction step. Recall that in [12] an upper bound on the time and space complexity of the algorithm from [9] is given, which is basically exponential w.r.t. the size of an input implicit generalization (cf. [12], Theorem 5.5). In fact, our algorithm from Sect. 5 also has an upper bound with a single exponentiality, no matter whether we apply this algorithm to a single implicit generalization or to a disjunction of implicit generalizations: In order to see this, recall from Theorem 2.1, that the number of constrained terms in the complement of a term $t\vartheta$ is quadratically bounded by the size of this term. Hence, also the number M_{ij} of constrained terms in the representation of the complement of $t_i\vartheta_{ij}$ w.r.t. t_i as well as the number N_k of constrained terms in the complement representation of

I_k is quadratically bounded by the size of $t_i \vartheta_{ij}$ and I_k, respectively. But then the number of possible values of i, K, A and B, that our algorithm from Sect. 5 has to inspect, is clearly exponentially bounded in the size of the original disjunction $\mathcal{I}$ of implicit generalizations.

In [15], the results on implicit generalizations of terms are extended to tuples of terms. Moreover, recall that the term tuples contained in a disjunction $\mathcal{I} = I_1 \vee \ldots \vee I_n$ of implicit generalizations with $I_j = t_j/t_{j1} \vee \ldots \vee t_{jm_j}$ can be represented by an equational formula of the form $\bigvee_{j=1}^{n} (\exists \boldsymbol{x}_j)(\forall \boldsymbol{y}_j)(\boldsymbol{z} = t_j \wedge t_{j1} \neq t_j \wedge \ldots \wedge t_{jm_j} \neq t_j)$ with free variables in $\boldsymbol{z}$, where $\boldsymbol{x}_j$ denotes the variables occurring in t_j and $\boldsymbol{y}_j$ denotes the variables occurring in $t_{j1} \vee \ldots \vee t_{jm_j}$. It is shown in [15], that only a slight extension of the transformation given in [10] is required so as to transform any equational formula into an equivalent one of this form. But then, an algorithm for the finite explicit representability problem of disjunctions of implicit generalizations immediately yields a decision procedure for the negation elimination problem of arbitrary equational formulae. In [3], a different decision procedure for the negation elimination problem is given by appropriately extending the reduction system from [2]. Even though this approach differs significantly from the method of [15], this algorithm also requires that certain terms have to be transformed w.r.t. terms that are themselves the result of a previous transformation step. Hence, analogously to the algorithm of [15], it does not seem as though there exists a singly exponential upper bound on the complexity of this algorithm even if it is only applied to equational formulae which correspond to a disjunction of implicit generalizations.

Several publications deal with the complexity of the emptiness and the finite explicit representability problem. In [6], [7] and [5], the coNP-completeness of the emptiness problem was proven. The coNP-hardness of the emptiness problem was also shown in [12], where the coNP-hardness of the finite explicit representability problem was then proven by reducing the emptiness problem to it. A coNP-membership proof of the finite explicit representability problem of a single implicit generalization was given in [11]. The coNP-membership in case of disjunctions of implicit generalizations has been an open question so far.

7   Conclusion

In this paper, we have revisited the finite explicit representability problem of implicit generalizations. We have provided an alternative algorithm for this decision problem which allowed us to prove the coNP-completeness in case of disjunctions of implicit generalizations. The most important aim for future research in this area is clearly the search for a more efficient algorithm. By our considerations from Sect. 6, our algorithm from Sect. 5 has a similar worst case complexity to the algorithm for single implicit generalizations from [9]. In contrast to the algorithms from [15] and [3], this upper bound on the complexity holds for our algorithm also in case of disjunctions of implicit generalizations. Hence, our algorithm can be seen as a step towards a more efficient algorithm for this decision problem, to which further steps should be added by future research.

References

[1] H.Comon, C.Delor: Equational Formulae with Membership Constraints, in Journal of Information and Computation, Vol 112, pp. 167-216 (1994).

[2] H. Comon, P. Lescanne: Equational Problems and Disunification, in Journal of Symbolic Computation, Vol 7, pp. 371-425 (1989).

[3] M.Fernández: Negation Elimination in Empty or Permutative Theories, in Journal of Symbolic Computation, Vol 26, pp. 97-133 (1998).

[4] G.Gottlob, R.Pichler: Working with ARMs: Complexity Results on Atomic Representations of Herbrand Models, in Proceedings of LICS'99, pp.306-315, IEEE Computer Society Press (1999).

[5] D.Kapur, P.Narendran, D.Rosenkrantz, H. Zhang: Sufficient-completeness, ground-reducibility and their complexity, in Acta Informatica, Vol 28, pp. 311-350 (1991).

[6] K.Kunen: Answer Sets and Negation as Failure, in Proceedings of the Fourth Int. Conf. on Logic Programming, Melbourne, pp. 219-228 (1987).

[7] G.Kuper, K.McAloon, K.Palem, K.Perry: Efficient Parallel Algorithms for Anti-Unification and Relative Complement, in Proceedings of LICS'88, pp. 112-120, IEEE Computer Society Press (1988).

[8] J.-L.Lassez, M.Maher, K.Marriott: Elimination of Negation in Term Algebras, in Proceedings of MFCS'91, LNCS 520, pp. 1-16, Springer (1991).

[9] J.-L.Lassez, K.Marriott: Explicit Representation of Terms defined by Counter Examples, in Journal of Automated Reasoning, Vol 3, pp. 301-317 (1987).

[10] M.Maher: Complete Axiomatizations of the Algebras of Finite, Rational and Infinite Trees, in Proceedings of LICS'88, pp. 348-357, IEEE Computer Society Press (1988).

[11] M.Maher, P.Stuckey: On Inductive Inference of Cyclic Structures, in Annals of Mathematics and Artificial Intelligence, Vol 15 No 2, pp. 167-208, (1995).

[12] K. Marriott: Finding Explicit Representations for Subsets of the Herbrand Universe, PhD Thesis, The University of Melbourne, Australia (1988).

[13] A.Martelli, U.Montanari: An efficient unification algorithm, in ACM Transactions on Programming Languages and Systems, Vol 4 No 2, pp. 258-282 (1982).

[14] R.Pichler: The Explicit Representability of Implicit Generalizations, full paper, available from the author (2000).

[15] M. Tajine: The negation elimination from syntactic equational formulas is decidable, in Proceedings of RTA'93, LNCS 690, pp.316-327 Springer (1993).

On the Word Problem for Combinators

Rick Statman*

Department of Mathematical Sciences
Carnegie Mellon University, Pittsburgh, PA 15213
statman@cs.cmu.edu

Abstract. In 1936 Alonzo Church observed that the "word problem" for combinators is undecidable. He used his student Kleene's representation of partial recursive functions as lambda terms. This illustrates very well the point that "word problems" are good problems in the sense that a solution either way - decidable or undecidable - can give useful information. In particular, this undecidability proof shows us how to program arbitrary partial recursive functions as combinators.

I never thought that this result was the end of the story for combinators. In particular, it leaves open the possibility that the unsolvable problem can be approximated by solvable ones. It also says nothing about word problems for interesting fragments i.e., sets of combinators not combinatorially complete.

Perhaps the most famous subproblem is the problem for S terms. Recently, Waldmann has made significant progress on this problem. Prior, we solved the word problem for the Lark, a relative of S. Similar solutions can be given for the Owl (S^*) and Turing's bird U. Familiar decidable fragments include linear combinators and various sorts of typed combinators. Here we would like to consider several fragments of much greater scope. We shall present several theorems and an open problem.

1   Introduction

In 1936 Alonzo Church [3] observed that the "word problem" for combinators is undecidable. He used his student Kleene's representation of partial recursive functions as lambda terms. This illustrates very well the point that "word problems" are good problems in the sense that a solution either way – decidable or undecidable – can give useful information. In particular, this undecidability proof shows us how to program arbitrary partial recursive functions as combinators.

I never thought that this result was the end of the story for combinators. In particular, it leaves open the possibility that the unsolvable problem can be approximated by solvable ones. It also says nothing about word problems for interesting fragments i.e., sets of combinators not combinatorially complete.

Perhaps the most famous subproblem is the problem for S terms. Recently, Waldmann has made significant progress on this problem [12]. In [8] we solved the word problem for the Lark, a relative of S. Similar solutions can be given

* This research supported in part by the National Science Foundation CCR-9624681

for the Owl (S^*) and Turing's bird U [7]. Familiar decidable fragments include linear combinators and various sorts of typed combinators. Here we would like to consider several fragments of much greater scope. We shall present several theorems and an open problem. We begin with a series of decidable word problems which approximate the undecidable general case.

Proper combinators P are given by reduction rules

$$Px(1)\ldots x(p) \to X$$

where X is an applicative combination of the indeterminates $x(1)\ldots x(p)$. P is said to be a compositor if X cannot be written without parentheses using the convention of association to the left. Equivalently, in the terminology of Craig's conditions [9], P has compositive effect. Otherwise, P is said to be a co-compositor. Examples of co-compositors are

$Cxyz \to xzy$ (The Cardinal)
$Kxy \to x$ (The Kestrel)
$Ix \to x$ (The Identity)
$Wxy \to xyy$ (The Warbler)

and the typical compositor is the Bluebird

$Bxyz \to x(yz)$.

Suppose that we fix a finite set of compositors Q with rules

$$Qx(1)\ldots x(q) \to X.$$

Consider the set of all applicative combinations of arbitrary co-compositors and these finitely many compositors, and select finitely many closed instances

$$QM(1)\ldots M(q) \to \left[\frac{M(1)}{x(1)}, \ \ldots, \ \frac{M(q)}{x(q)} \right] X$$

of the reduction rules for the compositors in this set. These finitely many instances together with the reduction rules for all co-compositors induce a reducibility relation $\twoheadrightarrow$ and a congruence relation $\twoheadleftrightarrow$.

Theorem: The relation $\twoheadleftrightarrow$ is decidable. Indeed, when finitely many co-compositors are selected in advance the problem is solvable in polynomial time.

Next we consider a tantalizing open problem an several closed variants.

In [9] proved that any basis of proper combinators must contain one of order > 2. This suggests that the word problem for all proper combinators of order < 3 might be decidable.

Problem: Determine whether the word problem for all proper combinators of order < 3 is decidable.

Slightly less is decidable. P is said to be hereditarily of order one (HOO) if P has reduction rule

$$Px \to X$$

where X is a normal applicative combination of x and previously defined members of HOO. Examples are

$Ix \to x$ (The Identity)
$Mx \to xx$ (The Mockingbird)
$K^*x \to I$ (The Identity once removed)
$C^{**} \to xI$ (The Cardinal twice removed)

Theorem: The word problem for HOO is decidable. Indeed, it is log-space complete for polynomial-time.

Slightly more is undecidable. P is said to be hereditarily of order two (HOT) if P has reduction rule

$$Pxy \to X$$

where X is a normal applicative combination of x, y, and previously defined members of HOT.

Examples:

$Lxy \to x(yy)$ (The Lark)
$Uxy \to y(xxy)$ (Turing's bird)
$Oxy \to y(xy)$ (The Owl)

Theorem: The word problem for HOT is undecidable. Indeed every partial recursive function, under an appropriate encoding, can be represented in HOT. Below we omit most proofs except for one or two lines indicating the direction of proof. Proof will be present in the final version of the paper.

2   Compositors and Co-compositors

Again, proper combinators P are given by reduction lines

$$Px(1)\ldots x(p) \to X \tag{1}$$

where X is a applicative combination of the indeterminates $x(1)\ldots x(p)$. Evidently, P is a co-compositor if and only if there is an integer r and a function $f : [1, r] \to [1, p]$ such that

$$X = x(f(1))\ldots x(f(r)).$$

Let COCO be the set of all co-compositors.

Suppose that we fix a finite set CO of compositors Q with rules

$$Qx(1)\ldots x(q) \to X. \tag{2}$$

Let COMB be the set of all applicative combinations of members of CO $\cup$ COCO. Select finitely many closed instances

$$QM(1)\ldots M(q) \to \left[\frac{M(1)}{x(1)}, \ldots, \frac{M(q)}{x(q)}\right] X \tag{3}$$

of the reduction rules for members of CO where $M(1), \ldots, M(q)$ members of COMB. These finitely many instances together with the reduction rules for all co-compositors induce a reducibility relation $\twoheadrightarrow$ and a congruence relation $\leftrightarrow\twoheadrightarrow$. More precisely, $\twoheadrightarrow$ is the monotone, transitive, reflexive closure of all COMB instances of (1) for P in COCO and the finitely many (3) for Q a member of CO. $\leftrightarrow\twoheadrightarrow$ is the congruence closure of $\twoheadrightarrow$.

Now the reducibility relation $\twoheadrightarrow$ need not be Church–Rosser since the instances (3) may not be closed under reducibility. For this reason we shall extend $\twoheadrightarrow$ to a Church–Rosser reducibility which generates the same $\leftrightarrow\twoheadrightarrow$. We define reductions $\to (n)$ by induction on n as follows.

Basis; $n = 0$. $\to (0)$ is $\to$ and $\twoheadrightarrow (0)$ is $\twoheadrightarrow$.

Induction Step; $n > 0$. We define $\to (n)$ to be $\to (n - 1)$ together with all instances (3) such that there exist $N(1), \ldots, N(q)$ satisfying

(i) $M(i) \twoheadrightarrow (n - 1)N(i)$ for $i == 1, \ldots, q$

(ii) $QN(1)\ldots N(q) \to (n - 1)\left[\frac{N(1)}{x(1)}, \ldots, \frac{N(q)}{x(q)}\right] X$.

$\twoheadrightarrow (n)$ is defined to be the monotone, transitive, reflexive closure of $\to (n)$.

Finally, we define $\to +$ to be the union of the $\to (n)$ and $\twoheadrightarrow +$ the union of the $\twoheadrightarrow (n)$. Clearly the congruence closure of $\twoheadrightarrow +$ is just $\leftrightarrow\twoheadrightarrow$ and $\twoheadrightarrow +$ is Church–Rosser. Given a finite $\twoheadrightarrow +$ reduction.

$$M(1) \twoheadrightarrow +M(2) \twoheadrightarrow + \ldots M(m - 1) \twoheadrightarrow +M(m) \tag{4}$$

where each step consists of contracting a single $\to +$ redex we assign a triple of integers with coordinates

(i) the least n such that (4) is an $\twoheadrightarrow (n)$ reduction

(ii) $m =$ the length of (4)

(iii) $| M(1) |=$ the length of $M(1)$

ordered lexicographically. We refer to the triple as the ordinal of (4). Evidently, the ordinal of (4) is less than omega cubed.

We define the positive subterms of a term M as follows. If $M = PM(1)\ldots M(p)$ then the positive subterms of M consist of M and all the positive subterms of $M(1)$ and $\ldots$ and $M(p)$.

We have the following:

Fact 1: If $M \twoheadrightarrow +N$ then every positive subterm of N is a $\twoheadrightarrow +$ reduct of a positive subterm of M or

(*) a positive subterm of one side or the other of one of the closed instances (3) for Q in CO.

Moreover, the ordinals of these reductions are not larger than the ordinal of the reduction $M \twoheadrightarrow +N$.

Proof: By induction on the ordinal of the reduction.

Now suppose that M is given and define a set of identities AXIOM(M) as follows. AXIOM(M) contains all the identities

(i) $QM(1)\ldots M(q) = [\frac{M(1)}{x(1)},\ldots,\frac{M(q)}{x(q)}]X$ for the selected close instances of (3) for the members Q of CO.

AXIOM(M) also contains all the identities

(ii) $PN(1)\ldots N(p) = N(f(1))\ldots N(f(p))$ for each member P of $COCO$ which appears in either M or one of the $M(j)$ in (i) and for each $N(1),\ldots,N(p)$ which are positive subterms of either M or (*).

Fact 2: If $M \twoheadrightarrow +N$ then AXIOM(M) $\vdash M = N$

Proof: By induction on the ordinal of the reduction $M \twoheadrightarrow +N$.

We obtain the following:

Proposition 1: $M \leftrightarrow\!\!\!\twoheadrightarrow N$ if and only if AXIOM(M) $\cup$ AXIOM(N) $\vdash M = N$

Proof: Clearly AXIOM(M) $\cup$ AXIOM(N) $\vdash M = N \Rightarrow M \leftrightarrow\!\!\!\twoheadrightarrow N$. Now suppose that $M \leftrightarrow\!\!\!\twoheadrightarrow N$. By the Church–Rosser theorem there exists L such that $M \twoheadrightarrow +L+ \twoheadleftarrow N$. By FACT 2, AXIOM(M) $\vdash M = L$ and AXIOM(N) $\vdash N = L$. Thus, AXIOM(M) $\cup$ AXIOM(N) $\vdash M = N$. ∎

Thus, we obtain the decidability proof by remarking that the problem of determining whether a finite list of closed identities implies a closed identity is decidable (the word problem for finitely presented algebras) [1].

3   Combinators Hereditarily of Order One

Members of HOO are atoms with associated reduction rules. These reduction rules generate a notion of reducibility $\twoheadrightarrow$ and a congruence $\leftrightarrow\!\!\!\twoheadrightarrow$. HOO and $\rightarrow$ are defined simultaneously by induction as follows.

If X is an applicative combination of x's then H defined by the reduction rule $Hx \rightarrow X$ belongs to HOO.

If X is a $\twoheadrightarrow$ normal combination of x's and previously defined members of HOO then H defined by the reduction rule $Hx \rightarrow X$ belongs to HOO.

We use H, J, L to range over HOO combinators, and M, N for HOO combinations below. We shall write $\lambda x.\ X$ for the member of HOO with reduction rule ending in X.

$\rightarrow$ is a regular left normal combinatory reduction system [5], so it satisfies the Church–Rosser and Standardization theorems. Clearly, any normal HOO combination belongs to HOO. If M is a HOO combination with no normal form we write $M = @$. This makes sense since the corresponding lambda term is an order zero unsolvable. More generally it is easy to see that $\longleftrightarrow$ coincides with beta conversion of the corresponding lambda terms.

We define the notion of $@$ normal form ($@$nf) as follows.

M is in $@$nf if $M = a$ HOO combinator H or $M = HJM(1)\ldots M(m)$ where $HJ = @$ and each $M(i)$ for $i = 1, \ldots, m$ is in $@$nf.

It is easy to see that $@$nf's always exist but they are hardly unique. The following relation $\rightarrowtail$ is useful for computing $@$nf's. $\rightarrowtail$ is the monotone closure of

$$HM \rightarrowtail [\tfrac{M}{x}]X \text{ if } M = @$$

$$HJ \rightarrowtail \begin{cases} L \text{ if } HJ \longleftrightarrow L \\[2mm] [\tfrac{J}{x}]\, X \text{ if } HJ = @ \end{cases}$$

for $H = \lambda x X$.

The relation is actually decidable; more about this below. A simple induction shows

Fact 1: $M \longleftrightarrow H \Rightarrow M \rightarrowtail H$.

We need the following notation. We write $M = M[M(1), \ldots, M(m)]$ if $M(1),\ldots, M(m)$ are disjoint occurrences of the corresponding HOO combinations in M.

Lemma 1: If $M = M[M(1), \ldots, M(m)]$, for $i = 1, \ldots, m$ we have $M(i)$ $\longleftrightarrow H(i)$, and $M \rightarrow N$ then we can write $N = N[N(1), \ldots, N(m)]$ with , for $i = 1, \ldots, m, N(i) \longleftrightarrow J(i)$ and $M[H(1), \ldots, H(m) \rightarrowtail N[J(1), \ldots, J(m)]$

Proof: By a case analysis of the redex position.

From this we obtain the following

Proposition 2: IF $M \rightarrowtail N$ and N is $@$ normal then $M \rightarrowtail N$.

[is the partial order on members of HOO defined by the cover relations
$J [H$ if $H = \lambda x J$
$\lambda x.X(i) [H$ for $i = 1, \ldots, n$ if $H = \lambda x.xX(1) \ldots X(n)$.
A sequence $H_- = H(1) \ldots H(t)$ of HOO combinators is said to be admissible if H_- is closed under [and $H(i) [H(j) \Rightarrow i < j$. Note that if H_- is admissible and $H(i)H(j) \longleftrightarrow H$ then H is a member of H_-.

We define the $n*n$ matrix MATRIX(H_) with entries in $\{1, \ldots, n, @\}$ by

$$\text{MATRIX(H\_)}(i, j) = \begin{cases} k \text{ if } H(i)H(j) = H(k) \\ \\ @ \text{ otherwise} \end{cases}$$

The procedure ( )@ is computed on H_{-} combinations as follows. $(H(i))@ = H(i)$

$(H(i)M(1)\ldots M(m))@ =$

$$\begin{cases} (H(j)\ M(2)\ldots M(m))@ & \text{if } H(i) = \lambda x.H(j) \\ & \text{or } (M(1))@ = H(k) \text{ and} \\ & H(i)H(k) \leftarrow\!\!\!\leftarrow\!\!\!\rightarrow\!\!\!\rightarrow H(j) \\ H(i)\ H(j)\ (M(2))@\ldots(M(m))@ & \text{if } (M(1))@ = H(j) \text{ and} \\ & H(i)\ H(j) = @ \\ \left[\frac{(M(1))@}{x}\right] X(M(2))@\ldots(M(m))@ & \text{if } M(1) = @ \text{ and} \\ & H(i) = \lambda x.\ X \text{ with } x \text{ in } X. \end{cases}$$

Although the output of ( )@ can be exponentially long in the input this is only because of repeated subterms. The procedure will run in time polynomial in the input and H_{-} if the output is coded as a system of assignment statements.

The relation $\mapsto$ is defined to be the monotone closure of

$$HJ \mapsto ([J/x]\,X)\,@$$

where $H = \lambda x.\ X$. $\mapsto$ is particularly useful in conversion between @ normal forms. Observe here that the congruence relation generated by $\mapsto$ restricted to admissible H_{-} can be presented as a finitely presented algebra and so, as above, it is decidable by [1].

FACT 2: If $M = @$ then $(MN)@ \leftarrow\!\!\!\leftarrow\!\!\!\rightarrow\!\!\!\rightarrow (M)@\ (N)@$.

FACT 3: If $H = \lambda x.\ X$ and $M = @$ then $(HM)@ \leftarrow\!\!\!\leftarrow\!\!\!\rightarrow\!\!\!\rightarrow ([M/x]\ X)\,@$.

Lemma 2: If $M \rightarrowtail N$ then $(M)@ \mapsto (N)@$.

Proof: By induction on M.

Proposition 3: If M and N are @ normal forms and $M \rightarrow\!\!\!\rightarrow N$ then

$$M_| \rightarrow\!\!\!\rightarrow N.$$

Corollary : If M and N are @ normal forms and $M \leftarrow\!\!\!\leftarrow\!\!\!\rightarrow\!\!\!\rightarrow N$ then there exists a @ normal form R such that $M \mapsto\!\!\!\rightarrow R \leftarrow\!\!\!\leftarrow_| N$.

We now suppose that MATRIX (H_{-}) is given and we wish to compute MATRIX
$(H_{-}H(n+1))$. Toward this end we define a procedure $( )$ which takes as an

input an $H\_H(n+1)$ combination and depends on $H\_$ and a parameter V, a subset of $\{1,\ldots,n+1\} \otimes \{n+1\}$ (here we suppose that MATRIX$(H\_)$ has been supplemented with values for pairs not in V).

Input: M
If $M = H(i)$ then return i else
If $M = H(i)M(1)\ldots M(m)$ then do

$\qquad$ If $H(i) = \lambda x.\, H(j)$ then return $\$\,(H(j)M(2)\ldots M(m))$ else

$\qquad$ set $h := \$(M(1))$ and

$\qquad$ If $h = (k,1)$ then return $(k,1)$ else if $h = k$ then

$\quad$ cases: (i,k) belongs to V return (i,k)

$\qquad\qquad\qquad i = n+1$ and $k < n+1$ let $H(n+1) = \lambda x.\, X(n+1)$

$\qquad\qquad\qquad$ set $h := \$([H(k)/x]X(n+1))$

$\qquad\qquad\qquad$ if $h = p$ then $\$\,(H(p)M(2)\ldots M(m))$ else

$\qquad\qquad\qquad\qquad\qquad$ return h

$\qquad\qquad (i,k)$ does not belong to V

$\qquad\qquad\qquad$ if MATRIX$(H\_)(i,k) = p$ then $(H(p)M(2)\ldots M(m))$

$\qquad\qquad\qquad$ else return (i,k).

Note that if the values $\$([H(k)/x]\,X(n+1))$ for $k = 1,\ldots,n$ have been pre-computed and store for look up then the procedure $\$(\ )$ runs in time polynomial in the input. $\$(\ )$ computes a first approximation to the head of a @nf for the input. It is used as follows. For $i = 1,\ldots,n+1$ set $h(i) = \$\,([H(n+1)/x]\,X(i))$. Define a graph $G(V)$ as follows. The points of $G(V)$ are the values $h(i)$ and the pairs $(i,n+1)$ in V. The edges are directed

$$(i, n+1) \longrightarrow h(i).$$

Given $(i,n+1)$ in V, $(i,n+1)$ begins a unique path which either cycles or termi-nates in a value outside of V. If this path cycles then $H(i)H(n+1) \longleftrightarrow\!\!\!\rightarrow$ @ as we shall see below. The path terminates in a pair (j,k) only if MATRIX$(H\_)(j,k) =$ @ so again $H(i)H(n+1) =$ @. Finally, if the path terminates in an integer k then the last edge in the path is

$$(j, n+1) \longrightarrow k$$

and we can conclude $H(j)H(n+1) \longleftrightarrow\!\!\!\rightarrow H(k)$. Thus, at least one new value can be added to the matrix and V decreased by at least one.

Lemma 3: If $[H(j)/x]\,X(i) \longrightarrow\!\!\!\rightarrow H(i)H(j)M(1)\ldots M(m)$ then $H(i)H(j) =$ @

Proof: By the standardization theorem.

Given admissible $H\_$, MATRIX$(H\_)$ can be computed recursively from the initial segments of $H\_$ in time polynomial in $H\_$.

Now suppose that we are given two HOO combinations together with the reduction rules for their atoms. Construct an admissible H_- containing these atoms.

This can be done in time polynomial in the input. Next compute MATRIX(H_-) as above. Using MATRIX(H_-) compute $(M)@$ and $(N)@$ as systems of assignment statements. Finally, add to these systems the equations $H(i)H(j) = ([H(j)/x]\ X(i))$ for each pair $H(i), H(j)$ in H_- (or rather the corresponding systems of assignment statements), and, using the algorithm for the word problem for finitely presented algebras [4], test whether $M = N$ is a consequence of these statements. In summary

Proposition 4: The word problem for HOO combinations can be solved in polynomial time.

It can be shown by encoding the circuit value problem that the word problem for HOO combinations is log space complete for polynomial time.

4 Combinators Hereditarily of Order Two

Members of HOT are atoms with associated reduction rules. These reduction rules generate a notion of reducibility $\twoheadrightarrow$ and a congruence $\leftrightarrow\!\!\!\!\rightarrow$. HOT and $\rightarrow$ are defined simultaneously by induction as follows.

If X is an applicative combination of x''s and y''s then H is defined by the reduction rule $Hxy \rightarrow X$ belongs to HOT.

If X is a $\twoheadrightarrow$ normal combination of x's, y's, and previously defined members of HOT then H defined by the reduction rule $Hxy \rightarrow X$ belongs to HOT.

We shall write $\lambda xy.\ X$ for the member of HOT with reduction rule ending in X. This notation induces a translation of HOT combinations into lambda terms. It will be convenient to refer to this translation below. In particular, we will say that the HOT combination M is strongly normal if it is in normal form and its translation has a beta normal form.

It can be shown that HOT is not finitely generated and therefore HOT is not combinatorially complete.

Simple data types can be encoded into HOT as follows.

Booleans:

$T = K$
$F = K^*$
IMP $= \lambda xy.\ xyT$
NEG $= (\lambda xy.\ yFT)T$

Integers:
$0 = K^*K^*$
SUCC $= \lambda xy.\ yF)T$
SUCC# $= \lambda ab.\ \text{SUCC}(ab)$

$n + 1 = $ SUCC n (Barendregt numerals)

PRED $= (\lambda xy.\ yF)T$

PRED# $= \lambda ab.\ abT0(abF)$

ZERO $= (\lambda xy.\ yT)T$

$sg = (\lambda xy.\ yT01)T$

$sg \sim= (\lambda xy.\ yT10)T$

Primitive recursive functions can be encoded into strongly normal HOT combinations as follows.

$+ \quad = (\lambda uv.\ v0UU(\lambda xy.\ yT0\ (\text{SUCC}\#(x(yF))))v)T$

$+\# = (\lambda uv.\ uv0UU(\lambda xy.\ yT0\ (\text{SUCC}\#(x(yF))))(uv)v$

$- \quad = (\lambda uv.\ vIUU(\lambda xy.\ yT0\ (\text{PRED}\#(x(yF))))v)T$

$* \quad = (\lambda uv.\ v0UU(\lambda xy.\ yT0(+\#(x(yF))))v)T$

so that we have

$$+n\ m \longrightarrow (\text{SUCC}^n)xm = \text{SUCC}^{(n+m)}0 = n + m$$

$$-nm \longrightarrow \begin{cases} (\text{PRED}^n)m \longrightarrow m - n \text{ if } m + 1 > n \\ 0 \qquad\qquad\qquad\qquad \text{if } n > m \end{cases}$$

$$*n\ m \longrightarrow (+m)^n\ 0 \longleftrightarrow n^*m.$$

Using similar techniques and constructions one can find strongly normal HOT combinations SQ,ROOT,D,QUADRES,PAIR and R such that

$D\ n\ m \longleftrightarrow |\ n - m\ |$

SQ $T\ n \longleftrightarrow n^2$

ROOT $T\ n \longleftrightarrow$ square root n rounded down

QUADRES $T\ n \longleftrightarrow n-(\text{square root } n \text{ rounded down})^2$

PAIR $n\ m \longleftrightarrow ((n + m)^2 + m)^2 + n$

$R\ T\ n \longleftrightarrow$ QUADRES T (ROOT $T\ n$).

Note that $R\ T$ (PAIR $n\ m$) $\longleftrightarrow m$ and QUADRES T (PAIR $n\ m$) $\longleftrightarrow n$.

Suppose now that M and N are strongly normal HOT combinations. Define

SUM$(M, N) = \lambda yx.\ x0 + (x0(Mx))(x0Nx))T$

COMP$(M, N) = (\lambda yx.\ x0M(x0Nx)T$

IT$(M) = (\lambda yx.\ x0UU(\lambda uv.\ vT0(y0M(u(vF))))x)T.$

Then

SUM$(M, N)n \longleftrightarrow +(Mn)(Nn)$

COMP$(M, N)\ n \longleftrightarrow M(Nn)$

IT$(M)\ n \longleftrightarrow M^n0$

Thus by [6] page 93 we have verified that

Proposition 5: Every primitive recursive unary function is representable by a HOT combination.

Partial recursive functions can be represented by HOT combinations as follows. Let t be the characteristic function of Kleene's T predicate i.e., $T(e, x, y) \Leftrightarrow t(e, x, y) = 0$ and define primitive recursive functions r and s by

$r(x) = x-$ (square root x rounded down)^2

$s(x) =$ square root x rounded down $-$ (square root (square root x rounded down) rounded down)^2.

The unary primitive recursive function $t(e, r(x)\ s(x))$ is represented by a strongly normal HOT combination $H(e)$. By [6] pages 83 and 84 we have $H(e)$ PAIR(n, m) $\longleftrightarrow\!\!\!\rightarrow t(e, n, m)$. Now define

$$G(e) = \lambda vu.\ uIUU(\lambda xy.\ yIH(e)yT(yIRy)(x(yI(\text{PAIR})(yILy)$$
$$((yI(\text{SUCC})(yIRy))))(uI(\text{PAIR})u0))T$$

and we obtain $G(e)n = \min\{m : t(e, n, m) = 0\}$. Thus, if F is a strongly normal HOT combination representing Kleene's result extracting function the term $\lambda yx.\ xIF(xIG((e)x))T$ is a strongly normal HOT combination which represents the partial recursive function $\{e\}$. Thus, we have

Proposition 6: Every partial recursive function is representable by a strongly normal HOT combination.

References

[1] Ackermann, Solvable Cases of the Decision Problem, North Holland, Amsterdam, 1954.

[2] Barendregt, The Lambda Calculus, North Holland, 1981.

[3] Church, A note on the Entscheidungsproblem, *JSL*, **1**, 40-41, 101-102.

[4] Kozen, The complexity of finitely presented algebras, *S. T. O. C.*, 1977.

[5] Klop, Combinatory Reduction systems, Math. Centrum, Amsterdam, 198 .

[6] Peter, Recursive Functions, Academic Press, 1967.

[7] Smullyan, To Mock a Mockingbird, Knopf, 1985, 244-246.

[8] Statman, The word problem for Smullyans's Lark combinator is decidable, *Journal of Symbolic Computatiom*, **7**, 1989, 103-112.

[9] On translating lambda terms into combinators, *LICS*, 1986, 37-383.

[10] Combinators hereditarily of order one, Tech. Report 88-32, Carnegie Mellon University, Department of Mathematics, 1988.

[11] Combinators hereditarily of order two, Tech. Report 88-33, Carnegie Mellon University, Department of Mathematics, 1988.

[12] Waldmann, How to decide S, Forschungergebnisse 97/03, Friedrich-Schiller-Universität Jena, Fakultät Mathematik, 1997.

An Algebra of Resolution

Georg Struth

Institut für Informatik, Albert-Ludwigs-Universität Freiburg
Georges-Köhler-Allee, D-79110 Freiburg i. Br., Germany
struth@informatik.uni-freiburg.de

Abstract . We propose an algebraic reconstruction of resolution as
Knuth–Bendix completion. The basic idea is to model propositional ordered Horn resolution and resolution as rewrite-based solutions to the
uniform word problems for semilattices and distributive lattices. Completion for non-symmetric transitive relations and a variant of symmetrization normalize and simplify the presentation. The procedural content
of resolution, its refutational completeness and redundancy elimination
techniques thereby reduce to standard algebraic completion techniques.
The reconstruction is analogous to that of the Buchberger algorithm by
equational completion.

1 Introduction

Is it a mere coincidence that the resolution procedure from computational logic,
the Knuth–Bendix procedure from universal algebra and the Buchberger algorithm from computer algebra look so similar? Already in the mid-eighties,
Buchberger suggested presenting the three procedures as instances of an *universal algorithm type* of completion [4]. In this spirit, the Buchberger algorithm
has been reconstructed in terms of Knuth–Bendix completion [5]. But is the
same program possible with resolution? Our main result is a positive answer
to this question. But before further explanations, let us turn to the Buchberger
algorithm.

Logically, the Buchberger algorithm solves the uniform word problem for
certain polynomial rings: Given a finite set P of equations (a presentation) and an
equation $s \approx t$ (over some set of constants) in the language of polynomial rings,
determine whether every polynomial ring that satisfies P also satisfies $s \approx t$.
For empty P, a canonical term rewrite system E for polynomial rings solves the
problem. Otherwise, P and E are completed; critical pairs between E and P are
eagerly determined. This *symmetrization* technique [9], which is not restricted
to polynomial rings, normalizes and simplifies P. Arbitrary equations between
polynomial ring expressions are rewritten as equations between polynomials.
Critical pairs of P (S-polynomials) are deduced modulo these normalizations and
certain redundant polynomials are deleted. P is thereby effectively transformed
into a Gröbner basis G , such that a polynomial equation $p_0 \approx 0$ is a consequence
of P when p_0 can be rewritten to 0 by G.

At first sight, ordered resolution may appear similar to Knuth–Bendix completion: certain consequences (determined by resolution) are added to an initial

L. Bachmair (Ed.): RTA 2000, LNCS 1833, pp. 214–228, 2000.
© Springer-Verlag Berlin Heidelberg 2000

clause set, certain redundant clauses are deleted and an ordering on clauses triggers these operations. But the intended reconstruction also leads to questions that are less obvious to answer: What is the word problem solved by resolution? Is resolution rule a critical pair computation? Is there any symmetrization? Our main result shows that (ground ordered) Horn resolution and resolution solve the uniform word problem for semilattices and distributive lattices via a variant of Knuth–Bendix completion for non-symmetric transitive relations [17]. The resolution rule is a lattice-theoretic critical pair computation and the initial clause set a presentation P in terms of inequalities. It is even equivalent to the distributivity axiom. Arbitrary lattice expressions are normalized via the respective non-symmetric counterparts of canonical systems [11] given by Levy and Agustí. Redundant inequalities are deleted in accordance with non-symmetric completion. Thereby P is effectively transformed into a resolution basis R that provides a certain normal form proof for precisely those inequalities that are consequences of P. As a special case, P is inconsistent or trivial iff $1 < 0$ is a consequence of P iff $1 < 0$ is in R.

The algebraic reconstruction of resolution provides a refined and uniform procedural view: First, the completion-based ordering constraints for resolution are weaker than the standard ones, where only maximal atoms are resolved. Only for boolean lattices (in analogy to the Buchberger algorithm), the stronger constraints also yield solutions to the uniform word problem. Otherwise, they are too restrictive for word problems, but still sufficient for deciding inconsistency. Second, we obtain refutational completeness of resolution as a corollary of correctness of non-symmetric completion. Finally, redundancy concepts for ordered resolution are inherited from non-symmetric completion.

The remainder is organized as follows. Section 2 introduces some rewriting and lattice theory basics. Section 3 sketches non-symmetric completion as far as needed for solving our word problems. Sections 4 and 5 specialize ground non-symmetric completion to solutions of uniform word problems for semilattices and distributive lattices. Section 6 discusses applications and extensions of these results: A strengthening of the ordering constraints to ordered resolution, the extension to boolean lattices and the lifting to the non-ground case. Section 7 contains a conclusion.

2   Preliminaries

Let $T_\Sigma(X)$ be a set of terms with signature Σ and variables in X. Terms are identified with $\Sigma \cup X$-labeled trees with nodes or *positions* in the monoid $\mathbb{N}^*$. ϵ denotes the root and $n.i$ the i-th child of node n. A variable is *linear* (*non-linear*) in a term, if it is the label of exactly one node (least two nodes). For a term t and a substitution σ, a *skeleton position* of $t\sigma$ with respect to t is a node labeled by a function or constant symbol in t. A *variable instance position* of $t\sigma$ with respect to t is a node $p.q$ such that p is a node labeled by a variable in t. $s[t]_p$ denotes replacement of the subterm $s|_p$ of a term s at position p by a term t.

A *quasi-ordered set* (quoset) $(P, <)$ is a set P endowed with a reflexive transitive relation $<$. A quoset is a *partially ordered set* (a poset), if $<$ is also anti-symmetric. Poset identify elements that are congruent modulo $\sim = (< \cap >)$ in the associated quoset.

A *join semilattice* is a quoset $(L, <)$ closed under least upper bounds (joins) for all pairs of elements. For all $x, y, z \in L$, the join $x \vee y$ of x and y satisfies $x \vee y < z$ iff $x < z$ and $y < z$. A *meet semilattice* is a quoset $(L, <)$ closed under greatest lower bounds (meets) for all pairs of elements. For all $x, y, z \in L$, the meet $x \wedge y$ of x and y satisfies $z < x \wedge y$ iff $z < x$ and $z < y$. Joins and meets are unique up to $\sim$. $\bigvee P$ and $\bigwedge P$ denote joins and meets of finite $P \subseteq L$.

A *lattice* is both a join and a meet semilattice. The signature $L = \{\vee, \wedge\}$ of the set $T_L(X)$ of lattice terms can be assumed variadic for flattening terms. A lattice $(L, <)$ is *distributive*, if for all $x, y, z \in L$ one of $x \wedge (y \vee z) < (x \wedge y) \vee (x \wedge z)$, $(x \vee y) \wedge (x \vee z) < x \vee (y \wedge z)$ and therefore the other one holds. The inequalities $x \wedge (y \vee z) > (x \vee y) \wedge (x \vee z)$ and $(x \vee y) \wedge (x \vee z) > x \vee (y \wedge z)$ hold in every lattice. Lattices are usually defined as partial orderings, but for comparing with resolution, quasi-orderings are more convenient.

A mapping f between quosets $A = (A, <_A)$ and $B = (B, <_B)$ is *monotonic*, if $x <_A y$ implies $f(x) <_B f(y)$, a *join homomorphism*, if $f(x \vee y) = f(x) \vee f(y)$ holds and a *meet homomorphism*, if $f(x \wedge y) = f(x) \wedge f(y)$ holds for all $x, y \in A$. Joins and meets are monotonic in both arguments.

A *presentation* P for a signature Σ of some algebra is a pair (G, R), where G is a set of constants and R a set of *defining identities* $s \approx t$ over *words* $s, t \in T_\Sigma(G)$. The *uniform word problem for* a class K of Σ-algebras is the following decision problem: Does $\text{Th}(K) \cup P \models s \approx t$ hold for the theory $\text{Th}(K)$ of K, for an arbitrary finite presentation $P = (G, R)$ and identity $s \approx t$ over $T_\Sigma(G)$? It is *solvable*, if there is an algorithm to decide the problem for arbitrary P and $s \approx t$. When K is axiomatized by a finite set E of identities, a solution exists iff the congruence defining the quotient algebra of the term algebra $T_\Sigma(G)$ by E and P can be effectively constructed, for instance by a canonical term rewrite system for E. We still say *uniform word problem*, when it is defined by inequalities.

3   Non-symmetric Rewriting and Completion

We presuppose the basic concepts and notation of equational rewriting, as given, for instance, in [7]. Non-symmetric rewriting is a technique for solving word or reachability problems for binary relations that are partitioned into an increasing part S and a decreasing part R with respect to some syntactic ordering $\prec$ on the universe of the relations, by search along normal-form paths [11, 15] (c.f. [16] for a brief introduction in the context of lattices). It is based on the following purely relational observations.

Lemma 1. *Let R and S be binary relations on some set A.*

(i) $S^*R^* \subseteq R^*S^*$ iff $(R \cup S)^* \subseteq R^*S^*$.
(ii) $SR \subseteq R^*S^*$ implies $(R \cup S)^* \subseteq R^*S^*$, if $(R \cup S^{-1})$ is well-founded.

Setting $S = R^{-1}$, lemma 1 (i) and (ii) specialize to the Church–Rosser theorem and Newman's lemma of equational rewriting. We visualize $R \cup S$ as a digraph with edges labeled by R or S and nodes by elements of A. We consider proofs or paths along the quasi-ordering $(R \cup S)^*$ induced by R and S. A *peak* now is a proof of the form SR; a *valley* a proof of the form R^*S^*.

Let R and S induce *rewrite relations* $\to_R$ and $\to_S$. A term s rewrites to a term t in one step at position p, written as $s \to_R^p t$ or $s \to_S^p t$, if there is some rewrite rule $l \to_R r \in R$ or $l \to_S r \in S$ such that $s|_p = l\sigma$ for some position p and substitution σ and $t = s[r\sigma]_p$. This presumes that all predecessor nodes of p are labeled by names of monotonic functions. With our notation notation, $\to_R$ increases and $\to_S$ increases from left to right in presence of the syntactic ordering $\prec$. This is consistent with the standard arrow notation for clauses. Replacement of peaks by valleys is now more complex than for equational rewriting. In general, a critical pair is a pair in SR that can possibly not be replaced by a valley. As in equational rewriting, there are critical pairs arising from skeleton positions. For $l_1 \to_R r_1$ and $l_2 \to_S r_2$, they are either of the form $(l_1\sigma[l_2\sigma]_p, r_1\sigma)$ for a most general unifier σ of r_2 and $l_1|_p$ or of the form $(l_2\sigma, r_2\sigma[r_1\sigma]_p)$ for a most general unifier σ of $r_2|_p$ and l_1. But now also critical pairs from variable instance positions of non-linear variables arise.

Example 1. Let $R = \{x \wedge x \to_R x\}$, $S = \{a \to_S b\}$. The peak $b \wedge a \to_S b \wedge b \to_R b$ cannot be replaced by a valley. Only a "backward" rewrite step, leading to $a \wedge a \to_S b \wedge a \to_S b \wedge b \to_R b$ yields a replacement via $a \wedge a \to_R a$.

Backward steps are unnecessary for linear variables. Then every peak can still be replaced by a valley. For variable critical pairs the position q of the variable instance position $p.q$ can in general not be bounded. These pairs can be represented, for instance, by context variables. The variable critical pair of example 1 then becomes $(C[a] \wedge C[b], C[b])$.

Non-symmetric rewriting extends to non-symmetric rewriting modulo a congruence induced by a set E of equations, as in the equational case [8]. In particular, for the associativity commutativity congruence (AC), the cliffs arising from ER- or SE-proofs can be eliminated by extended rules. The meet operation, for example, is associative and for all $r, s, t \in T_L(X)$. A rule $r \wedge s \to_R t$ or $r \to_S t \wedge s$ is extended to $(r \wedge s) \wedge x \to_R t \wedge x$ or $r \wedge x \to_S (t \wedge s) \wedge x$, where x is a fresh extension variable. Extending all rules in R and S, it then remains to replace syntactic unification by AC-unification. Consider [17] for concise definitions.

A non-symmetric completion procedure is a transformation on sets of inequalities that preserves the set of consequences modulo the theory of (monotonic) quasi-orderings. On success, its solution allows spanning every consequence by a valley. By well-foundedness of R and S^{-1}, valleys can be effectively constructed by searching for a common node of the finite acyclic R- and S^{-1}-digraphs. To this end, exactly the critical pairs as part of the transitive-monotonicity closure of the input set is computed. Additionally, certain redundant inequalities are eliminated on the fly to keep the solution set small. In contrast to the decision procedure, well-foundedness of $R \cup S^{-1}$ is needed for completion.

We consider the subcase of non-symmetric AC-completion, where extension variables are the only non-ground objects. In particular therefore, all terms are linear. Since the deviations from the equational case are small, we only give a sketch that can be completed, for instance, using [1]. Let $T_\Sigma(G)$ be a set of ground terms and V a set of extension variables. We use an AC-compatible reduction ordering $\prec$ that is total on $T_\Sigma(G)$, such that $f(s_1, \ldots, s_m) \prec f(t_1, \ldots, t_n)$, for an AC-symbol f, if $\{\!\{ s_1, \ldots, s_m \}\!\} \prec' \{\!\{ t_1, \ldots, t_n \}\!\}$, where $\{\!\{ . \}\!\}$ denotes a multiset and $\prec'$ the multiset extension of $\prec$ for all AC-symbols f [1].

We define a non-symmetric completion procedure C_{AC} for linear terms as a state transition system. States are triples (I, R, S) of a set I of (unordered) inequalities $l < r$ and two sets R and S of ordered inequalities $l \to_R r$ and $l \to_S r$ with $l \succ r$ and $l \prec r$. The initial state is $(I_0, \emptyset, \emptyset)$ with I_0 ground, that is without extended rules. The transitions are modeled by the following rules.

$$\frac{(I, R, S)}{(I \cup \{s < t\}, R, S)}, \qquad \text{(Deduce)}$$

if (s, t) is an AC-critical pair. This rule can also be written as an inference rule on inequalities, using AC-unifiers from a minimal complete set.

$$\frac{l_2 \to_S r_2 \quad l_1[r_2']_p \to_R r_1}{l_1\sigma[l_2\sigma]_p < r_1\sigma}, \qquad \frac{l_2 \to_S r_2[l_1']_p \quad l_1 \to_R r_1}{l_2\sigma < r_2\sigma[r_1\sigma]_p}.$$

$$\frac{(I \cup \{s < t\}, R, S)}{(I, R \cup \{s \to_R t\}, S)}, \qquad \frac{(I \cup \{s < t\}, R, S)}{(I, R, S \cup \{s \to_S t\})}, \qquad \text{(Orient)}$$

if $s \succ t$ and $s \prec t$, respectively.

$$\frac{(I, R, S)}{(I, R \cup \{s \to_R t\}, S)}, \qquad \frac{(I, R, S)}{(I, R, S \cup \{s \to_S t\})}, \qquad \text{(Extend)}$$

if $s \to_R t$ ($s \to_S t$) is an extension of a non-extended rule in R (S).

$$\frac{(I \cup \{s < t\}, R, S)}{(I, R, S)}, \qquad \frac{(I, R \cup \{s \to_R t\}, S)}{(I, R, S)}, \qquad \frac{(I, R, S \cup \{s \to_S t\})}{(I, R, S)},$$
$$\text{(Delete)}$$

if $s < t$, $s \to_R t$ or $s \to_S t$ can be replaced by a smaller proof defined as follows. In particular, every proof of $s < s$ can be replaced by the empty valley. We

---

[1] Such orderings exist [3, 6] even orienting the distributive laws in the desired direction. For our reconstruction of resolution, these orderings can be substantially simplified.

compare single proof steps (s, t) using (l, r) which is either an inequality in I, an (extended or non-extended) R or S rule or an AC-axiom, by the following tuples: by $\langle \{s, t\}, \perp, \perp \rangle$ for $l < r \in I$, $\langle \{s, t\}, \perp, r \rangle$ for $l \approx r$ an AC-axiom, $\langle \{s\}, l, r \rangle$ for $l \to_R r \in R$ non-extended, $\langle \{s\}, \perp, r \rangle$ for $l \to_R r \in R$ extended, $\langle \{t\}, r, l \rangle$ for $l \to_S r \in S$ non-extended and $\langle \{t\}, r, \perp \rangle$ for $l \to_S r \in S$ extended. Tuples are ordered lexicographically using the multiset extension of $\prec$ for the first component, the encompassment ordering $\lessdot^2$ for the second one and $\prec$ for the third one. Proofs are compared as multisets of one-step proofs by the multiset-extension of the one-step ordering. All these proof-orderings are denoted by $\prec$. They inherit well-foundedness from their components.

A run of C_{AC} *succeeds*, if its limit state is of the form $(\emptyset, R, S)$. It is *(weakly) fair*, if every continuously enabled transition is eventually executed. The pair (R, S) is a *normal system*, if $(R \cup S)^* \subseteq R^* S^*$, R and S^{-1} are well-founded and no element of R or S can be deleted.

Under the transition rules of C_{AC}, proofs can be successively replaced by simpler ones with respect to $\prec$.

Lemma 2. *For the completion procedure* C_{AC} *and the ordering* $\prec$ *on one-step and multi-step proofs.*

(i) *The transition rules of* C_{AC} *induce a well-founded proof transformation relation with respect to* $\prec$.

(ii) *The procedure effectively transforms every* $I \cup AC$-*proof into a valley.*

Lemma 2 leads to the following correctness property.

Theorem 1. *The limit state of a fair successful run of* C_{AC} *is a normal system.*

This holds in particular, when the procedure terminates. In opposition to equational completion, the DELETE rules of C_{AC} are search-based and the DEDUCE rule persists in the ground case. The former is the case, since $a < b$ and $a < c$ imply neither $b < c$ nor $c < b$, the latter, since the direction of arrows in the DEDUCE rule does not match with any DELETE-rule.

4   The Uniform Word Problem for Semilattices

We now consider the *symmetrization*, the interaction between a finite presentation in terms of semilattice inequalities $\bigwedge S < \bigwedge T$, where $S, T \subseteq G$ and G is finite, and a normal system for semilattices. We normalize the presentation and compute its critical pairs with the normal system for semilattices. Since the language of semilattices is very simple—join is the only operation symbol— the AC-ordering $\prec$ is very simple, too: we compare semilattice terms simply as multisets of generators.

Lemma 3 ([11]). *Given the AC-compatible ordering* $\prec$, *the following set* N_S *of rules is a normal system for semilattices.*

$$x \wedge y \to_R x, \qquad x \wedge y \to_R y, \qquad x \to_S x \wedge x, \qquad x \wedge y \to_S x \wedge x \wedge y.$$

---

^{2} $s \lessdot t$, if some subterm of t is an instance of s.

The last rule is extended, since meet is associative and commutative. The first three rules together imply that $x \wedge x < x$ and $x < x \wedge x$ hold. We can thus normalize with respect to idempotence and consider semilattice terms as sets and not as multisets. Sets and multisets are also the natural data-structures for clauses in resolution. Writing a clause as $\Gamma \longrightarrow \Delta$, where Γ and Δ are multisets of atoms, it is straightforward to show that $\longrightarrow$ is a quasi-ordering. Semilattice inequalities can still further be normalized, although not by means of rewriting.

Lemma 4. *Let s be a semilattice term and let $T \subset G$ be a finite set of generators. Then $s < \bigwedge T$ iff $s < t$ for all $t \in T$.*

The proof is immediate from the definition of meet. We can thus restrict our attention to uniform word problems defined by *Horn inequalities* of the form $\bigwedge S < t$, where $S \subseteq G$ and $t \in G$. Translation to the resolution context then yields Horn clauses. For our comparison with resolution, semilattices with a minimal element 0 and a maximal element 1 are especially important. Hence $1 \wedge s = s$ and $0 \wedge s = 0$ for all semilattice terms s. In particular $1 < 0$ holds only in the trivial semilattice consisting of one single element; hence it denotes triviality or inconsistency.

Lemma 5. *A finitely presented semilattice is finite and at most of size $2^{|G|}$.*

Of course at most $2^{|G|}$ different lattice terms can be built from G using the data-structure of sets. Consequently, each AC-extension of a ground rewrite rule from the presentation and each rule from the normal system in lemma 3 can be replaced—highly inefficiently—by a finite set of ground extensions. By lemma 4, extensions can be restricted to left-hand sides of inequalities. Also AC-matching and unification become very simple. Procedurally, these ground extensions should of course be lazily generated. It now remains to turn this discussion into an inference system.

Given the AC-ordering $\prec$ that is total on ground terms, we define a lattice-theoretic Horn ordered resolution calculus HOR as a specialization of C_{AC} by the following inference rules on semilattice inequalities. We first introduce some notation. Let s, s', t be semilattice terms; let $a, b, c, \ldots$ and $a_1, a_2, a_3, \ldots$ be generators. We use both transition and inference rules to obtain a format that is close to Horn resolution.

$$\frac{(I \cup \{s < a_1 \wedge \cdots \wedge a_n\}, R, S)}{(I \cup \{s < a_i : 1 \leq i \leq n\}, R, S)}, \qquad \text{(SPLITTING)}$$

$$\frac{s \wedge a \wedge a \rightarrow_R b}{s \wedge a < b}, \qquad \text{(IDEMPOTENCE)}$$

$$\frac{s \rightarrow_S a \qquad s' \wedge a \rightarrow_R b}{s \wedge s' < b}. \qquad \text{(RESOLUTION)}$$

The ORIENT and DELETE transition rules are inherited from C_{AC}. Special derived transition rules are

$$\frac{(I \cup \{s \wedge a < a\}, R, S)}{(I, R, S)} \qquad \text{(LB)}$$

and similarly with respect to R and S. We dispense with EXTEND rules because we can finitely represent them.

Proposition 1. HOR *is a normalizing specialization of* C_{AC} *for semilattices. Every fair run succeeds; its limit state is a normal system for the presentation modulo the theory of semilattices.*

Proof. Let $P = (G, R)$ be a finite presentation of a semilattice. By lemma 4, the SPLITTING rule is an equivalence transformation to Horn inequalities, whereas all other rules in HOR presuppose that format. Without loss of generality we assume that all inequalities in P are Horn.

We now compute the critical pairs between oriented variants of a Horn inequality $s < a$ of P with the rules of N_S from lemma 3. First, there are no critical pairs with the lower bound rules $x \wedge y \rightarrow_R x$ and $x \wedge y \rightarrow_R y$, since a is a generator and all variables linear. Second, we consider the critical pairs with $x \rightarrow_S x \wedge x$. For a rule $s \wedge a \wedge a \rightarrow_R b$, there is a critical pair $s \wedge a < b$. This leads to an instance of the IDEMPOTENCE rule. There is also a variable critical pair $C[s \wedge a \wedge a] < C[s \wedge a \wedge a] \wedge C[b]$, which can be finitely represented, according to lemma 5, by appropriate lattice terms. Using SPLITTING, all these critical pairs can be deleted in favor of $s \wedge a \wedge a \rightarrow_R b$. Third, we consider the critical pairs with $x \wedge y \rightarrow_S x \wedge x \wedge y$. Both the skeleton critical pair and the variable critical pair contribute nothing new.

We now consider the critical pairs between two members $s < a$ and $s' \wedge a < b$ of P. This leads immediately to the RESOLUTION rule, depicted in figure 1.

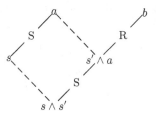

Figure1. Hasse diagram for Horn resolution

Algebraically, it can be derived using monotonicity of meet and transitivity. By our discussion following lemma 5, we can dispense with extended rules in favor of finite expansions of the presentation. But these only yield expressions that can be deleted.

We have now considered all critical pairs among elements of the presentation and between members of the presentation and the normal system for semilattices. We observe two facts. First, the Horn property is preserved by these DEDUCE steps. Second, all generators in the conclusions of these steps already appear in the premises. We can therefore apply the above analysis to an entire run of C_{AC}, not only to the presentation. Therefore, the C_{AC} rules specialize to HOR rules. Since ORIENT is trivial, it remains to consider LB. Obviously, every Horn inequality $s \wedge a < a$ or rule $s \wedge a \rightarrow_R a$ can be deleted in presence of $x \wedge y \rightarrow_R y$ from N_S. Also this rule can be recursively be applied at each stage of the process.

It is now straightforward to inspect that for all rules of HOR, the conclusion is smaller than the maximal premise. Therefore the associated proof transformation relation terminates. As a consequence of lemma 2 and 1 and of the fact that all ground rules can be oriented, we obtain the desired result, that is a normal system for the presentation. $\square$

Simple examples show that neither ground non-symmetric completion nor un-ordered ground Horn resolution always terminate, even in presence of fairness assumptions [17]. By lemma 5 and the fact that no rule of HOR introduces new variables or constants, however, HOR can from some stage on only produce in-equalities with multiple occurrences of some generators at the left-hand side. These can of course be deleted by IDEMPOTENCE. In order to express termina-tion modulo idempotence in a generic way, we call an inference *redundant*, if either some premise or the conclusion can be deleted.

Lemma 6. HOR *terminates (up to redundant inferences) for every presentation.*

Theorem 2. HOR *solves the uniform word problem for semilattices.*

Proof. For solving the uniform word problem, split a given equational presen-tation into inequalities and run HOR to obtain a normal system. Then split the word identity $\bigwedge S \approx \bigwedge T$ into word inequalities $\bigwedge S < t$ for all $t \in T$ and $\bigwedge T < s$ for all $s \in S$ and try to connect them by valleys using the normal sys-tem, according to the standard decision procedure of non-symmetric rewriting, as described in section 3. $\square$

The following inconsistency (or triviality) test is a corollary to theorem 2.

Corollary 1. *Let P be a presentation for a semilattice in which $1 < 0$ holds. Let 0 and 1 be least elements in the precedence for $\prec$. Then HOR derives $1 < 0$ from P.*

Proof. If $1 \leq 0$ is implied by P in the theory of semilattices, then there is a valley proving this inequality by theorem 2. If 1 and 0 are minimal in the precedence, then the only valley can be $1 \leq 0$ itself. Hence the inequality must be in the normal system of P. $\square$

Corollary 1 can be extended to an algebraic refutational completeness proof as follows: There is no meet homomorphism from a semilattice in which $1 < 0$ holds to the two element lattice **2**, since meet homomorphisms are required to map zero to zero and one to one and moreover they are monotonic. Thus there is no satisfying valuation for this presentation, that is the presentation is inconsistent.

Lemma 7. *Replace the* IDEMPOTENCE *inference rule by the transition rule*

$$\frac{(I \cup \{s \wedge a \wedge a < b\}, R, S)}{(I \cup \{s \wedge a < b\}, R, S)}$$

in HOR. *The new procedure still has the properties of proposition 1, lemma 6, theorem 2 and corollary 1. Thus* IDEMPOTENCE *can be applied as a simplification.*

Proof. Consider the inequality $s \wedge a \wedge a < b$ in a presentation P. By the instance $a \wedge a \to_R a$ of $x \wedge y \to_R x$ from N_S it rewrites to to $s \wedge a < b$. In general, this is not sound for inequalities, but since $x \to_S x \wedge x$ is also in N_S and therefore $x \wedge x < x$ and $x < x \wedge x$ hold, the step is an equivalence transformation. But then $s \wedge a \wedge a < b$ can be deleted. Again the argument can be extended on runs. This yields the above transition rule. $\square$

5   The Uniform Word Problem for Distributive Lattices

We now consider distributive lattices and apply the same technique as for semi-lattices. Now our signature consists of joins and meets.

Lemma 8 ([11]). *Given two AC-compatible orderings for R- and S-rules[3], the following set of rules is a normal system N_D for distributive lattices.*

$$x \vee x \to_R x \qquad x \vee x \vee y \to_R x \vee y \qquad x \wedge y \to_R x$$
$$x \wedge (y \vee z) \to_R (x \wedge y) \vee (x \wedge z) \qquad (x \wedge (y \vee z)) \wedge w \to_R ((x \wedge y) \vee (x \wedge z)) \wedge w$$
$$x \to_S x \wedge x \qquad x \wedge y \to_S x \wedge x \wedge y \qquad x \to_S x \vee y$$
$$(x \vee y) \wedge (x \vee z) \to_S x \vee (y \wedge z) \qquad ((x \vee y) \wedge (x \vee z)) \vee w \to_S (x \vee (y \wedge z)) \vee w$$

It allows the following normalization of lattice terms.

Lemma 9. *Let L be a distributive lattice, G a finite set of generators.*

(i) *Every lattice term t is equivalent in L to a join of meets of generators and to a meet of joins of generators from t.*
(ii) $\bigvee S < \bigwedge T$ *holds for $S, T \subset T_\Sigma(G)$, iff $s < t$ holds for all $s \in S$ and $t \in T$.*
(iii) *An inequality $s < t$ holds in L, iff there is a set of inequalities $\bigwedge S_i < \bigvee T_i$ that hold in L and the S_i (T_i) are sets of generators from s (t).*

Proof. (ad i) Since $x \wedge (y \vee z) = (x \wedge y) \vee (x \wedge z)$ and $x \vee (y \wedge z) = (x \vee y) \wedge (x \vee z)$ hold in a distributive lattice, distributing out terms according to the rules of the above rewrite system is an equivalence transformation.

(ad ii) By lemma 4 for meets and by duality for joins.

(ad iii) We put (i) and (ii) together. The transformation is effective. $\square$

We can thus restrict our attention to uniform word problems defined by inequalities of the form $\bigwedge S < \bigvee T$, where $S, T \subseteq G$. We write $n_{jm}(t)$ ($n_{mj}(t)$) for the normal form of t as a join of meets (meet of joins) of generators. They can be determined with only linear blowup in the size of terms [18]. Again we obtain the data-structure of sets and can dispense with extended rules.

---

[3] Remind that well-foundedness of the separate orderings is sufficient for the decision procedure, but not for completion.

Lemma 10. *A finitely presented distributive lattice is finite and at most of size* $2^{2^{|G|}}$.

Working with two separate well-founded orderings is not possible for non-symmetric completion. However, one single ordering, giving either join precedence over meet or conversely, would be sufficient if either the S-rules or the R-rules of N_D dealing with distributivity were discarded. The orientation of the other rules in N_D is invariant under this choice of the precedence. At the level of the following calculus, where all inequalities have the form $\bigwedge S < \bigvee T$, where $S, T \subseteq G$, the choice is even irrelevant, since joins and are always separated and never need to be compared. One can therefore treat S and T equally as multisets of generators.

We define a lattice-theoretic ordered resolution calculus OR for distributive lattices as a specialization of C_{AC} by the following inference rules on lattice inequalities. We first introduce some notation. Let u, v be lattice terms, let s, s' and $s_1, s_2, s_3, \dots$ be terms in the language of the meet semilattice and t, t' and $t_1, t_2, t_3, \dots$ be terms in the language of the join semilattice. We denote generators by $a, b, c, \dots$ or $a_1, a_2, a_3, \dots$ or $b_1, b_2, b_3, \dots$. We again distinguish between transition and inference rules.

$$\frac{(I \cup \{u < v\}, R, S)}{(I \cup \{n_{jm}(u) < n_{mj}(v)\}, R, S)}, \qquad \text{(DISTRIBUTE)}$$

$$\frac{(I \cup \{a_1 \vee \cdots \vee a_m < b_1 \wedge \cdots \wedge b_n\}, R, S)}{(I \cup \{a_i < b_j : 1 \le i \le m, 1 \le j \le n\}, R, S)}, \qquad \text{(SPLITTING)}$$

$$\frac{s \wedge a \wedge a \to_R t}{s \wedge a < t}, \qquad \frac{s \to_S t \vee a \vee a}{s < t \vee a}, \qquad \text{(IDEMPOTENCE)}$$

$$\frac{s \to_S t \vee a \qquad s' \wedge a \to_R t'}{s \wedge s' < t \vee t'}. \qquad \text{(RESOLUTION)}$$

The ORIENT and DELETE transition rules are inherited from C_{AC}. Special derived transition rules are

$$\frac{(I \cup \{s \wedge a < a \vee t\}, R, S)}{(I, R, S)} \qquad \text{(LB/UB)}$$

and similarly with respect to R and S.

Proposition 2. OR *is a normalizing specialization of* C_{AC} *for distributive lattices. Every fair run succeeds; its limit state is a normal system for the presentation modulo the theory of distributive lattices.*

Proof. The proof is similar to that of proposition 1. We assume a precedence where joins are bigger than meets. Again, the SPLITTING rule, which is defined in accordance with lemma 9, is used for preprocessing lattice terms. Here it is applied together with the DISTRIBUTE rule. Since this rule corresponds to an equivalence transformation, it is a simplification rule that allows us to discard the non-normalized inequality, like for idempotence in proposition 1. Most of the critical pairs have already been considered in the proof of proposition 1, up to lattice duality. It remains to consider the critical pairs among members of the presentation and between them and the distributivity laws. By lemma 10 we can again dispense with extended rules.

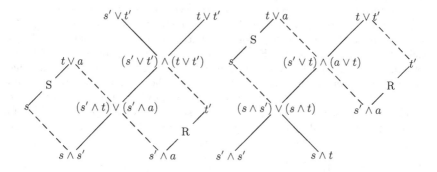

Figure2. Hasse diagrams for Resolution

There are no immediate critical pairs among members of the presentation, except when one is a Horn inequality or a dual Horn inequality $a \leq \bigvee T$ for $a \in G$ and $T \subseteq G$. However two members of the presentation can have critical pairs modulo distributivity in a particular case.

In order to work consistently with our choice of precedence, we must exclude the S-rules in N_D dealing with distributivity. We will argue that this restriction is justified. The only critical pair between an inequality $s \rightarrow_S t \vee u$ and the distributive inequality $x \wedge (y \vee z) \rightarrow_R (x \wedge y) \vee (x \wedge z)$ is $x \wedge s \rightarrow_S (x \wedge t) \vee (x \wedge u)$. By lemma 10, we can dispense with x again by instantiation with generators. In the special case that one of t and u, u say, is a generator a, an additional critical pair with an inequality $s' \wedge a \rightarrow_R t'$ may exist. Otherwise the above critical pair can be normalized and then deleted. The additional critical pair is $s \wedge s' \wedge x' < (s' \wedge x' \wedge t) \vee (x' \wedge t')$, where x has been instantiated by $s' \wedge x'$. Normalizing with respect to DISTRIBUTE, using SPLITTING and LB/UB, the pair simplifies to $s \wedge s' \wedge x' < t \vee t'$. Since every instance of this inequality can be deleted in favor of $s \wedge s' < t \vee t'$, only this last rule must be kept. This derivation of an instance of RESOLUTION is depicted in the left-hand diagram of figure 2. Algebraically, the dotted lines denote application of monotonicity of join and meet; the diagonal of the diagram shows the conclusion. Note that only the R-rule dealing with distributivity has been used.

Similarly to the preceding paragraph, using a precedence where meets are bigger than joins and excluding the R-rules in N_D dealing with distributivity, yields a critical pair $(x \vee s_1) \wedge (x \vee s_2) \rightarrow_R x \vee t$ from $s_1 \wedge s_2 \rightarrow_R t$ and $(x \vee y) \wedge (x \vee z) \rightarrow_S x \vee (y \wedge z)$. This situation is completely dual to the one above (by exchange of join and meet and inverting the inequalities). It leads to the situation depicted in the right-hand diagram of figure 2. Both derivations deduce the same conclusion from the same premises; nothing new is introduced.

This justifies discarding one variant of the distributivity rules. The RESOLUTION rule is the most general way in which two members of the presentation can overlap modulo distributivity. It covers the case where one of the premises is a Horn inequality or its dual.

The format of the presentation is preserved for all operations and no new generators are introduced. Therefore the argument can be extended from generators to the whole run of C_{AC}. Together with the results of proposition 1, this yields the inference rules of OR. For termination of the proof transformation it remains to show that every conclusion of a RESOLUTION rule is smaller than the maximal premise. This is not immediately the case, as the ordering constraints $t \succ s' \succ t' \succ s \succ a$ for the above rule show. Appropriate instances of the distributivity law and monotonicity must therefore be implicitly added for enforcing it. Lemma 2 and 1 and the fact that all ground rules can be oriented, then yield the desired result, that is a normal system for the presentation. $\square$

Our derivation of the resolution rule reveals the surprising fact that it is essentially an application of the distributivity law. In fact, a lattice $(L, <)$ is distributive iff $s < t \vee a$ and $s' \wedge a < t$ imply $s \wedge s' < t \vee t'$ for all $s, s', t, t', a \in L$. A similar characterization can be found in [12].

Lemma 11. OR *terminates (up to redundant inferences) for every presentation.*

Theorem 3. OR *solves the uniform word problem for distributive lattices.*

Corollary 2. *Let P be a presentation for a distributive lattice with 0 and 1 such that $1 < 0$ holds in that lattice. Let moreover 0 and 1 be the least elements of the precedence for $\prec$. Then $1 < 0$ is in the normal system derived from P.*

The proofs of lemma 11, theorem 3 and corollary 2 are similar to those of lemma 6, theorem 2 and corollary 1. Again, corollary 2 can be extended to an algebraic completeness proof.

Lemma 12. *Replace the* IDEMPOTENCE *inference rules by the transition rules*

$$\frac{(I \cup \{s \wedge a \wedge a < t\}, R, S)}{(I \cup \{s \wedge a < t\}, R, S)} \qquad \frac{(I \cup \{s < t \vee a \vee a\}, R, S)}{(I \cup \{s < t \vee a\}, R, S)}$$

in OR. *The new procedure still has the properties of proposition 2, lemma 11, theorem 3 and corollary 2. Thus* IDEMPOTENCE *can be applied as a simplification.*

The proof is similar to that of lemma 7.

6 Discussion

Our derivation of resolution calculi from non-symmetric rewriting and completion is quite fortunate for several reasons. First, semilattices and lattices are two of the few structures for which non-symmetric normal systems have so far been given. This has to do with the inherent weakness of non-symmetric rewriting, due to variable critical pairs. Even the derivation of these simple systems (which one might even be able to guess) has been based on ad hoc reasoning [11]; however a representation of general non-symmetric completion as a second-order procedure might be promising. Second, finiteness of finitely presented semilattices and distributive lattices lead to a simple circumvention of context variables, that otherwise might have caused difficulties.

The precise translation from lattice inequalities to clauses is straightforward; we leave it implicit. The calculi described so far are weaker than the standard ordered resolution calculi (c.f [10]), since we have not required that maximal terms are cut out. Only for boolean lattices this restriction still yields a solution to the uniform word problem, whereas for semilattices and distributive lattices, too few critical pairs are computed. A simple proof permutation argument (c.f. [17]) shows that inconsistency tests are still possible with the stronger constraints. Hence only for boolean lattices is the situation similar to the Buchberger algorithm. Algebraically, the ideals of boolean lattices and rings have similar properties, whereas those of distributive lattices and semilattices behave differently. Our methods are in particular important for the boolean case, where—in opposition to the distributive case—no equational canonical rewrite system exists [14].

Our calculi can be lifted to the non-ground case. There are no variable critical pairs; predicate symbols are interpreted as non-monotonic operations which block the application of rewrite steps below the lattice level. An ordered or unfailing version of non-symmetric completion should be used to obtain the ordering restrictions of resolution. Equational ordered or unfailing completion [13, 2, 1] is a semi-decision procedure for word problems. Critical pair computations are performed not on ordered equations, but on orderable instances. For many applications, unique normal forms only for ground instances of terms (which is weaker than for terms) suffices. For instance, all ground instances of the commutativity law are orderable, whereas the law itself is not. Consequently, ordered completion is based on syntactic unification and AC-compatible orderings are superfluous. However, a detailed comparison between the methods of non-ground ordered resolution and equational ordered or unfailing completion is left for future work.

7   Conclusion

We have reconstructed ground ordered resolution algebraically as a solution to lattice-theoretic uniform word problems based on a generalization of Knuth–Bendix completion to non-symmetric transitive relations. In fact, a variant of the resolution rule yields exactly the specific difference between lattices and distributive lattices. Our construction is analogous to the simulation of the Buchberger algorithm by equational completion. We essentially used normalization techniques to integrate the effect of distributive lattice theory in the presentation. The refutational completeness proof of resolution and redundancy elimination techniques can thus be reduced to those of non-symmetric rewriting. Beyond the results presented in this text, lifting of resolution shows interesting correspondences to unfailing completion and a more abstract ideal-theoretic comparison between our lattice word problem and the Buchberger algorithm could lead to structural insights.

References

[1] L. Bachmair. *Canonical Equational Proofs.* Birkhäuser, 1991.

[2] L. Bachmair, N. Dershowitz, and D.A. Plaisted. Completion without failure. In H. Aït-Kacy and M. Nivat, editors, *Resolution of Equations in Algebraic Structures,* volume 2: Rewriting Techniques, chapter 1, pages 1–30. Academic Press, 1989.

[3] L. Bachmair and D. Plaisted. Termination orderings for associative-commutative rewriting systems. *J. Symbolic Computation,* 1(4):329–349, 1985.

[4] B. Buchberger. Basic features and development of the critical-pair/completion procedure. In J.-P. Jouannaud, editor, *Rewriting Techniques and Applications, [1st International Conference],* volume 202 of *LNCS,* pages 1–45. Springer-Verlag, 1985.

[5] R. Bündgen. Buchberger's algorithm: The term rewriter's point of view. *Theoretical Computer Science,* 159(2):143–190, 1996.

[6] C. Delor and L. Puel. Extension of the associative path ordering to a chain of associative-commutative symbols. In C. Kirchner, editor, *Rewriting Techniques and Applications,* volume 690 of *LNCS,* pages 389–404. Springer-Verlag, 1993.

[7] N. Dershowitz and J.-P. Jouannaud. Rewrite systems. In J. van Leeuwen, editor, *Handbook of Theoretical Computer Science,* volume B: Formal Models and Semantics, chapter 6, pages 244–320. Elsevier, 1990.

[8] J.-P. Jouannaud and H. Kirchner. Completion of a set of rules modulo a set of equations. *SIAM J. Comput.,* 15:1155–1194, 1986.

[9] P. Le Chenadec. *Canonical Forms in Finitely Presented Algebras.* Research Notes in Theoretical Computer Science. Pitman [u.a.], London;New York;Toronto, 1986.

[10] A. Leitsch. *The Resolution Calculus.* Springer-Verlag, 1997.

[11] J. Levy and J. Agustí. Bi-rewrite systems. *J. Symbolic Computation,* 22:279–314, 1996.

[12] P. Lorenzen. Algebraische und logistische Untersuchungen über freie Verbände. *The Journal of Symbolic Logic,* 16(2):81–106, 1951.

[13] U. Martin and T. Nipkow. Ordered rewriting and confluence. In M. Stickel, editor, *10th International Conference on Automated Deduction,* volume 449 of *LNCS,* pages 366–380. Springer-Verlag, 1990.

[14] R. Socher-Ambrosius. Boolean algebra admits no convergent term rewriting system. In R.E. Book, editor, *4th International Conference on Rewriting Techniques and Applications,* volume 488 of *LNCS,* pages 264–274. Springer-Verlag, 1991.

[15] G. Struth. Non-symmetric rewriting. Technical Report MPI-I-96-2-004, Max-Planck-Institut für Informatik, Saarbrücken, 1996.

[16] G. Struth. On the word problem for free lattices. In H. Comon, editor, *Rewriting Techniques and Applications, 8th International Conference, RTA-97,* volume 1232 of *LNCS,* pages 128–141. Springer, 1997.

[17] G. Struth. *Canonical Transformations in Algebra, Universal Algebra and Logic.* PhD thesis, Institut für Informatik, Universität des Saarlandes, 1998.

[18] G.S. Tseitin. On the complexity of derivation in propositional calculus. In J. Siekmann and G. Wrightson, editors, *Automation of Reasoning,* volume 2, pages 466–483. Springer-Verlag, 1983. reprint.

Deriving Theory Superposition Calculi from Convergent Term Rewriting Systems[*]

Jürgen Stuber

Max-Planck-Institut für Informatik
Im Stadtwald, 66123 Saarbrücken, Germany
`juergen@mpi-sb.mpg.de`

Abstract. We show how to derive refutationally complete ground superposition calculi systematically from convergent term rewriting systems for equational theories, in order to make automated theorem proving in these theories more effective. In particular we consider abelian groups and commutative rings. These are difficult for automated theorem provers, since their axioms of associativity, commutativity, distributivity and the inverse law can generate many variations of the same equation. For these theories ordering restrictions can be strengthened so that inferences apply only to maximal summands, and superpositions into the inverse law that move summands from one side of an equation to the other can be replaced by an isolation rule that isolates the maximal terms on one side. Additional inferences arise from superpositions of extended clauses, but we can show that most of these are redundant. In particular, none are needed in the case of abelian groups, and at most one for any pair of ground clauses in the case of commutative rings.

1 Introduction

Automated theorem provers face problems when they are used on theories whose axioms generate large search spaces. Overwhelmed by a huge number of trivial consequences of each fact, they fail to prove even rather simple theorems.

Our goal in this work is to improve the methods for superposition theorem proving in the context of algebraic theories. To avoid a separate completeness proof for each theory and to gain a better understanding of the general mechanism we develop a framework that allows us to derive superposition calculi systematically from convergent term rewriting systems for algebraic theories. This framework consists of a parameterized superposition calculus, where the parameters are a term ordering, a simplification function and a symmetrization function. We assume certain properties of the parameters which allow us to prove refutational completeness of the parameterized calculus. For many important algebraic theories such as abelian groups, commutative rings, modules or algebras suitable parameters exist (Stuber 1999a). We use the theory of commutative

---

[*] Complete proofs can be found in
`http://www.mpi-sb.mpg.de/~juergen/publications/Stuber1999Diss.html`.

L. Bachmair (Ed.): RTA 2000, LNCS 1833, pp. 229–245, 2000.

rings as our main example, since it is important in many applications and allows us to demonstrate our approach to critical pairs of extensions. For less well-behaved algebraic theories, e.g. modules over rings with zero divisors, certain critical pairs in the symmetrization cannot be made to converge uniformly and must be considered explicitly; such theories would require an extension of our approach.

We arrive at calculi which are improved in several respects. First, stronger ordering restrictions require that inferences apply only to certain maximal subterms; in the case of commutative rings these are maximal summands within the top-level sum. Second, we replace some direct uses of axioms by macro inferences. Superposition inferences into theory equations, which would lead to many variants of a clause, are replaced by special simplification rules, such as isolation for the inverse law, and by introducing semantic matching into the superposition rule. We formalize this by associating to each original equation an extended set of term rewriting rules, called its *symmetrization*. By implicitly using these extensions for semantic matching, we avoid to explicitly add the corresponding extended clauses. Instead of superposition inferences between such explicitly represented extensions the calculi contain an extension superposition rule that is needed to accommodate critical peaks between the implicit extensions. Since the form of the symmetrization is known for any particular theory, we can derive redundancy criteria which dispose of all or most of these inferences. In the case of abelian groups no such inference is needed, while in the case of commutative rings at most one extension superposition inference is needed for any pair of ground clauses. The combination of stronger ordering restrictions, macro inferences and redundancy criteria promises to be much more efficient than a more general calculus applied to part of the axioms. For instance, in purely equational reasoning it has been demonstrated that special calculi can improve performance greatly (Zhang 1993, Marché 1996).

Our goal is to obtain refutationally complete calculi for arbitrary first-order formulas, without restrictions on the logical structure or the set of function symbols. Here we consider only the ground case, as we believe that lifting the general calculus would offer no substantial new insights. More is possible by considering specific theories separately. For instance, on the ground level Godoy and Nieuwenhuis (2000) use the same approach for the theory of abelian groups, but they represent nonground equations uniformly as $t \approx 0$. This way of abstracting away information about the maximal summand during lifting has the possible advantage of avoiding inferences with group axioms, which are replaced by abelian group unification. Lifting is a problem for *unshielded* variables, i.e. variables which do not appear below a free function symbol. In general these variables must be split into a sum of a maximal and a nonmaximal part, which gives rise to very prolific inferences. For certain cases these problematic variables can be eliminated (Waldmann 1998, Stuber 1998a).

If, on the other hand, we restrict the ground case for commutative rings further to the case of ground equations over a finite set of constants, then our calculus generates essentially the same inferences as the Gröbner base algorithm for polynomials over the integers (Kandri-Rody and Kapur 1988).

Our work builds on several strands of research: automated first-order theorem proving, term rewriting, and the theory of Gröbner bases.

We prove refutational completeness by showing that our calculi reduce minimal counterexamples, which is the standard method for completeness proofs of superposition calculi (Bachmair and Ganzinger 1998). Wertz (1992) is the first to build superposition calculi for theorem proving modulo E, and in particular modulo AC. He uses interpretations where transitivity holds universally but E holds only below a certain bound. In contrast to this, the approach of Bachmair and Ganzinger (1994a) to AC-superposition sacrifices universal validity of transitivity to get universal validity of AC. This allows them to handle AC-matching and AC-unification as black boxes, and in many important cases, for instance when simplifying by rewriting, the bound on transitivity is satisfied. We follow the second approach. Nieuwenhuis and Rubio (1997) and Vigneron (1994) consider superposition calculi modulo AC with constraints. Bachmair, Ganzinger and Stuber (1995) develop a calculus for commutative rings. Their proof technique was not strong enough to avoid certain shortcomings, namely the explicit representation of the symmetrization and the weaker notion of redundancy. Superposition calculi for cancellative abelian monoids require a notion of rewriting on equations instead of terms, since additive inverses are in general not available (Ganzinger and Waldmann 1996, Waldmann 1997). The special case of divisible torsion-free abelian groups allows us to eliminate unshielded variables, which avoids the most prolific inferences (Waldmann 1997, Waldmann 1998). Previously we have shown that our approach is compatible with constraints for the special case of integer modules (Stuber 1996, Stuber 1998a). We have also carried it out for commutative rings in the ground case (Stuber 1998b).

In term rewriting Marché (1996) builds a range of theories from AC to commutative rings into equational completion. He explicitly adds symmetrizations and is not concerned with redundancy criteria for extensions. Like standard completion it does not handle unorientable equations, hence inferences below variables are not needed. Using the Cime system for completion with built-in theories (Contejean and Marché 1996), Marché demonstrates that the special treatment of theories can reduce the number of inferences greatly and can lead to large speedups.

The notion of symmetrization originates from string rewriting systems for finitely presented groups Greendlinger (1960), and is generalized by Le Chenadec (1986) to various other theories.

Another strand of research leading to this work is concerned with Gröbner or standard bases for polynomial simplification (Buchberger 1970, Buchberger 1987, Becker and Weispfenning 1993, Bachmair and Tiwari 1997).

The relation between completion for term rewriting systems, which is the basis of our calculus, and Gröbner basis algorithms has been noticed by Buchberger and Loos (1983). Bündgen (1996) formalizes this by encoding Gröbner basis computation, including the computation in the base rings, in term rewriting systems. Bachmair and Ganzinger (1994b) use constraints to abstract away computations in the base ring.

As the term ordering for the case of abelian groups or commutative rings we can use an ordering based on the MAPO (Delor and Puel 1993), while modules and algebras require a TPO (Stuber 1999b).

2   Preliminaries

We assume familiarity with first-order logic and term rewriting systems (Baader and Nipkow 1998, Dershowitz and Jouannaud 1990), and in particular the case modulo E (Jouannaud and Kirchner 1986). We denote rewriting by $\Rightarrow$, syntactical equality by $\approx$, and the empty clause by $\perp$.

For a set R of rewrite rules $gnd(R)$ denotes the set of their ground instances. By $[\neg](s \approx t)$ we denote a literal that is either positive or negative; corresponding literals are of the same polarity.

3   The Term Rewriting System

We require that a theory is represented by a ground term rewriting system T that is convergent modulo an equational theory E. That is, T is terminating and Church-Rosser modulo E. Then for any equational proof $s \overset{*}{\Leftrightarrow}_{T \cup E} t$ there exists a valley proof $s \overset{*}{\Rightarrow}_T s' \overset{*}{\Leftrightarrow}_E t' \overset{*}{\Leftarrow}_T t$. To avoid explicitly mentioning E-matching everywhere we assume that it is included in T. That is, $T = \{l' \Rightarrow r \mid l' =_E l, \ l \Rightarrow r \in gnd(T')\}$ for some term rewriting system T'. We assume a fixed set of function symbols F. A function symbol f is *free* in T if there exists a possibly nonground term rewriting system $\widehat{T}$ such that $T = gnd(\widehat{T})$ and f does not occur in $\widehat{T}$. Function symbols which are not free are called *interpreted*. The set of interpreted function symbols is denoted by F_T. Terms with a free function symbol at the root position are called *atomic* and will be denoted by α. We let $T_1 = T \cup E \cup \mathrm{Eq}$, where the rules in T are understood as equations and Eq is the first-order axiomatization of equality for F. This is the logical contents of the theory, with equality made explicit.

4   The Termination Ordering

We require an E-compatible simplification ordering $\succ_T$ that is total up to E on ground terms such that $\succ_T$ contains the rewrite system T. We will usually omit the subscript T as the ordering used will be clear from the context. An atomic term α is called a *maximal atomic term* in s if $s = u[\alpha, \alpha_1, \ldots, \alpha_n]$ where $n \geq 0$, $\alpha_1, \ldots, \alpha_n$ are atomic, u contains only function symbols from F_T, and $\alpha \succeq \alpha_i$ for $i = 1 \ldots n$.

To extend the term ordering $\succ_T$ to literals and clauses we assign to each of these a *complexity* c. For literals we let

$$c(s \approx t) = \{\{s\}, \{t\}\} \tag{1}$$
$$c(s \not\approx t) = \{\{s, t\}\} \tag{2}$$

and $\succ$ on literals is the two-fold multiset extension of $\succ$ on terms applied to these complexities. This has the effect that the ordering on literals is the lexicographic combination of $\succ$ on the maximal term, the ordering $- \succ +$ on the polarity of the literal and $\succ$ on the minimal term. For a clause $C = L_1 \vee \ldots \vee L_n$ that is not an instance of transitivity we let

$$c(C) = \langle \{c(L_1), \ldots, c(L_n)\}, \emptyset \rangle \tag{3}$$

That is, the complexity of a nontransitivity clause is pair of the multiset of the complexities of its literals and the empty multiset. This uses the standard definition of the clause ordering by Bachmair and Ganzinger (1994a) in its first component. To extend it to transitivity we consider a ground instance

$$D = t_1 \not\approx s \vee s \not\approx t_2 \vee t_1 \approx t_2$$

and let

$$c(D) = \langle \{\{\{s\}\}\}, \{t_1, t_2\} \rangle.$$

We say that D has the *middle term* s and the *side terms* t_1 and t_2. Then the ordering on clauses is the lexicographic combination of the three-fold multiset extension of the term ordering and the multiset extension of the term ordering, applied to the complexities. By this definition transitivity instances with a middle term s are immediately below nontransitivity clauses with maximal term s in the term ordering. We call the middle term of transitivity instances and the maximal term of other clauses the *dominating term* of the clause, since it dominates the term ordering. It is possible to choose other well-orderings for the terms on the smaller side of literals (Bachmair and Ganzinger 1998) and for the side terms of a transitivity instance.

5 The Symmetrization Function

The symmetrization function is at the heart of our approach. It maps a single ground equation into a logically equivalent set of ground rewrite rules that behaves better in combination with the theory. That is, we consider terminating term rewriting systems of the form

$$T \cup \bigcup_{s \approx t} \mathcal{S}_T(s \approx t),$$

where the set of rules $\mathcal{S}_T(s \approx t)$ is designed such that $s \approx t$ becomes true and as much as possible of T and the equality axioms hold. It turns out that this works well for all axioms except transitivity, which for some theories causes problems due to nonconvergent peaks between extended rules.

We start by the notion of a set of rules being (strongly) symmetrized. Being symmetrized is a rather technical notion that is required by our general superposition calculus. It amounts to the convergence of all critical pairs that involve some rule from T or equation from E, and hence validity of the corresponding instances of transitivity. The notion of a strongly symmetrized set of rewrite rules becomes important when we later instantiate the general framework by specific theories. Strong symmetrization allows us to reduce equational proofs by normalizing the terms in the proof (Stuber 1997).

A set of rewrite rules S is *symmetrized* with respect to T modulo E if for all peaks $t_1 \Leftarrow_T t \Rightarrow_S t_2$ and for all cliffs $t_1 \Leftrightarrow_E t \Rightarrow_S t_2$ we have $t_1 \downarrow_{T \cup S} t_2$.

The set S is called *strongly symmetrized* with respect to T modulo E if S can be partitioned into sets S_i, $i \in I$, such that $T \cup S_i$ is convergent modulo E for all $i \in I$.

Proposition 1. *If a set of rewrite rules S is strongly symmetrized with respect to T modulo E then S is symmetrized with respect to T modulo E.*

Note that S being strongly symmetrized implies that peaks of the form $t_1 \Leftarrow_{S_i} t \Rightarrow_{S_i} t_2$ converge, which is not guaranteed if S is symmetrized but not strongly symmetrized. However, this is still much weaker than convergence, as peaks of the form $t_1 \Leftarrow_{S_i} t \Rightarrow_{S_j} t_2$ need not converge for $i \neq j$.

Our goal is to derive (strongly) symmetrized sets of rules directly for some given equation, so that the equation becomes true in the rewrite system. We break this into two steps. First the equation is brought into a certain theory-specific normal form by simplification, and then for any such equation a symmetrized set is obtained by applying a symmetrization function. For now we only assume that a set Norm_T of equations this normal form is given, and postpone the discussion of simplification. We continue by discussing symmetrization functions.

A *(strong) symmetrization function* $\mathcal{S}_T$ (for T) maps any equation $l \approx r$ in Norm_T to a (strongly) symmetrized set of rewrite rules $\mathcal{S}_T(l \approx r)$ such that

$$T_1 \cup \{l \approx r\} \models \mathcal{S}_T(l \approx r) \tag{4}$$

$$l \downarrow_{T \cup \mathcal{S}_T(l \approx r)} r \tag{5}$$

$$\mathcal{S}_T(l \approx r) \subseteq (\succ) \tag{6}$$

$$l' \succeq l \text{ for any } l' \Rightarrow r' \text{ in } \mathcal{S}_T(l \approx r) \tag{7}$$

(4) ensures soundness, (5) ensures that $l \approx r$ becomes true, (6) ensures termination, and (7) ensures that terms smaller than l cannot be rewritten by $\mathcal{S}_T(l \approx r)$. We call a rule $l' \Rightarrow r'$ in $\mathcal{S}_T(l \Rightarrow r) \setminus \{l \Rightarrow r\}$ an *extension* (of $l \Rightarrow r$). The symmetrization function is extended to sets of equations in Norm_T by

$$\mathcal{S}_T(R) = \bigcup_{l \approx r \in R} \mathcal{S}_T(l \approx r).$$

Assumption 2. *We assume from now on that $\mathcal{S}_T$ is a symmetrization function for T modulo E.*

To obtain a symmetrization function one starts with a set S containing a single rule in Norm_T and proceeds by adding to S critical pairs resulting from peaks of the form $t_1 \Leftarrow_T s \Rightarrow_S t_2$, in a way analogous to Knuth-Bendix completion. To obtain a strong symmetrization function one also has to consider critical peaks of the form $t_1 \Leftarrow_S s \Rightarrow_S t_2$. For the commutative theories that we consider here it turns out that the symmetrization function obtained by considering the first kind of peaks also makes the second kind converge. Thus the strong symmetrization property requires no extra effort in these cases. Without commutativity, however, an equation may have nontrivial overlaps with variants of itself. It is infeasible to derive a strong symmetrization function in that case, hence for instance Le Chenadec (1986) uses ordinary symmetrization for nonabelian groups.

6   Candidate Models

In this section we define a model functor I that maps any set N of ground clauses to an interpretation I_N. We show that I_N satisfies the theory and the equality axioms except for transitivity.

The construction of the interpretation extends the standard one by Bachmair and Ganzinger (1998) in several respects.

Firstly, rewriting is modulo E. Secondly, the built-in term rewriting system T is always included when constructing the interpretation. This ensures that these interpretations satisfy T. Thirdly, we have the additional restriction that a clause can be productive only if the rule it produces is in Norm_T. Finally, the term rewriting systems are built from symmetrizations of rules, which ensures that they are always symmetrized.

A ground clause $C \vee s \approx t$ is called *reductive* for $s \Rightarrow t$ if $s \approx t$ is strictly maximal in C and $s \succ t$. Only reductive clauses can contribute to an interpretation. Given a set N of ground clauses, we let N_C be the set of ground clauses in N which are smaller than C. For any set N of ground clauses we inductively define a set R_N of ground rules, a symmetrized set $S_N = \mathcal{S}_T(R_N)$ of ground rules, and the corresponding interpretation $I_N = (T \cup S_N)^{\Downarrow}$. Here $(T \cup S_N)^{\Downarrow}$ denotes the *valley closure* of $T \cup S_N$, i.e., the set $\{s \approx t \mid s \Downarrow_{T \cup S_N} t\}$. We may regard R, S and I as functions which map sets of clauses to sets of rewrite rules or equations. A rule $\{l \Rightarrow r\}$ is in R_N if there exists a clause $C = C' \vee l \approx r$ in N such that (i) C is false in I_{N_C}, (ii) C is reductive for $l \Rightarrow r$, (iii) $l \Rightarrow r$ is in Norm_T, (iv) l is irreducible by S_{N_C}, and (v) C' is false in $(T \cup S_{N_C} \cup \mathcal{S}_T(l \Rightarrow r))^{\Downarrow}$. In this case we say that C *produces* $l \Rightarrow r$ in R_N, or that C is *productive*. The set R_N is well-defined, since for any ground clause C only the interpretation for smaller clauses in N_C determines whether C produces a rule. Where N is clear from the context we write R_C for R_{N_C}, S_C for S_{N_C} and I_C for I_{N_C}.

Lemma 3. *Let $C = C' \vee l \approx r$ be a clause that produces $l \Rightarrow r$ in R_N. Then C' is false in I_N.*

We let $T_0 = \text{Refl} \cup \text{Symm} \cup \text{Mon} \cup E \cup T$.

Lemma 4. *Let N be a set of ground clauses. Then $I_N \models T_0$.*

It remains to consider instances of transitivity and clauses in N. These are in general not true in I_N. Validity of transitivity instances with middle term up to s in I_N is equivalent to the Church-Rosser-property of $T \cup S_N$ on terms up to s (Bachmair and Ganzinger 1994a). For commutative rings there are cases where two extended rules overlap in such a way that the resulting critical pair does not converge. Then $T \cup S_N$ is not Church-Rosser and transitivity does not hold. We say that a clause C in $\text{Trans} \cup N$ is a *counterexample* for I_N if C is false in I_N. Since the set of possible counterexamples is well-ordered by $\succ$, there is always a *minimal counterexample* if I_N is not a model of $N \cup T_1$.

7   Redundancy of Clauses and Simplification

We will need to refer to the specific construction of candidate models when we prove that certain clauses or inferences are redundant. In particular, we need that candidate models are built from (strongly) symmetrized sets of rewrite rules S_N, and we need to refer to the presence of certain rewrite rules in R_N. We achieve this by defining a special notion of consequence that takes into account only interpretations constructed by the model functor I, and by introducing a new atomic formula $s \Rightarrow t$ that is true in such an interpretation I_N whenever the rule $s \Rightarrow t$ is in R_N. Note that the R_N corresponding to I_N will always be known from the context via the set N of clauses. These atoms will be used only as unit clauses, and we will refer to them as rewrite rules. For sets of clauses or rewrite rules N_1 and N_2 we say that N_2 is an *I-consequence* of N_1, in symbols $N_1 \models_I N_2$, if $I_M \models N_1$ implies $I_M \models N_2$ for all sets of ground clauses M. Lemma 4 can then be rephrased as $\models_I T_0$.

Let C be some ground clause. We write Trans_C for the set of ground instances of transitivity in Trans which are smaller than C. The middle term of such an instance of transitivity is smaller than or equal to the dominating term of C. Then C is *redundant* (with respect to T) in a set of ground clauses N if

$$N_C \cup \text{Trans}_C \models_I C.$$

A (possibly nonground) clause is called *redundant* in a set of clauses N if all its ground instances are redundant in the set of ground instances of N. A clause is called *redundant* if it is redundant in $\emptyset$. A clause that is redundant in N cannot be the minimal counterexample for I_N, because some smaller clause in $N_C \cup \text{Trans}_C$ would have to be a counterexample for I_N as well. Note that we can use $N_C \cup \text{Trans}_C \cup T_0 \models C$ as a sufficient criterion for redundancy. This criterion corresponds to the notion of redundancy used by Bachmair and Ganzinger (1998).

Based on our notion of redundancy, we say that a ground clause D is a *simplification* (with respect to T) of a ground clause C if $\{C\} \cup T_1 \models D$ and C is redundant in $\{D\}$. That is, $C \succ D$, $\{C\} \cup T_1 \models D$, and $\{D\} \cup \text{Trans}_C \models_I C$.

Remember that the symmetrization function is only defined on equations in Norm_T. Equations not in this form need to be simplified before symmetrization can be applied. However, we want to restrict simplifications as much as possible, since ground simplifications become inferences when they are lifted. We formalize this by assuming that there exists a *simplification function* Simp_T which maps ground literals to sets of ground literals, such that L' is a simplification of L for all $L' \in \mathrm{Simp}_T(L)$. We say that Simp_T is *admissible* with respect to Norm_T if $\{L \mid \mathrm{Simp}_T(L) = \emptyset\} \subseteq \mathrm{Norm}_T$ where Norm_T is extended to literals in the obvious way. Since $\succ$ is well-founded, it suffices to nondeterministically apply simplifications in Simp_T to eventually reach a literal in Norm_T.

Assumption 5. *From now we assume* Simp_T *to be admissible with respect to* Norm_T.

The definition Simp_T imposes certain properties on Norm_T. Since any literal can be simplified to some literal in Norm_T, and since simplification preserves T_1-equivalence, for any literal there is a T_1-equivalent literal in Norm_T. This literal is in general not unique, since we want to avoid unnecessary reductions in right-hand sides of equations that would lead to inferences on nonmaximal summands. Moreover, for strong symmetrization functions the requirement that l is minimal among the left-hand sides of rules in $\mathcal{S}_T(l \Rightarrow r)$ translates into the requirement that equations in Norm_T are *left-minimal*. That is, l is minimal among the greater sides of T_1-equivalent equations. For if this were not the case, say there exists $l' \approx r'$ with $l' \succ r'$ and $l \succ l'$ then l' must be reducible by $\mathcal{S}_T(l \approx r)$, by some rule with left-hand side smaller than l.

8   The Inference System

We present a ground inference system that is based on the parameters introduced in the previous sections, namely the term rewriting system T, the ordering $\succ$, the set Norm_T, the symmetrization function $\mathcal{S}_T$ and the simplification function Simp_T.

We assume that in each ground clause a literal is selected; either some arbitrary negative literal, or a positive literal that is maximal in the entire clause. An *inference system* is a set of inferences. Each *inference* has a *main premise* C, *side premises* $C_1, \ldots, C_n$, and a conclusion D. These notions allow uniform definitions of reduction property and redundancy of inferences. Think of the main premise as the minimal counterexample, the side premises as productive clauses, and the conclusion as a smaller clause that is also false, but that is not in N. Thus, the main premise may either be a clause supposed to be from N, then we write

$$\frac{C_1 \quad \ldots \quad C_n \quad C}{D}$$

for the inference, with the main premise at the right. Or the main premise may be an instance of transitivity, then we omit it and write

$$\frac{C_1 \quad \ldots \quad C_n}{D},$$

for the inference. In this case we state the main premise explicitly. Instances of transitivity cannot be side premises. An inference is *strictly decreasing* if the conclusion is smaller than the main premise in the clause ordering. All inferences that we present are sound and strictly decreasing.

We let Sup_T be the set of the following inferences:

Let $l_1 \Rightarrow r_1$ and $l_2 \Rightarrow r_2$ be rules in Norm_T and $S_i = \mathcal{S}_T(l_i \Rightarrow r_i)$ for $i = 1, 2$. An *extension peak* between $l_1 \Rightarrow r_1$ and $l_2 \Rightarrow r_2$ with respect to T is a rewrite sequence

$$r_1' \Leftarrow_{S_1} l_1'[l_2'] \Rightarrow_{S_2} l_1'[r_2']$$

such that $l_i' \Rightarrow r_i'$ is a rule in S_i for $i = 1, 2$, l_1 is irreducible by $T \cup S_2$, and l_2 is irreducible by $T \cup S_1$.

T-*Extension Superposition* $\dfrac{l_1 \approx r_1 \vee C_1 \qquad l_2 \approx r_2 \vee C_2}{r_1' \approx l_1'[r_2'] \vee C_1 \vee C_2}$

if (i) $l_i \approx r_i$ is in Norm_T and selected in $l_i \approx r_i \vee C_i$ for $i = 1, 2$, and (ii) there exists an extension peak $r_1' \Leftarrow_{\mathcal{S}_T(l_1 \approx r_1)} l_1'[l_2'] \Rightarrow_{\mathcal{S}_T(l_2 \approx r_2)} l_1'[r_2']$ between $l_1 \Rightarrow r_1$ and $l_2 \Rightarrow r_2$.

The transitivity instance corresponding to the peak,

$$r_1' \not\approx l_1'[l_2'] \vee l_1'[l_2'] \not\approx l_1'[r_2'] \vee r_1' \approx l_1'[r_2'],$$

is the main premise of this inference. The explicit premises are side premises.

T-*Theory Simplification* $\dfrac{L \vee C}{L' \vee C}$

if (i) L is selected in $L \vee C$, and (ii) $L' \in \mathrm{Simp}_T(L)$.

T-*Reflexivity Resolution* $\dfrac{p \not\approx q \vee C}{C}$

if (i) $p \not\approx q$ is in Norm_T and selected in $p \not\approx q \vee C$, and (ii) $p =_E q$.

T-*Equality Factoring* $\dfrac{s \approx t \vee s' \approx t' \vee C}{t \not\approx t' \vee s' \approx t' \vee C}$

if (i) $s \approx t$ is in Norm_T and selected in $s \approx t \vee s' \approx t' \vee C$, and (ii) $s =_E s'$.

The single premise of Theory Simplification, Reflexivity Resolution and Equality Factoring is their main premise, they have no side premises.

T-*Superposition* $\dfrac{l \approx r \vee D \qquad [\neg](s[l''] \approx t) \vee C}{[\neg](s[r'] \approx t) \vee C \vee D}$

if (i) $l' \Rightarrow r'$ is in $\mathcal{S}_T(l \approx r)$, (ii) $l' =_E l''$, (iii) $[\neg](s[l''] \approx t)$ is selected in $[\neg](s[l''] \approx t) \vee C$ and in Norm_T, and (iv) $l \approx r$ is in Norm_T and selected in $l \approx r \vee D'$.

Superposition has the main premise $[\neg](s[l''] \approx t) \vee C$ and the side premise $l \approx r \vee D$.

An inference with main premise C, conclusion D and side premises $C_1, \ldots, C_n$, where $C_i = C_i' \vee l_i \approx r_i$ is reductive for $l_i \Rightarrow r_i$, is *redundant in* N if

$$N_C \cup \text{Trans}_C \cup \{l_i \Rightarrow r_i \mid i = 1, \ldots, n\} \cup \{\neg C_i' \mid i = 1, \ldots, n\} \models_I D.$$

An inference is called *redundant* if it is redundant in $\emptyset$. Here we exploit that side premises arise from productive clauses. Hence each C_i has the form $C_i' \vee l_i \approx r_i$ and is reductive for $l_i \Rightarrow r_i$ which is in Norm_T. We may assume that $l_i \Rightarrow r_i$ is in R_N and that C_i' is false in I_N.

Let C be the minimal counterexample for I_N and let π be an inference with main premise C, conclusion D and side premises $C_1, \ldots, C_k$ such that $C \succ D$, and each C_i is smaller than C, has the form $C_i = l_i \approx r_i \vee C_i'$ and is reductive for $l_i \Rightarrow r_i$. We say that π *reduces* C (with respect to I_N) if

$$I_N \models \neg D \wedge l_1 \Rightarrow r_1 \wedge \ldots \wedge l_k \Rightarrow r_k \wedge \neg C_1' \wedge \ldots \wedge \neg C_k'.$$

An inference system Calc has the *reduction property* (with respect to I) if Calc contains an inference that reduces C with respect to I_N for any set N of ground clauses such that I_N has the minimal counterexample $C \neq \bot$. A *refutation* of a clause set N in a calculus Calc is a sequence $C_1, \ldots, C_n$ of clauses such that each clause C_i is either from N or can be derived from clauses earlier in the sequence by an inference in Calc, and $C_n = \bot$. A calculus is *refutationally complete* for T_1 if for any set N of clauses such that $T_1 \cup N$ is inconsistent there exists a refutation of N in Calc. Calculi with the reduction property are refutationally complete (Bachmair and Ganzinger 1998).

Lemma 6 (Extension Superposition). *Let N be a set of ground clauses such that N does not contain the empty clause. Suppose that the minimal counterexample C for I_N is an instance of transitivity. Then Sup_T contains an Extension Superposition inference that reduces C.*

Proof: Let C be the minimal counterexample and let s be its middle term. Since C is minimal, instances of transitivity with smaller middle terms are true in I_N. This implies that $T \cup S_C$ is Church-Rosser modulo E below s, but that there exists some peak $t_1 \Leftarrow s \Rightarrow t_2$ such that t_1 and t_2 do not converge and $t_1 \approx t_2$ is false in I_N. As T is convergent and S_C is symmetrized modulo E with respect to T, all peaks involving T and all cliffs with E converge, so both rules used in the peak are from S_C. If the rewrite steps in the peak were in parallel positions of s then t_1 and t_2 would converge, which is not the case. Let $l_1' \Rightarrow r_1'$ and $l_2' \Rightarrow r_2'$ be the rules from S_C used in the peak. For $i = 1, 2$ the rule $l_i' \Rightarrow r_i'$ is from some symmetrization $\mathcal{S}_T(l_i \Rightarrow r_i)$ where $l_i \Rightarrow r_i$ is a rule in R_C that has been produced by some clause $C_i = l_i \approx r_i \vee C_i'$. If we suppose without loss of generality that $l_1 \succ l_2$ then l_1 is irreducible by $\mathcal{S}_T(l_2 \Rightarrow r_2)$ because this is a condition for C_1 being productive, and l_2 is irreducible by $\mathcal{S}_T(l_1 \Rightarrow r_1)$ because

$l_1 \succ l_2$ and l_2 is minimal among the left-hand sides in $\mathcal{S}_T(l_2 \Rightarrow r_2)$. Hence this is an extension peak of the form

$$t_1 = s[r_1'] \Leftarrow s[l_1'[l_2']] \Rightarrow s'[r_2'] = t_2.$$

Since C is the minimal counterexample, the context must be empty, and the peak has the form

$$r_1' \Leftarrow l_1'[l_2'] \Rightarrow l_1'[r_2'].$$

For such a peak Sup_T contains the Extension Superposition inference

$$\frac{l_1 \approx r_1 \vee C_1' \qquad l_2 \approx r_2 \vee C_2'}{r_1' \approx l_1'[r_2'] \vee C_1' \vee C_2'}$$

where C_i' is false in I_C and $l_i \Rightarrow r_i$ is in R_N for $i = 1, 2$. Since C_1', C_2' and $r_1' \approx l_1'[r_2']$ are false in I_N, the conclusion is false in I_N. Hence the inference reduces C. $\qquad\square$

For each of the other minimal counterexamples, which arise when some condition for productivity is violated, there is an inference that reduces it. Due to lack of space we have to skip the detailed proof; it is analogous to the case of standard superposition.

Theorem 7. Sup_T *has the reduction property.*

Putting things together and making all the conditions explicit we obtain the main theorem, stating that our method yields refutationally complete calculi:

Theorem 8. *Suppose that T is a ground term rewriting system that is confluent modulo E, $\succ$ is an E-compatible simplification ordering that is total up to E on ground terms, $T \subseteq (\succ)$, Simp_T is a simplification function, and $\mathcal{S}_T$ is a symmetrization function for T modulo E with respect to $\succeq$ such that Simp_T is admissible with respect to $\mathcal{S}_T$. Then Sup_T is refutationally complete for T_1.*

Variations are possible, for instance it is straightforward to replace equality factoring by factoring and merging paramodulation.

9   Extension Peaks Revisited

In the Extension Superposition rules stated above any extension peak between two rules leads to an inference, leading to a large or infinite number of inferences for any pair of clauses whose symmetrizations overlap. For specific theories we can do much better by exploiting the known structure of the symmetrizations. For the theories that we have considered either none or a single Extension Superposition inference suffices; all other such inferences are redundant. We call the extension peaks that give rise to these Extension Superposition inferences *critical extension peaks*. Furthermore, critical extension peaks are the only cause of transitivity counterexamples. We exploit this to relax the bound on transitivity somewhat, by bounding only subterms that occur as the middle terms in critical extension peaks. We need this extension in the case of commutative rings, where the bound is then on single summands instead of on the whole sum.

Let $S_i = \mathcal{S}_T(l_i \Rightarrow r_i)$ for $i = 1, 2$. Consider some extension peak $t_1 \Leftarrow_{S_1} s \Rightarrow_{S_2} t_2$ between rules $l_1 \Rightarrow r_1$ and $l_2 \Rightarrow r_2$. We call such an extension peak *redundant* in N if all Extension Superposition inferences that have the peak as their main premise are redundant in N. We call the extension peak *redundant* if it is redundant in $\emptyset$.

Lemma 9. *Let $t_1 \Leftarrow_{S_1} s \Rightarrow_{S_2} t_2$ be an extension peak between rules $l_1 \Rightarrow r_1$ and $l_2 \Rightarrow r_2$. The peak is redundant in N if and only if the extension superposition inference*

$$\frac{l_1 \approx r_1 \qquad l_2 \approx r_2}{t_1 \approx t_2}$$

with main premise $t_1 \not\succeq s \lor s \not\succeq t_2 \lor t_1 \approx t_2$ is redundant in N.

10   Commutative Rings

Due to space limitations we can only point out a few of the most important aspects for the case of commutative rings. We use the well-known convergent term rewriting system modulo AC for commutative rings of Peterson and Stickel (1981) and call it CR. $\mathrm{Norm}_{\mathrm{CR}}$ consists of equations of the form $n\phi \approx r$ where ϕ is irreducible with respect to CR, $\phi \succ r$, $n \geq 1$ and $\phi = \alpha_1 \cdots \alpha_k$ for $k \geq 0$, with $\phi = 1$ for $k = 0$. Here we use $n\phi$ as a shorthand for $\phi_1 + \cdots + \phi_n$ where $\phi_1 =_{\mathrm{AC}} \cdots =_{\mathrm{AC}} \phi_n$. For $n > 0$ we also use $(-n)\phi$ for $n(-\phi)$, and 0ϕ denotes just 0. A term $n\phi$ is called a *monomial*. We have the following symmetrization for commutative rings:

$$\mathcal{S}_{\mathrm{CR}}(\alpha \approx r) = \{\alpha \Rightarrow r\} \quad \text{if } \alpha \text{ is atomic or 1;}$$
$$\mathcal{S}_{\mathrm{CR}}(\phi \approx r) = \{\phi \Rightarrow r\}$$
$$\cup\, gnd(\{y \cdot \phi \Rightarrow y \cdot r\}) \quad \text{if } \phi \text{ is a proper product;}$$
$$\mathcal{S}_{\mathrm{CR}}(n\phi \approx r) = \{n\phi \Rightarrow r\}$$
$$\cup\, gnd(\{x + n\phi \Rightarrow x + r\})$$
$$\cup\, gnd(\{n(y \cdot \phi) \Rightarrow y \cdot r\})$$
$$\cup\, gnd(\{x + n(y \cdot \phi) \Rightarrow x + y \cdot r\})$$
$$\cup\, \{-\phi \Rightarrow (n-1)\phi - r\}$$
$$\cup\, gnd(\{-(y \cdot \phi) \Rightarrow (n-1)(y \cdot \phi) - (y \cdot r)\}) \quad \text{if } n \geq 2.$$

Extension peaks arise between extensions for multiplication. We obtain the following criterion, which leaves at most a single nonredundant extension peak for any pair of rules:

Theorem 10. *Let $n_i\phi_i \Rightarrow r_i$ be a rewrite rule in* Norm_{CR} *for* $i = 1, 2$, *and assume without loss of generality* $n_1 \geq n_2$.

These two rules have the single critical extension peak

$$r_1\psi_1 \Leftarrow n_1\phi \Rightarrow (n_1 - n_2)\phi + r_2\psi_2$$

where $\phi =_{AC} \text{lcm}(\phi_1, \phi_2) =_{AC} \phi_1\psi_1 =_{AC} \phi_2\psi_2$ *if (1)* $\psi_1 \neq 1$, *(2)* $n_1 > n_2$ *or* $\psi_2 \neq 1$, *and (3) either (a)* $n_1, n_2 \geq 2$, *(b)* $n_1 \geq 2$, $n_2 = 1$ *and* ϕ_2 *is a proper product with* $\gcd(\phi_1, \phi_2) \neq 1$, *or (c)* $n_1 = n_2 = 1$ *and* ϕ_1 *and* ϕ_2 *are proper products with* $\gcd(\phi_1, \phi_2) \neq 1$.

Otherwise there is no critical extension peak between these two rules.

Since only few instances of transitivity are not redundant, the truth of transitivity up to some bound implies that transitivity is true even up to the next nonredundant instance. Since by Theorem 10 nonredundant peaks have a single summand at the top, transitivity holds for all instances whose middle term contains only summands that do not exceed the bound individually, even if the sum as a whole is greater than the bound.

This extension is necessary to prove that isolation for commutative rings is compatible with the notion of redundancy, where isolation is the following inference:

CR-*Isolation* $$\dfrac{[\neg](m\phi_1 + s \approx n\phi_2 + t) \vee C}{[\neg]((m - n)\phi_1 \approx t - s) \vee C}$$

if (i) $\phi_1 =_{AC} \phi_2$, (ii) ϕ_1 is a product, (iii) ϕ_1 is irreducible with respect to CR, (iv) $m \geq n$, (v) $n \neq 0$ or $s \neq 0$, and (vi) $\phi_1 \succ s$ and $\phi_1 \succ t$.

Here the most difficult case arises when maximal summands in a negative literal must be cancelled. In this case we have to prove

$$\{m\phi + s \approx n\phi + t\} \cup \text{Trans}_D \models_I (m - 1)\phi + s \approx (n - 1)\phi + t,$$

where we must be careful not to exceed the bound D on transitivity, which is dominated by the maximal summand $m\phi$ or $n\phi$. Without this bound it would be sufficient to add $-\phi$ on both sides and to normalize with respect to the rewrite system CR for commutative rings. Since $-\phi$ exceeds the bound we have to be more careful. In the candidate model we have a rewrite proof

$$m\phi + s \overset{*}{\Rightarrow} n_0\phi + r_1 + s \Downarrow n_0\phi + r_2 + t \overset{*}{\Leftarrow} n\phi + t,$$

where reductions of maximal terms are done first and $n_0\phi$ is not reduced. We obtain rewrite proofs for (1) $r_1 + s \approx r_2 + t$ from the middle part and (2) $r_1 \approx (m - n)\phi + r_2$ from the peak formed by the reductions of the maximal terms. Using contexts $-r_1 + (m-1)\phi + [\,]$ on (1) and $-[\,] + (m-1)\phi + r_2 + t$ on (2) these can be combined to a suitable proof. The main idea to stay below the bound is to first reduce the maximal terms and then cancel the resulting smaller term r_1:

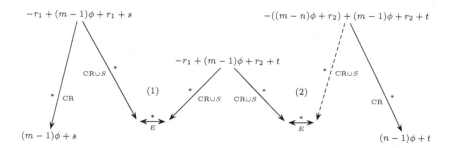

By normalizing each term in this proof with respect to CR we obtain a proof that contains $(m - 1)\phi$ as its maximal monomial, hence transitivity holds for all terms in the proof and we obtain a valley proof $p = (m - 1)\phi + s \Downarrow (n - 1)\phi + t$. Note that the normalization removes all occurrences of $-\phi$ in the dashed part of the proof.

Apart from Isolation the simplification function consists of rewriting with CR, which does not pose problems with respect to the bounded validity of transitivity. Simplification by rewriting may be limited such that rewriting always involves maximal summands, i.e., either reduce maximal summands, or cancel maximal summands by the inverse law. Other inferences like Superposition and Equality Factoring operate only on left-hand sides, which also restricts them to maximal terms. With these remarks the ground calculus for commutative rings can easily be derived from the general case.

11   Conclusion

We have presented an approach for constructing refutationally complete ground calculi for theories that can be represented by convergent term rewriting systems. This approach is suitable in particular for theories such as abelian groups or commutative rings. For commutative rings, overlaps of AC-extensions with respect to multiplication lead to problems with transitivity, and in turn to difficulties in the completeness proof. These were overcome by developing a redundancy criterion for these extension peaks, and by considering the effect of these redundancies on the validity of transitivity.

Acknowledgments

I thank Harald Ganzinger for his continued support, Uwe Waldmann for his valuable comments on parts of this work, and Hubert Baumeister, Fritz Eisenbrand, Patrick Maier and Georg Struth for interesting discussions on various aspects of this work.

References

BAADER, F. AND NIPKOW, T. (1998). *Term rewriting and all that*, Cambridge University Press, Cambridge, UK.

BACHMAIR, L. AND GANZINGER, H. (1994a). Associative-commutative superposition. In N. Dershowitz and N. Lindenstrauss (eds), *Proc. 4th Int. Workshop on Conditional and Typed Rewriting*, LNCS 968, Springer, Jerusalem, pp. 1–14.

BACHMAIR, L. AND GANZINGER, H. (1994b). Buchberger's algorithm: A constraint-based completion procedure. *Proc. 1st Int. Conf. on Constraints in Computational Logics*, LNCS 845, Springer, Munich, Germany, pp. 285–301.

BACHMAIR, L. AND GANZINGER, H. (1998). Equational reasoning in saturation-based theorem proving. *Automated Deduction - A Basis for Applications. Volume I*, Kluwer, Dordrecht, The Netherlands, chapter 11, pp. 353–397.

BACHMAIR, L. AND TIWARI, A. (1997). D-bases for polynomial ideals over commutative noetherian rings. In H. Comon (ed.), *8th Intl. Conf. on Rewriting Techniques and Applications*, LNCS 1103, Springer, Sitges, Spain, pp. 113–127.

BACHMAIR, L., GANZINGER, H. AND STUBER, J. (1995). Combining algebra and universal algebra in first-order theorem proving: The case of commutative rings. *Proc. 10th Workshop on Specification of Abstract Data Types*, LNCS 906, Springer, Santa Margherita, Italy, pp. 1–29.

BECKER, T. AND WEISPFENNING, V. (1993). *Gröbner Bases: A Computational Approach to Commutative Algebra*, Springer, Berlin.

BUCHBERGER, B. AND LOOS, R. (1983). Algebraic simplification. *Computer Algebra: Symbolic and Algebraic Computation*, 2nd edn, Springer, pp. 11–43.

BUCHBERGER, B. (1970). Ein algorithmisches Kriterium für die Lösbarkeit eines algebraischen Gleichungssystems. *Aequationes Mathematicae* **4**: 374–383.

BUCHBERGER, B. (1987). History and basic features of the critical pair/completion procedure. *Journal of Symbolic Computation* **3**: 3–38.

BÜNDGEN, R. (1996). Buchberger's algorithm: The term rewriter's point of view. *Theoretical Computer Science* **159**: 143–190.

CONTEJEAN, E. AND MARCHÉ, C. (1996). CiME: Completion modulo E. *Proc. 7th Int. Conf. on Rewriting Techniques and Applications*, LNCS 1103, Springer, New Brunswick, NJ, USA, pp. 18–32.

DELOR, C. AND PUEL, L. (1993). Extension of the associative path ordering to a chain of associative commutative symbols. *Proc. 5th Int. Conf. on Rewriting Techniques and Applications*, LNCS 690, Springer, pp. 389–404.

DERSHOWITZ, N. AND JOUANNAUD, J.-P. (1990). Rewrite systems. In J. van Leeuwen (ed.), *Handbook of Theoretical Computer Science: Formal Models and Semantics*, Vol. B, Elsevier/MIT Press, chapter 6, pp. 243–320.

GANZINGER, H. AND WALDMANN, U. (1996). Theorem proving in cancellative abelian monoids (extended abstract). *13th Int. Conf. on Automated Deduction*, LNAI 1104, Springer, New Brunswick, NJ, USA, pp. 388–402.

GODOY, G. AND NIEUWENHUIS, R. (2000). Paramodulation with built-in abelian groups. To appear in LICS 2000.

GREENDLINGER, M. (1960). Dehn's algorithm for the word problem. *Communications on Pure and Applied Mathematics* **13**: 67–83.

JOUANNAUD, J.-P. AND KIRCHNER, H. (1986). Completion of a set of rules modulo a set of equations. *SIAM Journal on Computing* **15**(4): 1155–1194.

KANDRI-RODY, A. AND KAPUR, D. (1988). Computing a Gröbner basis of a polynomial ideal over a Euclidean domain. *Journal of Symbolic Computation* **6**: 19–36.

LE CHENADEC, P. (1986). *Canonical Forms in Finitely Presented Algebras*, Pitman, London.

MARCHÉ, C. (1996). Normalised rewriting: an alternative to rewriting modulo a set of equations. *Journal of Symbolic Computation* **21**: 253–288.

NIEUWENHUIS, R. AND RUBIO, A. (1997). Paramodulation with built-in AC-theories and symbolic constraints. *Journal of Symbolic Computation* **23**: 1–21.

PETERSON, G. E. AND STICKEL, M. E. (1981). Complete sets of reductions for some equational theories. *Journal of the ACM* **28**(2): 233–264.

STUBER, J. (1996). Superposition theorem proving for abelian groups represented as integer modules. In H. Ganzinger (ed.), *Proc. 7th Int. Conf. on Rewriting Techniques and Applications*, LNCS 1103, Springer, New Brunswick, NJ, USA, pp. 33–47.

STUBER, J. (1997). Strong symmetrization, semi-compatibility of normalized rewriting and first-order theorem proving. In M. P. Bonacina and U. Furbach (eds), *Proc. Int. Workshop on First-Order Theorem Proving. RISC-Linz Report 97-50*, Research Institute for Symbolic Computation, Linz, Austria, pp. 125–129.

STUBER, J. (1998a). Superposition theorem proving for abelian groups represented as integer modules. *Theoretical Computer Science* **208**(1–2): 149–177.

STUBER, J. (1998b). Superposition theorem proving for commutative rings. In W. Bibel and P. H. Schmitt (eds), *Automated Deduction - A Basis for Applications. Volume III. Applications*, Kluwer, Dordrecht, The Netherlands, chapter 2, pp. 31–55.

STUBER, J. (1999a). *Superposition Theorem Proving for Commutative Algebraic Theories*, Dissertation, Technische Fakultät, Universität des Saarlandes, Saarbrücken. Submitted.

STUBER, J. (1999b). Theory path orderings. *Proc. 10th Int. Conf. on Rewriting Techniques and Applications (RTA-99)*, LNCS 1631, Springer, Trento, Italy, pp. 148–162.

VIGNERON, L. (1994). Associative-commutative deduction with constraints. *Proc. 12th Int. Conf. on Automated Deduction*, LNCS 814, Springer, Nancy, France, pp. 530–544.

WALDMANN, U. (1997). *Cancellative Abelian Monoids in Refutational Theorem Proving*, Dissertation, Universität des Saarlandes, Saarbrücken.

WALDMANN, U. (1998). Superposition for divisible torsion-free abelian groups. *Proc. 15th Int. Conf. on Automated Deduction (CADE-15)*, LNAI 1421, Springer, Townsville, Australia, pp. 144–159.

WERTZ, U. (1992). First-order theorem proving modulo equations. Technical Report MPI-I-92-216, Max-Planck-Institut für Informatik, Saarbrücken.

ZHANG, H. (1993). A case study of completion modulo distributivity and abelian groups. *Proc. 5th Int. Conf. on Rewriting Techniques and Applications*, LNCS 690, Springer, Montreal, pp. 32–46.

Right-Linear Finite Path Overlapping Term Rewriting Systems Effectively Preserve Recognizability

Toshinori Takai, Yuichi Kaji, and Hiroyuki Seki

Graduate School of Information Science
Nara Institute of Science and Technology
8916-5, Takayama, Ikoma, Nara, 630-0101, Japan
{toshin-t,kaji,seki}@is.aist-nara.ac.jp

Abstract. Right-linear finite path overlapping TRS are shown to effectively preserve recognizability. The class of right-linear finite path overlapping TRS properly includes the class of linear generalized semi-monadic TRS and the class of inverse left-linear growing TRS, which are known to effectively preserve recognizability. Approximations by inverse right-linear finite path overlapping TRS are also discussed.

1   Introduction

Much effort has been devoted to finding subclasses of TRSs which have reasonable computational power and for which important problems are decidable and, if possible, efficiently solvable. Tree automata inherit many favorable properties of finite-state automata on strings[5]. For a tree automaton M, let $L(M)$ be the set of terms accepted by M. A set T of terms is *recognizable* if there is a tree automaton M with $T = L(M)$. The class of recognizable sets is closed under boolean operations (union, intersection and complementation), and the emptiness problem is decidable for a recognizable set. If TRSs and recognizable sets of terms can be related appropriately, then the favorable properties of recognizable sets help us solve some problems in TRSs.

Two different directions for relating TRS and recognizable sets exist. One direction is the study of a TRS which effectively preserves recognizability[2, 6, 7, 8, 11, 13]. For a TRS R and a set L of terms, define $R^*(L) = \{t \mid \exists s \in L$ s.t. $s \to_R^* t\}$. A TRS R is said to *effectively preserve recognizability* if, for any tree automaton M, $R^*(L(M))$ is also recognizable and a tree automaton M_* such that $R^*(L(M)) = L(M_*)$ can be effectively constructed. Joinability, reachability and local confluence are decidable for a TRS which effectively preserves recognizability[7, 8]. Since it is undecidable whether a given TRS effectively preserves recognizability or not[6], decidable subclasses of TRSs which effectively preserve recognizability have been investigated. Such classes include ground TRS[1], right-linear monadic TRS[13], linear semi-monadic TRS[2] and linear generalized semi-monadic TRS[8]. Another direction of the study for relating TRS and recognizable sets is to find a class of TRS R such that the set

L. Bachmair (Ed.): RTA 2000, LNCS 1833, pp. 246–260, 2000.

$R^{-1*}(L(M)) = \{t \mid \exists s \in L(M) \text{ s.t. } t \to_R^* s\}$ is recognizable for any tree automaton M[4, 10, 12]. A linear growing TRS[10] has this property, and later, the result was extended to left-linear growing TRS[12]. Obviously, if a TRS R has this property, then $R^{-1} = \{l \to r \mid r \to l \in R\}$ preserves recognizability, and vice versa. A TRS R is (right-)linear semi-monadic if and only if R^{-1} is (left-)linear growing except that the variable restriction (l is not a variable and $Var(r) \subseteq Var(l)$ for each $l \to r \in R$) is dropped in [10, 12].

In this paper, a new class of TRSs, right-linear *finite path overlapping* TRS is proposed. A TRS in the class effectively preserves recognizability (Section 4), and the class properly includes known decidable classes of TRSs which effectively preserve recognizability (Section 3). Section 5 discusses approximations by inverse right-linear finite path overlapping TRS.

2   Preliminaries

2.1   Term Rewriting Systems

We use the usual notions for terms, substitutions, etc (see [3] for details). Let $\mathcal{F}$ be a finite set of *function symbols* and $\mathcal{X}$ an enumerable set of *variables*. The *arity* of $f \in \mathcal{F}$ is denoted by $a(f)$. $\mathcal{F}$ is called a *signature*. The set of $\mathcal{F}$-*terms*, or simply *terms*, defined in the usual way, is denoted by $\mathcal{T}(\mathcal{F}, \mathcal{X})$. The set of variables occurring in t is denoted by $Var(t)$. A term t is *ground* if $Var(t) = \emptyset$. The set of ground $\mathcal{F}$-terms is denoted by $\mathcal{G}(\mathcal{F})$. A term is *linear* if no variable occurs more than once in the term. A *substitution* σ is a mapping from $\mathcal{X}$ to $\mathcal{T}(\mathcal{F}, \mathcal{X})$, and written as $\sigma = \{x_1 \mapsto t_1, \ldots, x_n \mapsto t_n\}$ where t_i with $1 \leq i \leq n$ is a term which substitutes for the variable x_i. The term obtained by applying a substitution σ to a term t is written as $t\sigma$. $t\sigma$ is called an *instance* of t and t is said to *subsume* $t\sigma$. An *occurrence* (or *position*) in a term t is defined as a sequence of positive integers as usual, and the set of all the occurrences in a term t is denoted by $\mathcal{O}cc(t)$. Let λ denote the empty occurrence. If an occurrence o_1 is a prefix (resp. proper prefix) of o_2, then we write $o_1 \preceq o_2$ (resp. $o_1 \prec o_2$). Two occurrences o_1 and o_2 are *disjoint* if neither $o_1 \preceq o_2$ nor $o_2 \preceq o_1$. A subterm of t at an occurrence o is denoted by t/o. t/o is said to occur at *depth* $|o|$. The *depth* of a term t is $\max\{|o| \mid o \in \mathcal{O}cc(t)\}$. If a term t is obtained from a term t' by replacing the subterms of t' at occurrences $o_1, \ldots, o_m$ ($o_i \in \mathcal{O}cc(t')$, o_i and o_j are disjoint if $i \neq j$) with terms $t_1, \ldots, t_m$, respectively, then we write $t = t'[o_i \leftarrow t_i \mid 1 \leq i \leq m]$.

A *rewrite rule* is an ordered pair of terms, written as $l \to r$. The variable restriction ($Var(r) \subseteq Var(l)$ and l is not a variable) is not assumed unless stated otherwise. A *term rewriting system* (*TRS*) is a finite set of rewrite rules. For terms t, t' and a TRS R, we write $t \to_R t'$ if there exist an occurrence $o \in \mathcal{O}cc(t)$, a substitution σ and a rewrite rule $l \to r \in R$ such that $t/o = l\sigma$ and $t' = t[o \leftarrow r\sigma]$. Define $\to_R^*$ (resp. $\leftrightarrow_R^*$) to be the reflexive and transitive (resp. the reflexive, symmetric and transitive) closure of $\to_R$. The subscript R of $\to_R$ is omitted if R is clear from the context. A *redex* (*in* R) is an instance of l for some $l \to r \in R$. A *normal form* (in R) is a term which has no redex as its subterm. Let NF_R denote the set of all ground normal forms in R. A rewrite rule $l \to r$ is *left-linear*(resp. *right-linear*) if l is linear (resp. r is linear). A rewrite rule is *linear* if it is left-linear and right-linear. A TRS R is *left-linear* (resp. *right-linear, linear*) if every rule in R is left-linear (resp. right-linear, linear). A TRS R is *orthogonal* if R is left-linear and has no critical pair. For a TRS R, let R^{-1} denote $\{l \to r \mid r \to l \in R\}$. For a class C of TRSs, let C^{-1} denote the class of TRSs $\{R \mid R^{-1} \in C\}$.

Reachability, joinability, unifiability, unifier, most general unifier, confluence, local confluence are defined in the usual way. For a TRS R, two terms t_1 and t_2 are *R-unifiable* if there exists a substitution σ such that $t_1\sigma \leftrightarrow^*_R t_2\sigma$.

2.2   Tree Automata

A *tree automaton*[5] is defined by a 4-tuple $M = (\mathcal{F}, \mathcal{Q}, \mathcal{Q}_{final}, \Delta)$ where $\mathcal{F}$ is a signature, $\mathcal{Q}$ is a finite set of states, $\mathcal{Q}_{final} \subseteq \mathcal{Q}$ is a set of final states, and Δ is a finite set of transition rules of the form $f(q_1, \ldots, q_n) \rightharpoonup q$ where $f \in \mathcal{F}$, $a(f) = n$, and $q_1, \ldots, q_n, q \in \mathcal{Q}$ or of the form $q' \rightharpoonup q$ where $q, q' \in \mathcal{Q}$. The latter one is called an *ε-rule*. Consider the set of ground terms $\mathcal{G}(\mathcal{F} \cup \mathcal{Q})$ where $a(q) = 0$ for $q \in \mathcal{Q}$. A *move* of a tree automaton can be regarded as a rewrite relation on $\mathcal{G}(\mathcal{F} \cup \mathcal{Q})$ by regarding transition rules in Δ as rewrite rules on $\mathcal{G}(\mathcal{F} \cup \mathcal{Q})$. For terms t and t' in $\mathcal{G}(\mathcal{F} \cup \mathcal{Q})$, we write $t \vdash_M t'$ if and only if $t \to_\Delta t'$. The reflexive and transitive closure of $\vdash_M$ is denoted by $\vdash^*_M$. For a tree automaton M and $t \in \mathcal{G}(\mathcal{F})$, if $t \vdash^*_M q_f$ for a final state $q_f \in \mathcal{Q}_{final}$, then we say t is *accepted* by M. The set of ground terms accepted by M is denoted by $L(M)$. A set T of ground terms is *recognizable* if there is a tree automaton M such that $T = L(M)$. Also let $L(M(q)) = \{t \mid t \vdash^*_M q\}$ for a state q of M. Recognizable sets inherit some useful properties of regular (string) languages[5].

Lemma 1. *The class of recognizable sets is effectively closed under union, intersection and complementation. For a recognizable set L, the following problems are decidable. (1) Does a given ground term t belong to L? (2) Is L empty?* ☐

3   TRS which Preserves Recognizability

3.1   Definition and Known Results

For a TRS R and a set T of ground terms, define $R^*(T) = \{t \mid \exists s \in T \text{ s.t. } s \to^*_R t\}$. A TRS R is said to *effectively preserve recognizability* if, for any tree automaton M, the set $R^*(L(M))$ is also recognizable and we can effectively construct a tree automaton which accepts $R^*(L(M))$. In this paper, the class of TRSs which effectively preserve recognizability is written as EPR-TRS.

Theorem 1. *If a TRS R belongs to EPR-TRS, then the reachability relation and the joinability relation for R are decidable[7]. It is also decidable whether R is locally confluent or not[8].* ☐

Theorem 2. *For a confluent $R \in EPR\text{-}TRS$ and linear terms t_1 and t_2 with $Var(t_1) \cap Var(t_2) = \emptyset$, it is decidable whether t_1 and t_2 are R-unifiable or not.*

Proof. Since R is confluent, t_1 and t_2 are R-unifiable if and only if there exist a ground substitution σ and a term v such that $t_1\sigma \to_R^* v$ and $t_2\sigma \to_R^* v$. For a term t, let $I(t)$ denote the set of ground instances of t, i.e., $I(t) = \{t' \mid \exists\text{ground substitution } \sigma \text{ with } t' = t\sigma\}$. Then t_1 and t_2 are R-unifiable if and only if

$$R^*(I(t_1)) \cap R^*(I(t_2)) \neq \emptyset \tag{3.1}$$

since $Var(t_1) \cap Var(t_2) = \emptyset$. It is easy to see that $I(t)$ is recognizable for any linear term t. Thus $R^*(I(t_1))$ and $R^*(I(t_2))$ are recognizable since $R \in EPR\text{-}TRS$. By Lemma 1, the condition (3.1) is decidable. □

Unfortunately it is undecidable whether a given TRS belongs to EPR-TRS or not[6]. Therefore decidable subclasses of EPR-TRS have been proposed, for example, ground TRS by Brained[1], right-linear monadic TRS (RLM-TRS) by Salomaa[13], linear semi-monadic TRS (LSM-TRS) by Coquidé et al.[2], and linear generalized semi-monadic TRS (LGSM-TRS) by Gyenizse and Vágvölgyi[8]. Note that these papers assume the variable restriction.

Theorem 3. *ground TRS $\subset$ RLM-TRS $\subset$ EPR-TRS, and ground TRS $\subset$ LSM-TRS $\subset$ LGSM-TRS $\subset$ EPR-TRS.* □

Similar discussions can be found in [10], [4] and [12]. A TRS R (without the variable restriction) is *growing* if all variables in $Var(l) \cap Var(r)$ occur at depth 0 or 1 in l for every rewrite rule $l \to r$ in R[10]. Nagaya and Toyama[12] showed that for each left-linear growing TRS (LLG-TRS) R, R^{-1} effectively preserves recognizability. Note that if a TRS R satisfies the variable restriction then R is (linear, right-linear) semi-monadic if and only if R^{-1} is (linear, left-linear) growing and the left-hand side of every rewrite rule in R is not a constant. LLG-TRS^{-1} properly includes both of RLM-TRS and LSM-TRS and is incomparable with LGSM-TRS.

3.2   Finite Path Overlapping TRS

A new class of TRS named *finite path overlapping TRS* (*FPO-TRS*) is proposed in this section without assuming the variable restriction. As we will show in 3.3, the class of right-linear FPO-TRS properly includes the class of right-linear generalized semi-monadic TRS and LLG-TRS^{-1}. It is also shown in Section 4 that a right-linear FPO-TRS (without the variable restriction) effectively preserves recognizability. To the authors' knowledge, the proposed class is the largest decidable subclass of EPR-TRS. To define the class, some additional definitions are necessary. We say that a term s *sticks out of* t if t is not a variable and there is a variable occurrence γ ($\neq \lambda$) of t such that

1. for any occurrence o with $\lambda \preceq o \prec \gamma$, we have $o \in Occ(s)$ and the function symbol of s at o and the function symbol of t at o are the same, and

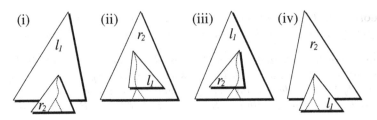

Fig. 1. The sticking out relations of rewrite rules.

2. $\gamma \in Occ(s)$ and s/γ is not a ground term.

When the occurrence γ is of interest, we say that s sticks out of t at γ. If s sticks out of t at γ and s/γ is not a variable (i.e. s/γ is a non-ground and non-variable term), then s is said to *properly stick out of* t (at γ). For example, a term $f(g(x), a)$ sticks out of $f(g(y), b)$ at the occurrence $1 \cdot 1$, and $f(g(g(x)), a)$ properly sticks out of $f(g(y), b)$ at the occurrence $1 \cdot 1$. A *finite path overlapping TRS (FPO-TRS)* is a TRS R such that the following *sticking-out graph* of R does not have a cycle of weight one or more.

Definition 1. *The sticking-out graph of a TRS R is a directed graph $G = (V, E)$ where $V = R$ (i.e. the vertices are the rewrite rules in R) and E is defined as follows. Let v_1 and v_2 be (possibly identical) vertices which correspond to rewrite rules $l_1 \rightarrow r_1$ and $l_2 \rightarrow r_2$, respectively. Replace each variable in $Var(r_i) \setminus Var(l_i)$ with a fresh constant, say $\Diamond$, for $i = 1, 2$.*

(i) If r_2 properly sticks out of a subterm of l_1, then E contains an edge from v_2 to v_1 with weight one.

(ii) If a subterm of r_2 properly sticks out of l_1, then E contains an edge from v_2 to v_1 with weight one.

(iii) If a subterm of l_1 sticks out of r_2, then E contains an edge from v_2 to v_1 with weight zero.

(iv) If l_1 sticks out of a subterm of r_2, then E contains an edge from v_2 to v_1 with weight zero.

The four cases are illustrated in Fig. 1.   □

Example 1. Let $R_1 = \{p_1 \colon f(x, a) \rightarrow f(h(y), x), \ p_2 \colon g(y) \rightarrow f(g(y), b)\}$. Fig. 2 shows the sticking-out graph of R_1. The right-hand side of p_2 properly sticks out of the left-hand side of p_1 at the occurrence 1, and hence there is an edge of weight one from p_2 to p_1. The sticking-out graph also has an self-looping edge of weight zero at p_2 since the left-hand side $g(y)$ of p_2 sticks out of $f(g(y), b)/1 = g(y)$. Since the variable y in p_1 is replaced with a constant $\Diamond$, the right-hand side of p_1 does not stick out of its left-hand side. There is no other edge since there is no other sticking-out relation between subterms of these rewrite rules. The

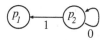

Fig. 2. The sticking-out graph of R_1.

sticking-out graph has a cycle of weight zero, but does not have a cycle of weight one or more, and hence R is finite path overlapping.

Let $R_2 = \{f(x) \rightarrow g(f(g(x)))\}$. The subterm $f(g(x))$ of the right-hand side of the (unique) rewrite rule properly sticks out of its left-hand side, as in condition (ii) of the definition of sticking-out graph. The sticking-out graph of R_2 consists of one vertex and one cycle with weight one. Therefore, R_2 is not finite path overlapping. Note that $R_2 \notin$ EPR-TRS since $R_2^*(\{f(a)\}) = \{g^n(f(g^n(a))) \mid n \geq 0\}$ is not recognizable. □

Remark that the sticking-out graph is effectively constructible for a given TRS R, and hence it is decidable whether a given TRS R is finite path overlapping or not (in $O(m^2 n^2)$ time where m is the maximum size of a term in R and n is the number of rules in R).

3.3 Inclusion Relation

Although a generalized semi-monadic TRS (GSM-TRS) was originally defined with the variable restriction in [8], we define GSM-TRS without the variable restriction to treat growing TRS, GSM-TRS and FPO-TRS in a uniform way.

A TRS R is *generalized semi-monadic* if the following condition holds for any pair of (possibly the same) rewrite rules $l_1 \rightarrow r_1$ and $l_2 \rightarrow r_2$ in R. For $i = 1, 2$, each variable in $Var(r_i) \setminus Var(l_i)$ is replaced with a fresh constant. For any occurrences $\alpha \in Occ(l_1)$ and $\beta \in Occ(r_2)$ such that $\alpha = \lambda$ or $\beta = \lambda$ and for any term l_3 which subsumes l_1/α, if r_2/β and l_3 are unifiable, then

1. l_1/α is a variable, or
2. for any $\gamma \in Occ(l_3)$ such that $l_1/\alpha \cdot \gamma$ is a variable, $(l_3/\gamma)\sigma$ is a variable or a ground term where σ is the most general unifier of r_2/β and l_3.

Lemma 2. *A TRS R is generalized semi-monadic if and only if the sticking-out graph of R has no edge with weight one.*

Proof. We show the only if part by contradiction. The if part can be shown in a similar way. Assume that R is a GSM-TRS and contains rules $l_1 \rightarrow r_1$ and $l_2 \rightarrow r_2$ (each variable in $Var(r_i) \setminus Var(l_i)$ has been replaced with a constant $\diamond$ for $i = 1, 2$) which satisfy condition (i) of the definition of sticking-out graph. In this case, there is an occurrence $\alpha \in Occ(l_1)$ such that r_2 properly sticks out of l_1/α. Let γ be the variable occurrence of l_1/α at which r_2 properly sticks out of l_1/α, then $l_1/\alpha \cdot \gamma$ is a variable and r_2/γ is a non-ground and non-variable term. Let l_3 be the term which satisfies the following conditions.

- *For an occurrence o with $\lambda \preceq o \prec \gamma$, l_3 and l_1/α have the same symbol at o,*
- *a variable, say x_o, occurs at an occurrence o such that o is disjoint to γ and o is written as $o' \cdot i$ with $o' \prec \gamma$ and*
- *a variable x_γ occurs at γ.*

It is easily understood that l_3 subsumes l_1/α and that l_3 and r_2 are unifiable by an mgu σ, which replaces x_γ by r_2/γ. Now we have $(l_3/\gamma)\sigma = r_2/\gamma$, which is neither a variable nor a ground term by assumption. This concludes that R is not a GSM-TRS. In a similar way, we can show that if any pair of rules in R satisfies condition (ii) of the definition of sticking-out graph, then R is not a GSM-TRS. □

Theorem 4. *The class of right-linear FPO-TRS properly includes the class of right-linear GSM-TRS.*

Proof. The class of right-linear FPO-TRS includes the class of right-linear GSM-TRS by Lemma 2. TRS R_1 in Example 1 is right-linear FPO but not GSM. If we take $l_1 = f(x, a)$, $r_2 = f(g(y), b)$, $\alpha = \beta = \lambda$ and $l_3 = f(x, z)$, then r_2 and l_3 are unifiable by an mgu $\sigma = \{x \mapsto g(y), z \mapsto b\}$. Let $\gamma = 1$, then $l_1/\alpha \cdot \gamma = l_1/1$ is a variable x and $(l_1/\alpha \cdot \gamma)\sigma = g(y)$. Therefore R_1 is not a GSM-TRS. □

4   A Right-Linear Finite Path Overlapping TRS Effectively Preserves Recognizability

4.1   Construction of Tree Automata

In this subsection, we will present a procedure which takes a right-linear TRS R and a tree automaton M as an input and constructs a tree automaton M_* such that $L(M_*) = R^*(L(M))$ if the procedure halts. In the next subsection, it is shown that if R is a right-linear FPO-TRS, then the procedure always halts. This concludes that right-linear FPO-TRS $\subseteq$ EPR-TRS. The procedure is an extension of the procedure to solve a semantic unification problem presented in [11]. In [11], rewrite rules are restricted so that variables appearing in the left-hand side more than once do not occur in the right-hand side. The restriction can be dropped in the following way. In the construction of M_*, a term is used as a state of the tree automaton. To deal with non-left-linear TRS, we need to construct a kind of product automata whose states are Cartesian products of terms. To represent such Cartesian products and usual first-order terms in a uniform way, we introduce a packed state. Intuitively, a packed state is an extension of a first-order term such that a finite set of terms, rather than a single term, occurs at a subterm occurrence. For a signature $\mathcal{F}$ and a finite set $\mathcal{Q}$, the set of *packed states*, denoted $\mathcal{P}_{\mathcal{F},\mathcal{Q}}$, is defined as follows.

- $\{q\} \in \mathcal{P}_{\mathcal{F},\mathcal{Q}}$ for any $q \in \mathcal{Q}$.
- If $p_1, p_2 \in \mathcal{P}_{\mathcal{F},\mathcal{Q}}$, then $p_1 \cup p_2 \in \mathcal{P}_{\mathcal{F},\mathcal{Q}}$.
- If $f \in \mathcal{F}$ and $p_1, \ldots, p_{a(f)} \in \mathcal{P}_{\mathcal{F},\mathcal{Q}}$, then $\{f(p_1, \ldots, p_{a(f)})\} \in \mathcal{P}_{\mathcal{F},\mathcal{Q}}$.

For the readability, a packed state $\{t_1, \ldots, t_n\}$ is written as $\langle t_1, \ldots, t_n \rangle$. For example, let $\mathcal{F} = \{f, g\}$ with $a(f) = 2$ and $a(g) = 1$ and $\mathcal{Q} = \{q_1, q_2\}$. We can easily verify that $\langle f(\langle q_1 \rangle, \langle q_2 \rangle), g(\langle g(\langle q_1 \rangle), \langle q_2 \rangle \rangle) \rangle$ belongs to $\mathcal{P}_{\mathcal{F}, \mathcal{Q}}$.

Procedure 1.

Input: a right-linear TRS R and a tree automaton $M = (\mathcal{F}, \mathcal{Q}, \mathcal{Q}_{final}, \Delta)$.
Output: a tree automaton M_* such that $L(M_*) = R^*(L(M))$.

Procedure 1 has a loop-structure to expand the set $\mathcal{Q}$ of states and the set Δ of transition rules of M. For a nonnegative integer k let M_k be the tree automaton which is obtained from M by executing the loop k times. We write $\vdash_k$ for $\vdash_{M_k}$. Observe that if $t \in R^*(L(M))$ and $t \rightarrow_R t'$ then $t' \in R^*(L(M))$ by the definition of $R^*(\cdot)$. Hence, if M_k accepts t, then we construct M_{k+1} so that it accepts t' by adding transition rules which simulates a rewrite step $t \rightarrow_R t'$. In general, if $t \rightarrow_R t'$ by applying a rule $l \rightarrow r$ and $t \vdash_k^* p$, where p is a state of M_k, then we add transition rules which make $t' \vdash_{k+1}^* p$ possible.

Step 1. Add a new state q_{any} to $\mathcal{Q}$ and add a transition rule $f(q_{any}, \ldots, q_{any}) \rightarrow q_{any}$ to Δ for each f in $\mathcal{F}$. Obviously, $t \vdash_M^* q_{any}$ for any $t \in \mathcal{G}(\mathcal{F})$. Let $M_0 = (\mathcal{F}, \mathcal{Q}_0, \mathcal{Q}_{final}^0, \Delta_0)$ be a "packed" version of M where $\mathcal{Q}_0 = \{\langle q \rangle \mid q \in \mathcal{Q}\} \subseteq \mathcal{P}_{\mathcal{F}, \mathcal{Q}}$, $\mathcal{Q}_{final}^0 = \{\langle q \rangle \mid q \in \mathcal{Q}_{final}\}$, and $\Delta_0 = \{f(\langle q_1 \rangle, \ldots, \langle q_n \rangle) \rightarrow \langle q \rangle \mid f(q_1, \ldots, q_n) \rightarrow q \in \Delta\} \cup \{\langle q' \rangle \rightarrow \langle q \rangle \mid q' \rightarrow q \in \Delta\}$.
Step 2. Let $k := 0$. This k is used as a loop counter of the procedure.
Step 3. Let $\mathcal{Q}_{k+1} := \mathcal{Q}_k$ and $\Delta_{k+1} := \Delta_k$.
Step 4. The set of transition rules is modified in this step. Let $l \rightarrow r$ be a rewrite rule in R. It is assumed that l has $m(\geq 0)$ variables $x_1, \ldots, x_m$ and the variable x_i has γ_i occurrences $o_i^j \in \mathcal{O}cc(l)$ $(1 \leq j \leq \gamma_i)$ in l. If there are states p and p_i^j $(1 \leq i \leq m, 1 \leq j \leq \gamma_i)$ in $\mathcal{Q}_k$ such that

$$l[o_i^j \leftarrow p_i^j \mid 1 \leq i \leq m, 1 \leq j \leq \gamma_i] \vdash_k^* p \tag{4.1}$$

and

$$L(M_k(p_i^1)) \cap \cdots \cap L(M_k(p_i^{\gamma_i})) \neq \emptyset \tag{4.2}$$

for $1 \leq i \leq m$, then add

$$p_i = \bigcup_{1 \leq j \leq \gamma_i} p_i^j \quad (1 \leq i \leq m) \tag{4.3}$$

to $\mathcal{Q}_{k+1}$ as new states, let $\rho = \{x_i \mapsto p_i \mid 1 \leq i \leq m\} \cup \{x \mapsto \langle q_{any} \rangle \mid x \in \mathcal{V}ar(r) \setminus \mathcal{V}ar(l)\}$, and do the following (a) and (b).
 (a) Add $\langle r\rho \rangle \rightarrow p$ to Δ_{k+1}. If a move of the tree automaton is caused by this rule, then the move is called a *rewriting move* of degree $k + 1$.
 (b) Execute **ADDTRANS**($\langle r\rho \rangle$). In **ADDTRANS**($\langle r\rho \rangle$), new states and transition rules are defined so that $r\rho \vdash_{k+1}^* \langle r\rho \rangle$.
Simultaneously execute this Step 4 for every rewrite rule and every tuple of states that satisfy conditions (4.1) and (4.2).
Step 5. Continue the loop until $\Delta_{k+1} = \Delta_k$. If $\Delta_{k+1} \neq \Delta_k$, then $k := k + 1$ and go to Step 3.

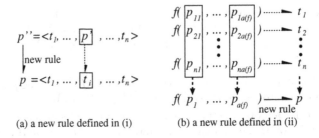

$p'' = <t_1, \ldots, \boxed{p}, \ldots, t_n>$

$\Big\vert$ new rule

$p = <t_1, \ldots, \boxed{t_i}, \ldots, t_n>$

(a) a new rule defined in (i)

(b) a new rule defined in (ii)

Fig. 3. The new rules introduced by **ADDTRANS**.

Step 6. Output M_k as M_*.   □

Procedure 2. [ADDTRANS] This procedure takes a packed state p as an input. If p has already been defined as a state, then the procedure defines no transitions. Otherwise, the procedure first defines p as a new state of $\mathcal{Q}_{k+1}$ and also defines transition rules as follows. It is required that if $p = \langle t_1, \ldots, t_n \rangle$ $(n \geq 2)$, then each $\langle t_i \rangle$ has been defined as a state.

Case 1. If $p = \langle c \rangle$ with c a constant, then define $c \rightharpoonup \langle c \rangle$ as a transition rule.

Case 2. If $p = \langle f(p_1, \ldots, p_{a(f)}) \rangle$ with $f \in \mathcal{F}$, then define $f(p_1, \ldots, p_{a(f)}) \rightharpoonup p$ as a transition rule and execute **ADDTRANS**(p_i) for $1 \leq i \leq a(f)$.

Case 3. If $p = \langle t_1, \ldots, t_n \rangle$ $(n \geq 2)$, then do the following (i) and (ii).

 (i) For each transition rule of the form $p' \rightharpoonup \langle t_i \rangle$ $(p' \in \mathcal{Q}_k, 1 \leq i \leq n)$, define a new ε-rule $p'' \rightharpoonup p$ and execute **ADDTRANS**(p'') where p'' is the state defined as $p'' = (p \setminus \langle t_i \rangle) \cup p'$ (see Fig. 3(a)).

 (ii) If there is a function symbol f and, for each i with $1 \leq i \leq n$, there are states p_{ij} with $1 \leq j \leq a(f)$ such that $f(p_{i1}, \ldots, p_{ia(f)}) \rightharpoonup \langle t_i \rangle$, then define a new rule $f(p_1, \ldots, p_{a(f)}) \rightharpoonup p$ and execute **ADDTRANS**(p_j) for $1 \leq j \leq a(f)$ where $p_j = p_{1j} \cup \cdots \cup p_{nj}$ (see Fig. 3(b)).   □

Example 2. Let $M = (\mathcal{F}, \mathcal{Q}, \mathcal{Q}_{final}, \Delta)$ be a tree automaton where $\mathcal{F} = \{f, g, h, c\}$ with $a(f) = 2$, $a(g) = a(h) = 1$ and $a(c) = 0$, $\mathcal{Q} = \{q_0, q_1, q_2\}$, $\mathcal{Q}_{final} = \{q_2\}$ and Δ consists of the following transition rules:

$$c \rightharpoonup q_0, \quad h(q_0) \rightharpoonup q_1, h(q_1) \rightharpoonup q_0,$$
$$f(q_0, q_0) \rightharpoonup q_2, f(q_1, q_0) \rightharpoonup q_2.$$

It can be easily verified that $L(M) = \{f(h^m(c), h^{2n}(c)) \mid m, n \geq 0\}$. Let $R = \{f(x, x) \to g(x), g(x) \to x\}$. R is a right-linear FPO-TRS. We apply Procedure 1 to M and R. Consider the rewrite rule $f(x, x) \to g(x)$ in Step 4 for $M_0(k = 0)$. Since a move $f(\langle q_0 \rangle, \langle q_0 \rangle) \vdash_0 \langle q_2 \rangle$ is possible, new transition rules

$$\langle g(\langle q_0 \rangle) \rangle \rightharpoonup \langle q_2 \rangle \text{ and } g(\langle q_0 \rangle) \rightharpoonup \langle g(\langle q_0 \rangle) \rangle$$

are added to Δ_1. The latter rule is added in Case 2 of **ADDTRANS**$(\langle g(\langle q_0 \rangle) \rangle)$. Next, consider the rewrite rule $g(x) \to x$ in Step 4 for M_1 ($k = 1$). Since $g(\langle q_0 \rangle) \vdash_1 \langle g(\langle q_0 \rangle) \rangle \vdash_1 \langle q_2 \rangle$, $\langle q_0 \rangle \dashrightarrow \langle q_2 \rangle$ is added to Δ_2. Thus we obtain $h(h(c)) \vdash_2^* \langle q_0 \rangle \vdash_2 \langle q_2 \rangle \in \mathcal{Q}_{final}^0$ and hence $h(h(c)) \in L(M_2)$. We can verify that $M_3 = M_2$ ($= M_*$) and $L(M_*) = R^*(L(M)) = \{g(h^{2n}(c)) \mid n \geq 0\} \cup \{h^{2n}(c) \mid n \geq 0\} \cup L(M)$. □

The following lemma states a basic property of a packed state, which is used for the proof of Lemma 4.

Lemma 3. *Let* $M_k = (\mathcal{F}, \mathcal{Q}_k, \mathcal{Q}_{final}^0, \Delta_k)$ *be the tree automaton constructed in Procedure 1* ($k \geq 1$). *Then, for each state* $\langle t_1, \ldots, t_n \rangle \in \mathcal{Q}_k$, $L(M_k(\langle t_1, \ldots, t_n \rangle))$ $= L(M_k(\langle t_1 \rangle)) \cap \cdots \cap L(M_k(\langle t_n \rangle))$. □

From the construction of the tree automata in Procedure 1, the inclusion hierarchy $L(M) = L(M_0) \subseteq L(M_1) \subseteq \cdots$ holds. Procedure 1 has the following soundness and completeness property. See [14] for the detailed proof.

Lemma 4. *For a ground term* s, $s \in R^*(L(M))$ *if and only if there is an integer* k *such that* $s \in L(M_k)$.

Proof. (If part) The following claim can be shown.

> **Claim A** *For a ground term* s, *and states* $p, p' \in \mathcal{Q}_k$, *if there is a sequence of moves* $s \vdash_k^* p' \vdash_k p$ *such that*
> *(i)* $p' \vdash_k p$ *is a rewriting move at the root occurrence, and*
> *(ii) there is no rewriting move in* $s \vdash_k^* p'$,
> *then there is a term* s' *such that* $s' \to_R s$ *and* $s' \vdash_{k-1}^* p$.

By claim A, we can show by induction on k that for a ground term s and a state p, if $s \vdash_k^* p$, then there is a ground term u such that $u \to_R^* s$ and $u \vdash_0^* p$. If the state p is in $\mathcal{Q}_{final}$, then this implies that if $s \in L(M_k)$ then $s \in R^*(L(M))$.

(Only if part) It can be shown that if $s' \to_R^* s$ with $s' \in L(M_0)$, then there is an integer k such that $s \in L(M_k)$ by induction on the length of the derivation $s' \to_R^* s$. □

The following theorem is obtained directly from Lemma 4.

Theorem 5. *For a right-linear TRS R, if Procedure 1 halts then* $L(M_*) = R^*(L(M))$. □

4.2 Termination of Procedure 1

We show that if a right-linear FPO-TRS is given to Procedure 1, then there is an upper-bound limit on the number of states which are newly defined. Once the set of states saturates, then the set of transition rules also saturates and the procedure halts. First, as a measure of the size of a state, we introduce the concept of the *layer* of a packed state. Intuitively, the number of layers of a packed state is the number of right-hand sides of rewrite rules which are used for defining the state. For a packed state $p \in \mathcal{Q}_k$, define the number of *layers* of p, denoted layer(p), as follows; (1) if $p \in \mathcal{Q}_0$ or $p = \langle t \rangle$ with t a ground subterm of a rewrite rule in R, then layer(p) = 0, (2) if $p = p_1 \cup p_2$, then layer(p) = max{layer(p_1), layer(p_2)}, and (3) if $p = \langle (r/o)\sigma \rangle$ with $l \to r \in R$, $o \in \mathcal{O}cc(r)$, $\mathcal{V}ar(r/o) = \{x_1, \ldots, x_n\}$ and $\sigma = \{x_i \mapsto p_i \mid 1 \leq i \leq n\}$, then layer($p$) = 1 + max{layer($p_i$) | $1 \leq i \leq n$}. Remark that layer(p) is not defined for all packed states, but all packed states in Procedure 1 are of the form (1), (2) or (3). Also remark that layer(p) is not always uniquely determined by this definition. If different values are defined as layer(p), then we choose the minimum among the values as layer(p). We note that if $x_i \in \mathcal{V}ar(r) \setminus \mathcal{V}ar(l)$, then $p_i = \langle q_{any} \rangle$ and layer(p_i) = 0. This means that variables which occurs in the right-hand side only are ignored for defining the number of layers.

Example 3. Consider the states of the tree automata in Example 2. Let $l \to r = f(x,x) \to g(x) \in R$, $o = \lambda$ and $\sigma = \{x \mapsto \langle q_0 \rangle\}$ in the above definition (3). Then, $p = \langle (r/o)\sigma \rangle = \langle g(\langle q_0 \rangle) \rangle$ and layer(p) = layer($\langle q_0 \rangle$) + 1 = 1.   □

Lemma 5. *For any non-negative integer j, the number of packed states which have j or less layers is finite.*
Sketch of proof. The lemma can be shown by induction on j.   □

In the following, it is shown that if R is a right-linear FPO-TRS, then layer(p) $\leq |R|$ for any state p defined by Procedure 1 where $|R|$ is the number of rewrite rules in R. An outline of the proof is as follows. First we associate each rule in R with a non-negative integer called a *rank*. If R is finite path overlapping, then the rank is well-defined and is less than $|R|$. Next, it is shown that if a rule with rank j is used in Step 4 of Procedure 1, then layer(p) $\leq j+1$ for any state p defined in the same step. The *rank* of a rule in R is defined based on the sticking-out graph $G = (V, E)$ of R. Let v be the vertex of G which corresponds to a rewrite rule $l \to r$ in R. The *rank* of $l \to r$ is the maximum weight of a path to v from any vertex in V. If R is finite path overlapping, then the rank of any rewrite rule is a non-negative integer less than $|R|$. For R_1 in Example 1, the ranks of p_1 and p_2 are one and zero, respectively, since there is an edge with weight one from p_2 to p_1.

Lemma 6. *Let $l \to r$ be a rewrite rule and $\rho = \{x_i \mapsto p_i \mid 1 \leq i \leq m\} \cup \{x \mapsto \langle q_{any} \rangle \mid x \in \mathcal{V}ar(r) \setminus \mathcal{V}ar(l)\}$ be a substitution which are used in Step 4 of Procedure 1. If the rank of $l \to r$ is j, then layer(p_i) $\leq j$ for each $1 \leq i \leq m$.* □

Before presenting a proof of the lemma, we first see how the number of layers of the state changes by a move of the tree automaton. A move of the tree automaton is either an ε-move or a non-ε-move. Transition rules used for ε-moves are ε-transition rules of the original tree automaton M, or rules for rewriting moves defined in Step 4(a) of Procedure 1, or the rules defined in Case 3(i) of **ADDTRANS** procedure. In all three cases, it can be shown that the number of layers in a state which is associated with the head does not increase. Transition rules used for non-ε-moves are non-ε-transition rules of M, or the rules defined

number of layers

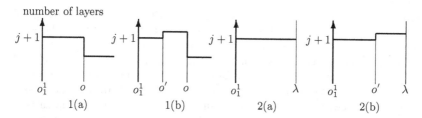

Fig. 4. The number of layers of a state of M_k in the sequence (4.1).

in Cases 1, 2 or 3(ii) of **ADDTRANS**. In all cases, the number of layers in a state is increased by one or not changed by the move.

Proof of Lemma 6. The proof is by induction on the loop variable k of Procedure 1. When $k = 0$, every state belongs to Q_0 and layer(p_i) = 0 for $1 \leq i \leq n$, and the lemma holds for any j. Assume that the lemma holds for $k \leq n - 1$, and consider the case with $k = n$. The inductive part is shown by contradiction. Without loss of generality, let p_1 be a state such that layer(p_1) $\geq j + 1$. Since $p_1 = \bigcup_{1 \leq l \leq \gamma_1} p_1^l$, we can assume p_1^1 is the state such that layer(p_1^1) $\geq j + 1$ without loss of generality. Let us observe how the number of layers of the state changes as the head of M_k moves from o_1^1 to the root in the sequence (4.1) of moves. There are four different cases.

1. The number of layers decreases at a certain occurrence. Let o be the innermost occurrence among such occurrences. There are two different subcases:
   (a) The number of layers does not increase at any o' with $o \prec o' \prec o_1^1$.
   (b) There is an occurrence o' with $o \prec o' \prec o_1^1$ such that the number of layers increases at o'.
2. The number of layers does not decrease at all. There are two subcases:
   (a) The number of layers does not increase at any o' with $\lambda \prec o' \prec o_1^1$.
   (b) There is an occurrence o' with $\lambda \prec o' \prec o_1^1$ such that the number of layers increases at o'.

These four cases are illustrated in Fig. 4.

Assume that the number of layers changes as in case 1(a) above. In this case we can derive a contradiction as follows (See [15] for the precise and more formal proof). From the observation before this proof, we know that the number of layers decreases at o only if an ε-move occurs at o. We can furthermore show that this ε-move is caused by a transition rule for rewriting moves. Let $l' \to r'$ be the rewrite rule used for defining this transition rule in Step 4 of Procedure 1. Then, the state just before the ε-move occurs at o can be written as $\langle r' \rho' \rangle$. Remark that layer($\langle r' \rho' \rangle$) = layer(p_1^1) $\geq j + 1$ since the number of layers changes as in case 1(a). This implies that the substitution ρ' replaces a variable in r' with a state which has j or more layers. Therefore, by using the inductive hypothesis, the rule $l' \to r'$ must have rank j or more. On the other hand, the fact that the number of layers does not increase at o' with $o \prec o' \prec o_1^1$ implies that r'

properly sticks out of l/o as in condition (i) of the definition of sticking-out graph (Definition 1). Hence, the rank of $l \to r$ must be larger than the rank of $l' \to r'$, and consequently must be $j + 1$ or more, a contradiction.

For case 2(a), it can be shown that there is a rewrite rule $l' \to r'$ with rank j or more, and a subterm of r' sticks out of l as in condition (ii) of the definition of the sticking-out graph. For cases 1(b) and 2(b), it can be shown that there is a rewrite rule $l' \to r'$ with rank $j + 1$ or more, and the rule satisfies conditions (iii) and (iv) of the definition of sticking-out graph, respectively. In either case, a contradiction is derived. Hence, the inductive part is shown and the proof completes.   □

For a right-linear FPO-TRS R, the rank of every rule is less than $|R|$ and hence the number of layers of any packed state is $|R|$ or less by Lemma 6. By Lemma 5, the number of packed states is finite and the following theorem holds.

Theorem 6. *Procedure 1 halts for a right-linear FPO-TRS.*   □

In general, the running time of Procedure 1 is exponential to both of the size of a TRS R and the size of a tree automaton M.

Corollary 1. *LLG-TRS^{-1} $\subset$ RLGSM-TRS $\subset$ RLFPO-TRS $\subset$ EPR-TRS, where RLGSM-TRS and RLFPO-TRS are the classes of the right-linear GSM-TRS and the right-linear FPO-TRS, respectively.*

Proof. LLG-TRS^{-1} $\subset$ RLGSM-TRS can easily be shown by definition. RLGSM-TRS $\subset$ RLFPO-TRS is by Theorem 4. RLFPO-TRS $\subset$ EPR-TRS is by Theorems 5 and 6.   □

5   Decidable Approximations

In this section, we investigate decidable approximations of TRS along the lines of [4, 10, 12]. A TRS R' is an approximation of a TRS R if $\to_R^* \subseteq \to_{R'}^*$ and $\mathrm{NF}_R = \mathrm{NF}_{R'}$. An approximation mapping α is a mapping from TRSs to TRSs such that $\alpha(R)$ is an approximation of R for any TRS R. For a class C of TRSs, a C approximation mapping is an approximation mapping such that $\alpha(R) \in C$ for every TRS R.

Jacquemard[10] introduced a linear growing approximation mapping. Later Nagaya and Toyama[12] introduced a better approximation called a left-linear growing approximation mapping and presented decidable results on them. An RLFPO-TRS^{-1} approximation mapping α is such that for a TRS R, α replaces some variables in the right-hand side r_2 of a rewrite rule $l_2 \to r_2$ in R^{-1} with a new variable which is not in $Var(l_2)$, so that r_2 cannot contribute to an edge in the sticking-out graph of $\alpha(R^{-1})$. For example, replacing variable x with x' in the right-hand side of the rule in R_2 of Example 1 yields an RLFPO-TRS^{-1} approximation of R_2^{-1}. The following results are a generalization of [12].

Let α be an approximation mapping and Ω be a fresh constant. A redex at an occurrence p in $t \in \mathcal{G}(\mathcal{F})$ is α-needed if there exists no $s \in \mathrm{NF}_R$ such that $t[p \leftarrow \Omega] \rightarrow^*_{\alpha(R)} s$ and s contains no Ω. This definition is due to [4]. If R is orthogonal, then every α-needed redex is a needed redex in the sense of Huet and Lévy [9]. Let CBN-NF$_\alpha = \{R \mid$ every term $t \notin$ NF$_R$ has an α-needed redex $\}$. By Theorems 15 and 29 in [4] and Lemma 1 of this paper, the following theorem holds.

Theorem 7. *Let R be a left-linear TRS and α be an EPR-TRS^{-1} approximation mapping. Then the following problems are decidable. (1) Is a given redex in a given term α-needed? (2) Is R in CBN-NF$_\alpha$?* □

Corollary 2. *Let R be an orthogonal TRS in EPR-TRS^{-1} which satisfies the variable restriction such that l is not a variable and $Var(r) \subseteq Var(l)$ for every $l \rightarrow r \in R$.*

(1) Every term $t \notin$ NF$_R$ has a needed redex.
(2) It is decidable whether a given redex in a given term is needed. □

To conclude this section, we provide an orthogonal TRS R in FPO-TRS^{-1} such that there exists no left-linear growing approximation mapping β which satisfies $R \in$ CBN-NF$_\beta$.

Example 4. Let $R = \{g(h(x)) \rightarrow f(x, x, x)\} \cup R'$ be an orthogonal TRS where R' consists of the following five rewrite rules:

$$f(a, b, x) \rightarrow a, f(b, x, a) \rightarrow a, f(x, a, b) \rightarrow a,$$
$$f(a, a, a) \rightarrow a, f(b, b, b) \rightarrow b.$$

It can be easily verified that R is in FPO-TRS^{-1}. Every term $t \notin$ NF$_R$ has a needed redex in R by Corollary 1 and Corollary 2 (1). On the other hand, a left-linear growing approximation mapping β should be $\beta(R) = \{g(h(y)) \rightarrow f(x, x, x)\} \cup R'$ for some variable $y \neq x$. Consider a term $t = f(g(h(a)), g(h(a)), g(h(a)))$. Obviously, $g(h(a)) \rightarrow^*_{\beta(R)} a$ and $g(h(a)) \rightarrow^*_{\beta(R)} b$. Hence, t has no β-needed redex. Thus, $R \notin$ CBN-NF$_\beta$. □

6 Conclusion

A new class of TRS named finite path overlapping TRS (RLFPO-TRS) is proposed. It is shown that an RLFPO-TRS effectively preserves recognizability, and that the class properly includes known decidable classes of TRSs which effectively preserve recognizability. Approximations by the proposed class are also discussed. RLFPO-TRS does not include simple EPR-TRSs such that $R = \{f(x) \rightarrow f(f(x))\}$. To construct a tree automaton M_* which accepts $R^*(L(M))$ for a given tree automaton M, we might need an operation which "merges" equivalent states of a tree automaton.

Acknowledgments

The authors express their thanks to Mr. Kouji Kitaoka for his significant contribution in the early stage of this research. They also would like to thank the anonymous referees for their carefully reading the paper and giving valuable comments, and Prof. Dee A. Worman for her reading and editing the paper.

References

[1] Brained, W.S.: "Tree generating regular systems," Inform. and Control, **14**, pp. 217–231, 1969.

[2] Coquidé, J.L., Dauchet, M., Gilleron, R. and Vágvölgyi, S.: "Bottom-up tree pushdown automata: classification and connection with rewrite systems," Theoretical Computer Science, **127**, pp. 69–98, 1994.

[3] Dershowitz, N. and Jouannaud, J.-P.: "Rewrite Systems," *Handbook of Theoretical Computer Science, Vol. B, Formal Models and Semantics*, pp. 243–320, Elsevier Science Publishers, 1990.

[4] Durand, I. and Middeldorp, A.: "Decidable call by need computations in term rewriting (extended abstract)," Proc. of CADE-14, North Queensland, Australia, LNAI **1249**, pp. 4–18, 1997.

[5] Gécseq, F. and Steinby, M.: *Tree Automata*, Académiai Kiadó, 1984.

[6] Gilleron, R.: "Decision problems for term rewriting systems and recognizable tree languages," Proc. of STACS'91, Hamburg, Germany, LNCS **480**, pp. 148–159, 1991.

[7] Gilleron, R. and Tison, S.: "Regular tree languages and rewrite systems," Fundamenta Informaticae, **24**, pp. 157–175, 1995.

[8] Gyenizse, P. and Vágvölgyi, S.: "Linear generalized semi-monadic rewrite systems effectively preserve recognizability," Theoretical Computer Science, **194**, pp. 87–122, 1998.

[9] Huet, G. and Lévy, J.-J.: "Computations in Orthogonal Rewriting Systems," Chapters I and II, in *Computational Logic, Essays in Honor of Alan Robinson*, MIT Press, pp. 396–443, 1991.

[10] Jacquemard, F.: "Decidable approximations of term rewriting systems," Proc. of RTA96, New Brunswick, NJ, LNCS **1103**, pp. 362–376, 1996.

[11] Kaji, Y., Fujiwara, T. and Kasami, T.: "Solving a unification problem under constrained substitutions using tree automata," J. of Symbolic Computation, **23**, 1, pp. 79–117, 1997.

[12] Nagaya, T. and Toyama, Y.: "Decidability for left-linear growing term rewriting systems," Proc. of RTA99, Trento, Italy, LNCS **1631**, pp. 256–270, 1999.

[13] Salomaa, K.: "Deterministic tree pushdown automata and monadic tree rewriting systems," J. Comput. System Sci., **37**, pp. 367–394, 1988.

[14] Takai, T., Kaji, Y., Tanaka, T. and Seki, H.: "A procedure for solving an order-sorted unification problem – extension for left-nonlinear system," *Technical Report of NAIST*, NAIST-IS-TR98011, 1998. available at http://www.aist-nara.ac.jp/

[15] Takai, T., Kaji, Y. and Seki, H.: "A sufficient condition for the termination of the procedure for solving an order-sorted unification problem," *Technical Report of NAIST*, NAIST-IS-TR99010, 1999. available at http://www.aist-nara.ac.jp/

System Description:
The Dependency Pair Method

Thomas Arts

Computer Science Laboratory, Ericsson Utvecklings AB
Box 1505, 125 25 Älvsjö, Sweden
thomas@cslab.ericsson.se
http://www.ericsson.se/cslab/~thomas

1 Introduction

The dependency pair method refers to the approach for proving (innermost) termination of term rewriting systems by showing that no infinite chain of so-called dependency pairs exists (for an overview of the method see [AG00]). The method generates inequalities that should be satisfied by a suitable well-founded ordering. Well-known techniques for searching simplification orderings (such as path orderings or polynomial interpretations) may be used to find such an ordering. The key point is that, even if the TRS is not simply terminating, the dependency pair method often generates a set of inequalities that can be satisfied by a simplification ordering and herewith can prove termination of the TRS.

This paper describes a tool that implements the dependency pair approach with its most recent additions and refinements, such as modularity results that can effectively be used on larger TRSs [AG98] and operations on the dependency pairs such as narrowing, rewriting and instantiation of pairs. These refinements all increase the power of the method and turned out to be useful for larger examples from a verification case study [GA00].

The tool is described from the user's point of view via the interface (Sect. 2) and from an implementor's point of view via the strategies that are used to deal with the occurring complexity problems (Sect. 3). In order to demonstrate the concept, some standard solutions for finding an ordering satisfying the generated inequalities (e.g. lexicographic path ordering [DF85, BN98]) are implemented as well, but the open architecture of the tool encourages that these inequalities are solved by separate programs.

2 Interface

From the perspective of the user, the tool consists of three different layers. First, the user has the possibility to load a certain TRS, and to indicate whether to prove termination or innermost termination (Fig. 1). Another alternative is to prove termination by proving innermost termination, but that alternative is only valid when the TRS is of a certain form, such as being non-overlapping [Gra95]. After having chosen between the two kinds of termination, the dependency pairs

L. Bachmair (Ed.): RTA 2000, LNCS 1833, pp. 261–264, 2000.

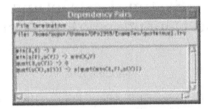

Fig. 1. Loading a TRS and choosing the kind of termination to be proved.

are displayed and if innermost termination is the goal, the usable rules are computed (Fig. 2). Now the user may choose to manipulate the dependency pairs (as far as the method allows this), by rewriting, narrowing or instantiating them one or many times. Consecutively, one can compute the dependency graph and its cycles, which results in a new window for every cycle, in which the same procedure can be repeated. Eventually one can activate the third layer by gen-

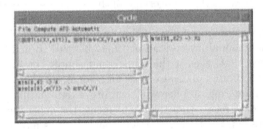

Fig. 2. Dealing with one cycle in the graph, its usable rules and the chosen rule(s) for the AFS.

erating the inequalities, which results in several windows, all containing a set of inequalities. Solving one such set, for example by the lexicographic path ordering, is equivalent with proving that for the originating cycle no infinite chain can exist. At the moment of writing, the tool supports the lexicographic path ordering in two variations and can use the POLO system [G95] for polynomial interpretations. Implementations of other standard orderings can be accessed via the 'export' function of the tool (writing the inequalities in a file). After exporting and solving the inequalities remotely, the user may claim the inequalities to be solved.

In order to use a weakly monotonic variation of the ordering solving the inequalities, before generating them one may use an argument filtering TRS (AFS) to eliminate certain arguments from function symbols.

When all the cycles have been shown to correspond to only finite chains of dependency pairs, the proof is finished and can be saved in a file. Alternatively to the manual approach, one can start a strategy which tries to perform the proof steps automatically.

3   Strategies

The tool allows the implementation of arbitrary strategies and supports at the moment one fully automatic strategy. The strategy can be seen as a standard way of pressing the buttons of the tool. As such, the strategy can be invoked within any cycle, starting to apply the strategy to the dependency pairs in that cycle. By evaluating the strategy, basically the following happens:

- It is checked whether the inequalities resulting from the dependency pairs and rules can be solved by any of the implemented orderings (without regarding additional AFS rules). If several sets of inequalities are generated, they are examined in parallel.
- All pairs that can be rewritten, are rewritten a fixed number of times. All pairs that can be narrowed are narrowed (but only one step) and if a possible instantiation of a pair makes the graph change, such an instantiation is performed. Hereafter the connected components of the cycle are re-computed and it is checked whether a cycle/component remains at this point.
- If the re-computation transforms one component into several new ones, the strategy is repeated for every component (this is performed in parallel).
- If the re-computation results in only one connected component, then all AFSs are generated and all sets of inequalities w.r.t. these AFSs are examined by the implemented orderings until one of the sets is satisfiable.
- If none of the sets is satisfiable, the cycles of the component are computed and the strategy is recursively applied to all these cycles.

The aim of this strategy is too postpone the most costly computations as long as possible.

The complexity of the application is determined by the exponential growth of sets of inequalities when we consider all argument filtering TRSs and the choice of one strict inequality in every set of generated inequalities. For every cycle in the graph we have to check whether a weakly monotonic ordering exists that satisfies a certain set of inequalities. Such a set is constructed by demanding left-hand sides of both dependency pairs and (usable) rules to be greater or equal to the corresponding right-hand sides. Only for one of the dependency pairs the inequality needs to be strict. Therefore, as many sets as dependency pairs are generated, where in each set a different dependency pair is chosen to correspond to a strict inequality. For every set, all possible AFSs are generated, the inequalities are normalized with respect to one AFS and, for example, the lexicographic path ordering is applied to the normalized inequalities.

In order to implement an efficient strategy, instead of creating all the cycles, the tool is more conservative and restricts to computing the connected components of the graph; such a component can consist of many cycles. Generating inequalities from a component is, pragmatically, performed by choosing all dependency pair related inequalities strict. Only when the connected component cannot be divided after operations on the dependency pairs, nor can the inequalities resulting from applying all possible AFSs be solved, it is decided to compute the cycles within the component.

The strategy is not guaranteed to terminate, but a time-out mechanism is provided and in addition the user can manually stop the evaluation of the strategy by pressing a button provided by the interface.

4   Conclusion

This implementation of the dependency pair method is a useful tool for those that try to use the method on examples that require many manipulations of the dependency pairs. In manual 'mode' one has the advantage of an accurate and quick experimentation technique. Finding the proofs automatically works fine for smaller examples and parts of other examples, but more research is necessary for finding a better way to check satisfiability by weakly monotonic orderings.

For a statement about performance a measure on number of lines in a TRS or number of function symbols times their arity is insufficient, since it does not relate to the number of dependency pairs, cycles, necessary narrowing steps and such. By unavailability of a satisfactory measure, only examples may demonstrate the usefulness of the application. Compared to techniques that search for simplification orderings compatible with the TRS, hardly any overhead is generated by using such techniques in the setting of the dependency pair method, whereas the former techniques become successfully applicable to many more TRSs. Compared to the prototype in Cime 2.0 [Cime], the application described in this paper contains the latest refinements of the dependency pair approach and is more user-friendly.

The tool is available at `http://www.ericsson.se/cslab/~thomas/deppairs`.

References

[AG98]   T. Arts and J. Giesl. Modularity of termination using dependency pairs. In *Proc. of RTA-98*, *LNCS* 1379, pages 226–240, Tsukuba, Japan, March/April 1998. Springer Verlag, Berlin.

[AG00]   T. Arts and J. Giesl. Termination of term rewriting using dependency pairs. *Theoretical Computer Science* 236:133-178, 2000.

[BN98]   F. Baader and T. Nipkow. Term Rewriting and All That. Cambridge University Press, 1998.

[Cime]   CiME 2 (1999). Pre-release at `http://www.lri.fr/~demons/cime-2.0.html`.

[DF85]   D. Detlefs and R. Forgaard. A Procedure for Automatically Proving Termination of a Set of Rewrite Rules. In *Proc. of RTA-85*, *LNCS* 202, pages 255–270, Dijon, France, May 1985, Springer Verlag, Berlin.

[G95]   J. Giesl. Generating polynomial orderings for termination proofs. In *Proc. of RTA-95*, *LNCS* 914, pages 426–431, Kaiserslautern, Germany, 1995. Springer Verlag, Berlin.

[GA00]   J. Giesl and T. Arts. Verification of Erlang Processes by Dependency Pairs *Applicable Algebra in Engineering, Communication, and Computing*, to appear. Preliminary version appeared as Technical Report IBN 00/52, Darmstadt University of Technology, Germany.
`http://www.inferenzsysteme.informatik.tu-darmstadt.de/~reports/notes/ibn-00-52.ps`.

[Gra95]   B. Gramlich. Abstract relations between restricted termination and confluence properties of rewrite systems. *Fundamenta Informaticae*, 24:3–23, 1995.

REM (Reduce Elan Machine): Core of the New ELAN Compiler

Pierre-Etienne Moreau

INRIA & Bouygues Telecom (France)
`moreau@loria.fr`

1 Introduction

ELAN is a powerful language and environment for specifying and prototyping deduction systems in a language based on rewrite rules controlled by strategies. It offers a natural and simple logical framework for the combination of the computation and deduction paradigms. It supports the design of theorem provers, logic programming languages, constraint solvers and decision procedures.

ELAN takes from functional programming the concept of abstract data types and the function evaluation principle based on rewriting. In ELAN, a program is a set of labelled conditional rewrite rules $\ell : l \Rightarrow r$ if c. Informally, rewriting a ground term t consists of selecting a rule whose left-hand side (also called pattern) matches the current term (t), or a subterm (t_ω), computing a substitution σ that gives the instantiation of rule variables ($l\sigma = t_\omega$), and applying it to the right-hand side to build the reduced term (when instantiated conditions are satisfied). In general the normalisation of a term may not terminate, or terminate with different results corresponding to different selected rules, selected sub-terms and non-unicity of the substitution σ (in Associative and Commutative (AC) theories for example). So, evaluation by rewriting is essentially non-deterministic and backtracking may be needed to generate all results.

One of the main originalities of the ELAN language is to provide AC operators allowing a simpler and more concise specification, and a strategy language allowing the programmer to specify the control on rule applications. This is in contrast to many existing rewriting-based languages where the term reduction strategy is hard-wired and not accessible to the designer of an application.

The strategy language offers primitives for sequential composition, iteration, deterministic and non-deterministic choices of elementary strategies that are labelled rules. From these primitives, more complex strategies can be expressed. In addition, the user can introduce new strategy operators and define them by rewrite rules. Evaluation of strategy application is itself based on rewriting. Moreover, it should be emphasised that ELAN has logical foundations based on rewriting logic [4] and detailed in [1]. So the simple and well-known paradigm of rewriting provides both the evaluation mechanism of the language and the logical framework in which deduction systems can be expressed and combined.

L. Bachmair (Ed.): RTA 2000, LNCS 1833, pp. 265–269, 2000.
© Springer-Verlag Berlin Heidelberg 2000

The full **ELAN** system (interpreter, compiler, standard library, examples, applications and documentation) is available through the **ELAN** web page[1].

2   Main Design

The **ELAN** system is built around the **ELAN** front-end and the **ELAN** compiler. The front-end accepts parametrized modules with user defined mix-fix syntax. The mix-fix parser consists of a lex/yacc based parser for the fixed syntax and a context free parser based on the Earley algorithm. The front-end also contains a powerful preprocessor able to automatically generate signatures, rules and strategies. The corresponding intermediate representation is then transformed and optimised (common subexpressions are factorised, complex strategy constructions are translated into rewrite rules and basic strategy operators and then specialised by partial evaluation,*etc.*). This "normalised" representation is called Reduce Elan Format (REF).

In the design of the new **ELAN** compiler, we have attempted to make it as system independent as possible by modularizing its main components: the Reduce Elan Machine (**REM**), written in **Java**, reads a REF program and generates **C** modules; the non-deterministic library deals with choice points and backtracking; the runtime library defines term representation, memory management, internal data structure (bitmask, bipartite graph, Diophantine equation, *etc.*), and built-in operators (term comparison, string, identifier, boolean and integer public interfaces, *etc.*).

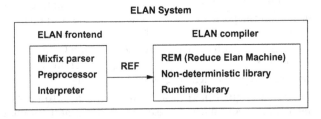

The key point is that the **REM** Compiler is completely independent of the **ELAN** syntax and is reusable with different systems and formalisms. Basically, a REF program is a text based easily parsable prefix format which contains the mix-fix signature definition and some associated properties (built-in, AC, *etc.*), a list of conditional rewrite rules and a list of strategies. The only real dependence between those components is that an operator defined as built-in in the REF program should be implemented in a module of the runtime library. To illustrate this aspect, we successfully used the **REM** Compiler to make executable an ASF+SDF [3] specification. The only thing to do was to translate an ASF program to REF (thanks to the μASF format) and extend the runtime library

---

in order to integrate the `cons` and the built-in list operations used by the list-matching algorithm (written in ASF itself). This experiment makes the REM Compiler a good candidate to make executable a sub-language of CASL[2].

3   Implementation and Optimisations

Each component of the ELAN Compiler is highly modular and has been designed to achieve efficiency through both high-level optimisations such as running a deterministic analysis algorithm before compiling rules and strategy expressions, and also low-level optimisations such as representing built-in data-types by native C integers for examples. For lack of space, we mention here only a few of the key optimisations.

Many-to-One Pattern Matching. In the syntactic case as in the AC case, the major source of efficiency is the use of compiled deterministic discrimination trees. This makes the pre-selection of all applicable rules to be linear wrt. the size of the term to reduce. This pre-selection is useful when applying non-deterministic strategies and when building bipartite graphs needed in the AC case.

Compact Bipartite Graph. The second source of efficiency of the AC matching algorithm is the definition of restricted classes of patterns for which a refined data structure of compact bipartite graph is used. This approach allows encoding, in a single data structure, all matching problems relative to a set of rewrite rules and improves the rule selection process (by minimising the bipartite graph construction cost, the number of matching attempts and the memory allocation).

Deterministic Analysis. The third optimisation that is crucial to the compilation scheme is the definition of an analysis algorithm which statically detects deterministic computations. With this approach, the search space size, the memory usage, the number of necessary choice points, and the time spent in backtracking and memory management can be considerably reduced. We can also benefit from the deterministic analysis to improve the efficiency of AC-matching and to detect some non-terminating strategies. In practice, this analysis often reduces the number of simultaneous active choice points (and the needed memory) to a constant and increases the efficiency from a factor 3 to 30 depending on the input specification (for instance, 3 for a Knuth-Bendix completion procedure, 6 for the N-queens problem, and 30 for the Fibonacci function). The interested reader is invited to read [2] for more details.

Data Structures and Memory Management. The runtime library relies on a Mark & Sweep garbage collector and an efficient allocation implementation using free lists. Practical studies have shown that recycling data structures (such as

---

[2] developed by the ESPRIT Working Group CoFI (Common Framework Initiative for Algebraic Specification and Development).

bipartite graphs) and integrating a generational copy collector can considerably reduce the time spent in memory management.

Independently, all data structures have been carefully designed: terms can be shared, AC-terms are maintained in Ordered Canonical Form (flattened, ordered and identical subterms are merged using multiplicities), and built-ins are implemented by native types and are not "wrapped" by artificial constructors. Internal operations are also optimised: destructive updates are performed when possible (unshared term not involved in non-deterministic computations) and bitmasks (intensively used by matching algorithms) are represented by 32-bits or 64-bits integers when possible, *etc.*

All these optimisations are essential to get an efficient implementation and are fully detailed in [5, 2]. Naturally, they can be re-used and applied to any similar language implementation.

The ELAN Compiler is available as part of the current ELAN 3.4 distribution and includes a friendly user interface. The generated programs and C-code are completely independent from the environment and can be edited, modified and integrated as components in any complex system.

4   Applications and Experimental Results

ELAN is an attractive framework for building advanced applications and formal tools. Among those, let us mention for instance the design of rules and strategies for constraint satisfaction problems, theorem proving tools in first-order logic with equality, the combination of unification algorithms and of decision procedures in various equational theories and the design of a tree automata library used to prove non-trivial protocol properties.

By compiling most significant ELAN programs (which involve many non-deterministic computations and a lot of backtracking), we note that the resulting executable specifications simulate the application of 1 to 2 million (complex conditional) rewrite rules per second on standard Intel or Alpha hardware (from 30,000 to 100,000 pure AC rewrite steps per second up to 15 millions for very simple examples such as Fibonacci numbers). Rather than claiming that the ELAN Compiler is equivalent or (often) faster than most comparable implementations (such as ASF+SDF, Brute, CafeOBJ, CiME, Epic, Maude, OBJ, RRL, Smaran, Tram, *etc.*), the main result of this work is that highly optimised (semi)-compilation techniques (ASF+SDF, ELAN and Maude) may promote rewrite rule based languages at the level of the best functional or logical language implementations. In order to support our intuition, we compare on two standard benchmarks the performance of the ELAN system[3] to the Objective Caml functional programming system and the GNU Prolog logic programming system. We also compare ELAN with RRL, OBJ and Brute on three pure AC benchmarks (see [5] for more details). The experimental results are given in the table below:

---

[3] On a Sun Enterprise running Solaris 5.6.

(time in second)	OCaml	GNU Prolog	Elan	RRL	OBJ	Brute
Fibonacci(35)	1.80	-	2.97	-	-	-
Nqueens(12)	19.0	57.0	73.0	-	-	-
Prop	-	-	0.43	> 24h	1164	1.78
Bool3	-	-	0.18	> 4h	> 24h	2.25
Sum100	-	-	1.32	-	> 24h	6.25

Acknowledgements

I sincerely thank Peter Borovanský, Horatiu Cirstea and Hélène Kirchner for helpful discussions and comments.

References

[1] P. Borovanský, C. Kirchner, H. Kirchner, P.-E. Moreau, and C. Ringeissen. An overview of ELAN. In *Proceedings of the 2nd International Workshop on Rewriting Logic and its Applications*, volume 15. Electronic Notes in Theoretical Computer Science, 1998.

[2] H. Kirchner and P.-E. Moreau. Non-deterministic computations in ELAN. In *Recent Developements in Algebraic Specification Techniques, 13th WADT'98*, volume 1589 of *Lecture Notes in Computer Science*, pages 168–182. Springer-Verlag, 1998.

[3] P. Klint. A meta-environment for generating programming environments. *ACM Transactions on Software Engineering and Methodology*, 2:176–201, 1993.

[4] J. Meseguer. Conditional rewriting logic as a unified model of concurrency. *Theoretical Computer Science*, 96(1):73–155, 1992.

[5] P.-E. Moreau and H. Kirchner. A compiler for rewrite programs in associative-commutative theories. In *"Principles of Declarative Programming"*, number 1490 in Lecture Notes in Computer Science, pages 230–249. Springer-Verlag, 1998.

TALP: A Tool for the Termination Analysis of Logic Programs

Enno Ohlebusch[1], Claus Claves[1], and Claude Marché[2]

[1] Faculty of Technology, University of Bielefeld
P.O. Box 10 01 31, 33501 Bielefeld, Germany
{enno,cclaves}@techfak.uni-bielefeld.de
[2] LRI, CNRS URA 410
Bât. 490, Université de Paris-Sud, Centre d'Orsay
91405 Orsay Cedex, France
marche@lri.fr

1   Introduction

In the last decade, the automatic termination analysis of logic programs has been receiving increasing attention. Among other methods, techniques have been proposed that transform a well-moded logic program into a term rewriting system (TRS) so that termination of the TRS implies termination of the logic program under Prolog's selection rule. In [Ohl99] it has been shown that the two-stage transformation obtained by combining the transformations of [GW93] into deterministic conditional TRSs (CTRSs) with a further transformation into TRSs [CR93] yields the transformation proposed in [AZ96], and that these three transformations are equally powerful. In most cases simplification orderings are not sufficient to prove termination of the TRSs obtained by the two-stage transformation. However, if one uses the dependency pair method [AG00] in combination with polynomial interpretations instead, then most of the examples described in the literature can automatically be proven terminating. Based on these observations, we have implemented a tool for proving termination of logic programs automatically. This tool consists of a front-end which implements the two-stage transformation and a back-end, the CiME system [CiM], for proving termination of the generated TRS. Experiments show that our tool can compete with other tools [DSV99] based on sophisticated norm-based approaches.

2   The Front-End

TALP takes a Prolog program and a query as input and proceeds in four steps:

1. The Prolog program is translated into a logic program $\mathcal{P}$. In this process, clauses with if-then-structures, disjunctions, or negated atoms are translated into new clauses. For instance, the clause $A \leftarrow B, \text{not } C, D$ is replaced with the clauses $A \leftarrow B, C, \text{fail}$ and $A \leftarrow B, D$. Cuts are ignored.

L. Bachmair (Ed.): RTA 2000, LNCS 1833, pp. 270–273, 2000.

2. The query determines which of the arguments in its predicates are used as input and output, respectively. According to this information, the tool tries to generate a *moding* for the logic program such that the program is *well-moded*. If this step is successfully completed, the logic program will be transformed into a TRS as follows.

3. Every atom $A = p(t_1, \ldots, t_n)$ with input positions $i_1, \ldots, i_k$ and output positions $i_{k+1}, \ldots, i_n$ associates with a rewrite rule

$$\rho(A) = p_{in}(t_{i_1}, \ldots, t_{i_k}) \rightarrow p_{out}(t_{i_{k+1}}, \ldots, t_{i_n})$$

and every program clause $C = A \leftarrow B_1, \ldots, B_m$ is transformed into a conditional rewrite rule $\rho(C) = \rho(A) \Leftarrow \rho(B_1), \ldots, \rho(B_m)$. The CTRS $\mathcal{R}_{\mathcal{P}} = \{\rho(C) \mid C \in \mathcal{P}\}$ obtained in this way is deterministic because the logic program $\mathcal{P}$ is well-moded.

4. Every rule $l \rightarrow r \Leftarrow c \in \mathcal{R}_{\mathcal{P}}$ with n conditions in c is transformed into $n+1$ unconditional rewrite rules (cf. Sect. 4):

$$U(l \rightarrow r \Leftarrow c) = \begin{cases} \{l \rightarrow r\}, & \text{if } c \text{ is empty} \\ \{l \rightarrow u(s, \vec{x})\} \cup U(u(t, \vec{x}) \rightarrow r \Leftarrow c'), & \text{if } c = s \rightarrow t, c' \\ \quad \text{where } u \text{ is a fresh function symbol and} \\ \quad \vec{x} = \mathcal{V}ar(l) \cap (\mathcal{V}ar(t) \cup \mathcal{V}ar(c') \cup \mathcal{V}ar(r)) \end{cases}$$

3   The Back-End C*i*ME

The back-end tries to prove termination of the rewrite systems generated by the front-end. This back-end is a pre-release of version 2 of the rewrite tool C*i*ME [CM96] which is available as an alpha version [CiM]. TALP uses one specific method available in C*i*ME for proving termination: the dependency pair method [AG00] in combination with polynomial interpretations [Gie95]. To be precise, C*i*ME performs the following steps:

1. It computes the *estimated dependency graph* of the rewrite system.
2. From the cycles in that graph, it computes a set of constraints of the form $t_1 > t_2$ or $t_1 \geq t_2$, that have to be satisfied by a *weakly* monotonic reduction ordering.

The next goal is to find such an ordering, which is done as follows [CMT99]:

3. With each symbol f in the signature, say of arity n, it associates a parametric polynomial interpretation of the simple linear form $P_f(x_1, \ldots, x_n) = a_1 x_1 + \cdots + a_n x_n + c$.
4. Every constraint is translated into constraints on polynomials, and then into non-linear Diophantine constraints over the a_i's and c's, by means of some (incomplete) positiveness criteria [HJ98].
5. The Diophantine constraints are solved for variables in the interval $[0; B]$, where B is a bound for coefficients given by the user, by using *finite domain* constraint solving techniques [BC93].

4   Example

If TALP gets the following Prolog program with query `flat(in, out)` as input

```
flat(niltree, nil).
flat(tree(X, niltree, T), cons(X, L)) :- flat(T, L).
flat(tree(X, tree(Y, T1, T2), T3), L) :-
  flat(tree(Y, T1, tree(X, T2, T3)), L).
```

then the first transformation yields the CTRS

$$\mathsf{flat_{in}}(\mathsf{niltree}) \to \mathsf{flat_{out}}(\mathsf{nil})$$
$$\mathsf{flat_{in}}(\mathsf{tree}(x, \mathsf{niltree}, t)) \to \mathsf{flat_{out}}(\mathsf{cons}(x, l)) \Leftarrow \mathsf{flat_{in}}(t) \to \mathsf{flat_{out}}(l)$$
$$\mathsf{flat_{in}}(\mathsf{tree}(x, \mathsf{tree}(y, t_1, t_2), t_3)) \to \mathsf{flat_{out}}(l) \Leftarrow$$
$$\mathsf{flat_{in}}(\mathsf{tree}(y, t_1, \mathsf{tree}(x, t_2, t_3))) \to \mathsf{flat_{out}}(l)$$

and the second transformation yields the unconditional TRS

$$\mathsf{flat_{in}}(\mathsf{niltree}) \to \mathsf{flat_{out}}(\mathsf{nil})$$
$$\mathsf{flat_{in}}(\mathsf{tree}(x, \mathsf{niltree}, t)) \to \mathsf{u_1}(\mathsf{flat_{in}}(t), x)$$
$$\mathsf{u_1}(\mathsf{flat_{out}}(l), x) \to \mathsf{flat_{out}}(\mathsf{cons}(x, l))$$
$$\mathsf{flat_{in}}(\mathsf{tree}(x, \mathsf{tree}(y, t_1, t_2), t_3)) \to \mathsf{u_2}(\mathsf{flat_{in}}(\mathsf{tree}(y, t_1, \mathsf{tree}(x, t_2, t_3))))$$
$$\mathsf{u_2}(\mathsf{flat_{out}}(l)) \to \mathsf{flat_{out}}(l)$$

Subsequently CiME is asked to find a linear polynomial interpretation with coefficients in the interval $[0; 2]$. It generates the following interpretation

$$
\begin{array}{lll}
[\![\mathsf{nil}]\!] = 0 & [\![\mathsf{flat_{out}}]\!](x_0) = 0 & [\![\mathsf{u_1}]\!](x_0, x_1) = 0 \\
[\![\mathsf{niltree}]\!] = 0 & [\![\mathsf{u_2}]\!](x_0) = 0 & [\![\mathsf{tree}]\!](x_0, x_1, x_2) = x_2 + 2x_1 + 1 \\
[\![\mathsf{flat_{in}}]\!](x_0) = 0 & [\![\mathsf{cons}]\!](x_0, x_1) = 0 & [\![\mathsf{FLAT_{in}}]\!](x_0) = x_0
\end{array}
$$

and the induced polynomial ordering satisfies all constraints obtained from the cycles in the estimated dependency graph.

5   Experimental Results

As in [DSV99], we have tested the TALP system on well-known examples (the benchmarks are collected in [LS97]). A Web interface for TALP is available at `http://bibiserv.techfak.uni-bielefeld.de/talp/`. Overall, our results are comparable to those reported in [DSV99] but there are examples for which TALP succeeds and other tools don't (e.g. the example in Sect. 4 and Example 2.3.1 in [Plü90]) and vice versa. During our experiments we made the following observations. For all those examples for which CiME was able to find a termination proof, it was also able to generate a suitable linear polynomial interpretation with coefficients in the interval $[0; 2]$. The restriction to linear polynomial interpretations seems to be a very good heuristic because whenever searching for a linear interpretation fails, then searching for a more general one, like simple-mixed, does not succeed either. Table 1 contains the execution times of the back-end for finding a termination proof on a Sun SPARCstation 10.

Program	Query	Time	Ref.
permutation	perm(i,o)	0.64 sec.	[Plü90] 1.2
transitivity	p(i,o)	0.19 sec.	[Plü90] 2.3.1
quicksort	qsort(i,o)	1.73 sec.	[Plü90] 6.1.1
mult	mult(i,i,o)	0.22 sec.	[Plü90] 7.2.9
mergesort	mergesort(i,o)	3.03 sec.	[Plü90] 8.2.1a
flat	flat(i,o)	2.19 sec.	[AZ96]

Table 1. Benchmarks

References

[AG00]   T. Arts and J. Giesl. Termination of term rewriting using dependency pairs. *Theoretical Computer Science*, 2000. To appear.

[AZ96]   T. Arts and H. Zantema. Termination of logic programs using semantic unification. In *Proc. 5th Int. Workshop on Logic Program Synthesis and Transformation*, volume 1048 of *Lecture Notes in Computer Science* pages 219–233, 1996. Springer-Verlag.

[BC93]   F. Benhamou and A. Colmerauer, editors. *Constraint Logic Programming: selected research*. MIT Press, 1993.

[CiM]   CiME 2. Prerelease available at http://www.lri.fr/~demons/cime-2.0.html.

[CM96]   E. Contejean and C. Marché. CiME: Completion Modulo E. *Proc. 7th Int. Conference on Rewriting Techniques and Applications*, volume 1103 of *Lecture Notes in Computer Science*, pages 416–419, July 1996. Springer-Verlag. System description available at http://www.lri.fr/~demons/cime.html.

[CMT99]   E. Contejean, C. Marché, A.-P. Tomás, and X. Urbain. Solving termination constraints via finite domain polynomial interpretations. Draft, 1999.

[CR93]   M. Chtourou and M. Rusinowitch. Méthode transformationnelle pour la preuve de terminaison des programmes logiques. Unpublished manuscript, 1993.

[DSV99]   S. Decorte, D. De Schreye, and H. Vandecasteele. Constraint-based termination analysis of logic programs. *ACM TOPLAS*, 1999. To appear.

[Gie95]   J. Giesl. Generating polynomial orderings for termination proofs. In *Proc. 6th Int. Conference on Rewriting Techniques and Applications*, volume 914 of *Lecture Notes in Computer Science*, April 1995. Springer-Verlag.

[GW93]   H. Ganzinger and U. Waldmann. Termination proofs of well-moded logic programs via conditional rewrite systems. In *Proc. 3rd Int. Workshop on Conditional Term Rewriting Systems*, volume 656 of *Lecture Notes in Computer Science*, pages 113–127, Berlin, 1993. Springer-Verlag.

[HJ98]   Hoon Hong and Dalibor Jakus. Testing positiveness of polynomials. *Journal of Automated Reasoning*, 21(1):23–38, August 1998.

[LS97]   N. Lindenstrauss and Y. Sagiv. Automatic termination analysis of logic programs (with detailed experimental results). Technical report, Hebrew University, Jerusalem, 1997.

[Ohl99]   E. Ohlebusch. Transforming conditional rewrite systems with extra variables into unconditional systems. In *Proc. 6th Int. Conference on Logic for Programming and Automated Reasoning*, volume 1705 of *Lecture Notes in Artificial Intelligence*, pages 111–130, 1999. Springer-Verlag.

[Plü90]   L. Plümer. *Termination Proofs for Logic Programs*, volume 446 of *Lecture Notes in Artificial Intelligence*. Springer-Verlag, Berlin, 1990.

Author Index

Lecture Notes in Computer Science

For information about Vols. 1–1760
please contact your bookseller or Springer-Verlag

Vol. 1803: S. Cagnoni et al. (Eds.), Real-World Applications and Evolutionary Computing. Proceedings, 2000. XII, 396 pages. 2000.

Vol. 1805: T. Terano, H. Liu, A.L.P. Chen (Eds.), Knowledge Discovery and Data Mining. Proceedings, 2000. XIV, 460 pages. 2000. (Subseries LNAI).

Vol. 1806: W. van der Aalst, J. Desel, A. Oberweis (Eds.), Business Process Management. VIII, 391 pages. 2000.

Vol. 1807: B. Preneel (Ed.), Advances in Cryptology – EUROCRYPT 2000. Proceedings, 2000. XVIII, 608 pages. 2000.

Vol. 1810: R.López de Mántaras, E. Plaza (Eds.), Machine Learning: ECML 2000. Proceedings, 2000. XII, 460 pages. 2000. (Subseries LNAI).

Vol. 1811: S.W. Lee, H.. Bülthoff, T. Poggio (Eds.), Biologically Motivated Computer Vision. Proceedings, 2000. XIV, 656 pages. 2000.

Vol. 1813: P.L. Lanzi, W. Stolzmann, S.W. Wilson (Eds.), Learning Classifier Systems. X, 349 pages. 2000. (Subseries LNAI).

Vol. 1815: G. Pujolle, H. Perros, S. Fdida, U. Körner, I. Stavrakakis (Eds.), Networking 2000 – Broadband Communications, High Performance Networking, and Performance of Communication Networks. Proceedings, 2000. XX, 981 pages. 2000.

Vol. 1816: T. Rus (Ed.), Algebraic Methodology and Software Technology. Proceedings, 2000. XI, 545 pages. 2000.

Vol. 1817: A. Bossi (Ed.), Logic-Based Program Synthesis and Transformation. Proceedings, 1999. VIII, 313 pages. 2000.

Vol. 1818: C.G. Omidyar (Ed.), Mobile and Wireless Communications Networks. Proceedings, 2000. VIII, 187 pages. 2000.

Vol. 1819: W. Jonker (Ed.), Databases in Telecommunications. Proceedings, 1999. X, 208 pages. 2000.

Vol. 1821: R. Loganantharaj, G. Palm, M. Ali (Eds.), Intelligent Problem Solving. Proceedings, 2000. XVII, 751 pages. 2000. (Subseries LNAI).

Vol. 1822: H.H. Hamilton, Advances in Artificial Intelligence. Proceedings, 2000. XII, 450 pages. 2000. (Subseries LNAI).

Vol. 1823: M. Bubak, H. Afsarmanesh, R. Williams, B. Hertzberger (Eds.), High Performance Computing and Networking. Proceedings, 2000. XVIII, 719 pages. 2000.

Vol. 1824: J. Palsberg (Ed.), Static Analysis. Proceedings, 2000. VIII, 433 pages. 2000.

Vol. 1825: M. Nielsen, D. Simpson (Eds.), Application and Theory of Petri Nets 2000. Proceedings, 2000. XI, 485 pages. 2000.

Vol. 1826: W. Cazzola, R.J. Stroud, F. Tisato (Eds.), Reflection and Software Engineering. X, 229 pages. 2000.

Vol. 1830: P. Kropf, G. Babin, J. Plaice, H. Unger (Eds.), Distributed Communities on the Web. Proceedings, 2000. X, 203 pages. 2000.

Vol. 1831: D. McAllester (Ed.), Automated Deduction – CADE-17. Proceedings, 2000. XIII, 519 pages. 2000. (Subseries LNAI).

Vol. 1832: B. Lings, K. Jeffery (Eds.), Advances in Databases. Proceedings, 2000. X, 227 pages. 2000.

Vol. 1833: L. Bachmair (Ed.), Rewriting Techniques and Applications. Proceedings, 2000. X, 275 pages. 2000.

Vol. 1834: J.-C. Heudin (Ed.), Virtual Worlds. Proceedings, 2000. XI, 314 pages. 2000. (Subseries LNAI).

Vol. 1835: D. N. Christodoulakis (Ed.), Natural Language Processing – NLP 2000. Proceedings, 2000. XII, 438 pages. 2000. (Subseries LNAI).

Vol. 1837: R. Backhouse, J. Nuno Oliveira (Eds.), Mathematics of Program Construction. Proceedings, 2000. IX, 257 pages. 2000.

Vol. 1838: W. Bosma (Ed.), Algorithmic Number Theory. Proceedings, 2000. IX, 615 pages. 2000.

Vol. 1839: G. Gauthier, C. Frasson, K. VanLehn (Eds.), Intelligent Tutoring Systems. Proceedings, 2000. XIX, 675 pages. 2000.

Vol. 1840: F. Bomarius, M. Oivo (Eds.), Product Focused Software Process Improvement. Proceedings, 2000. XI, 426 pages. 2000.

Vol. 1841: E. Dawson, A. Clark, C. Boyd (Eds.), Information Security and Privacy. Proceedings, 2000. XII, 488 pages. 2000.

Vol. 1842: D. Vernon (Ed.), Computer Vision – ECCV 2000. Part I. Proceedings, 2000. XVIII, 953 pages. 2000.

Vol. 1843: D. Vernon (Ed.), Computer Vision – ECCV 2000. Part II. Proceedings, 2000. XVIII, 881 pages. 2000.

Vol. 1844: W.B. Frakes (Ed.), Software Reuse: Advances in Software Reusability. Proceedings, 2000. XI, 450 pages. 2000.

Vol. 1845: H.B. Keller, E. Plöderer (Eds.), Reliable Software Technologies Ada-Europe 2000. Proceedings, 2000. XIII, 304 pages. 2000.

Vol. 1846: H. Lu, A. Zhou (Eds.), Web-Age Information Management. Proceedings, 2000. XIII, 462 pages. 2000.

Vol. 1847: R. Dyckhoff (Ed.), Automated Reasoning with Analytic Tableaux and Related Methods. Proceedings, 2000. X, 441 pages. 2000. (Subseries LNAI).

Vol. 1848: R. Giancarlo, D. Sankoff (Eds.), Combinatorial Pattern Matching. Proceedings, 2000. XI, 423 pages. 2000.

Vol. 1849: C. Freksa, W. Brauer, C. Habel, K.F. Wender (Eds.), Spatial Cognition II. XI, 420 pages. 2000. (Subseries LNAI).

Vol. 1850: E. Bertino (Ed.), ECOOP 2000 – Object-Oriented Programming. Proceedings, 2000. XIII, 493 pages. 2000.

Vol. 1851: M.M. Halldórsson (Ed.), Algorithm Theory – SWAT 2000. Proceedings, 2000. XI, 564 pages. 2000.

Vol. 1853: U. Montanari, J.D.P. Rolim, E. Welzl (Eds.), Automata, Languages and Programming. Proceedings, 2000. XVI, 941 pages. 2000.

Vol. 1855: E.A. Emerson, A.P. Sistla (Eds.), Computer Aided Verification. Proceedings, 2000. X, 582 pages. 2000.

Vol. 1857: J. Kittler, F. Roli (Eds.), Multiple Classifier Systems. Proceedings, 2000. XII, 404 pages. 2000.

Vol. 1860: M. Klusch, L. Kerschberg (Eds.), Cooperative Information Agents IV. Proceedings, 2000. XI, 285 pages. 2000. (Subseries LNAI).